Optical Materials
and Applications

OPTICAL SCIENCE AND ENGINEERING

Founding Editor
Brian J. Thompson
University of Rochester
Rochester, New York

RECENTLY PUBLISHED

Please visit our website **www.crcpress.com** *for a full list of titles*

Optical Materials and Applications

Edited by Moriaki Wakaki

CRC Press
Taylor & Francis Group
Boca Raton London New York

CRC Press is an imprint of the
Taylor & Francis Group, an **informa** business

CRC Press
Taylor & Francis Group
6000 Broken Sound Parkway NW, Suite 300
Boca Raton, FL 33487-2742

First issued in paperback 2019

© 2013 by Taylor & Francis Group, LLC
CRC Press is an imprint of Taylor & Francis Group, an Informa business

No claim to original U.S. Government works

ISBN-13: 978-0-8247-2759-8 (hbk)
ISBN-13: 978-1-138-37261-0 (pbk)

Library of Congress Cataloging-in-Publication Data

Optical materials and applications / editor, Moriaki Wakaki.
     p. cm. -- (Optical science and engineering)
    Summary: "This book brings together and couples results from electromagnetic, mechanics, thermodynamics, solid state physics, chemistry, optics, and materials science to provide a coherent picture of materials science, in particular their optical properties. It discusses the underlying physical basis of the each property, including constituents, bonding, and structural order. The book also discusses units and unit conversions, provides simple models for estimating properties or extrapolating data as well as examples of how to fit data using these models. Two chapters are completely devoted to practical problems in optical systems"-- Provided by publisher.
    Includes bibliographical references and index.
    ISBN 978-0-8247-2759-8 (hardback)
    1. Optical materials. I. Wakaki, Moriaki.

QC374.O668 2012
621.36--dc23
                                                     2012017895

Visit the Taylor & Francis Web site at
http://www.taylorandfrancis.com

and the CRC Press Web site at
http://www.crcpress.com

Contents

Preface

Light propagates in free space with the velocity of 2.997925 m/s^{-1} as a wave spatially and a collection of photons with an energy of $\varepsilon = h\nu$. Various kinds of optical materials or media are developed to utilize the fundamental natural phenomena of light. Many types of refractive and reflective optical systems consisting of lenses and mirrors have been developed for the simple manipulation of the spatial distribution of light. To design the optical system with a specific character, the materials to conform the system must have desired optical spectra for transmission and reflection. Also to design a certain active element, desired absorption and emission spectra are required. A fundamental review of typical active materials such as LASER is given in Chapter 4, "Materials for Solid-State Laser," by T. Itatani. These optical properties are realized by utilizing and designing the dielectric dispersion of each material.

As the first step, we should understand the basic optical properties of typical materials. The definition of optical material has been expanded largely due to the rapid expansion of optoelectronic applications caused by the progress of information technology (IT). Various types of optical windows are the typical optical materials, and these are classified as a passive element. Many types of optical components for dispersion control such as prisms, gratings, and filters and for phase control such as wave plates are needed to form advanced optical systems. A review of the fundamental optical properties of solids is given in Chapter 1, "Basic Theory on Optical Properties of Solids," by M. Wakaki and K. Kudo. A review of fundamental optical materials is given in Chapter 2, "Optical Materials for Ultraviolet, Visible, and Infrared," by T. Arai and M. Wakaki in the linear regime and a review extended to the nonlinear regime is given in Chapter 3, "Materials for Nonlinear Optics," by T. Arai and M. Wakaki.

Bulk materials are the basic material and offer basic information on the materials for the design of various modifications. Optical thin films are the beginning of artificial components. Artificial structures with subwavelength dimensions have been recently formed with physical (vacuum evaporation) or chemical techniques to control the phase of the light wave. A fundamental review of optical thin films is presented in Chapter 6, "Materials for Optical Thin Films," by T. Shibuya. Starting from a one-dimensional (1D) structure typically applied as an interference filter, nanoprocess to two-dimensional (2D) and three-dimensional (3D) structures to control the dispersion of the material has been developed. A review of such typical systems is presented in Chapter 7, "Materials for Nanophotonics," by K. Asakawa. Finally, the progress of IT greatly depends on the progress of the optical waveguide, which gives the optical systems compact and high-efficiency characteristics. A review of such an optical waveguide is given in Chapter 5, "Materials for Optical Waveguides," by B. P. Pal.

The editor of this book, M. Wakaki, appreciates each author's contribution to the basic ideas for beginners in this field and advanced topics for active researchers in the field, and also expresses great thanks to T. Shibuya for assisting in the arrangement of the manuscripts. The editor also expresses great thanks to the promotional and management editors of CRC Press for encouraging authors to finish the manuscripts.

Finally, I feel indebted to my wife, T. Wakaki, and my sons for giving me the time to do this work.

M. Wakaki
At old and stimulating city, Edinburgh

The Editor

Moriaki Wakaki is a professor in the Department of Optical and Imaging Science and Technology, School of Engineering, Tokai University, Japan. He is a chief professor of electro photo optics, Graduate School of Engineering, Tokai University. He is the author or coauthor of over 100 scientific papers. He was the editor-in-chief of the *Journal of Advanced Science* from 1996 to 2000, the newsletter of the Thermoelectric Conversion Research Society of Japan from 1997 to 2000, and *Oyokougaku (Applied Optics* in Japanese) from 2001 to 2003. Wakaki is a member of the Optical Society of America, the International Society for Optical Engineering, the Optical Society of Japan, the Laser Society of Japan, the Japan Spectroscopic Society, the Japan Society of Applied Physics, the Physical Society of Japan, the Surface Science of Japan, the Thermoelectric Society of Japan, the Carbon Society of Japan, the Japan Society of Plastic Surgery, and the Society of Advanced Science. He received MSc and DSc degrees in physics from Tokyo Educational University in 1972 and 1975, respectively. His current areas of research are solid state physics relating mainly to complex materials dispersed with nanoparticles, optoelectronics research, laser medicine and biophotonics, infrared astronomy relating mainly to the development of spectroscopic measuring systems and far-infrared detectors.

Wakaki has published the following works in English and Japanese: *Rediscover Optics* (Optronics Co. Press, Tokyo, 1997, in Japanese), *Introduction to Optical Engineering* (Jikkyou Syuppan Press, Tokyo, 1998, in Japanese), *Fundamentals of Quantum Optics* (Gendai Kougakusya Press, Tokyo, 1998, in Japanese), *Environment Conscious Material* (Nikka Giren Press, Tokyo, 2002, in Japanese), *Introduction to Wave Optics* (Jikkyou Syuppan Press, Tokyo, 2004, in Japanese), and *Physical Properties and Data of Optical Materials* (CRC Press, Boca Raton, Florida, 2007).

Contributors

Toshihiro Arai
Deceased
University of Tsukuba
Tsukuba, Ibaraki, Japan

Kiyoshi Asakawa
University of Tsukuba
Tsukuba, Ibaraki, Japan

Taro Itatani
Advanced Industrial Science and
    Technology (AIST)
Tsukuba, Ibaraki, Japan

Keiei Kudo
Deceased
University of Tsukuba
Tsukuba, Ibaraki, Japan

Bishnu P. Pal
Physics Department
Indian Institute of Technology
Delhi, India

Takehisa Shibuya
Department of Optical and Imaging
    Science and Technology
School of Engineering
Tokai University
Hiratsuka, Kanagawa, Japan

Moriaki Wakaki
Department of Optical and Imaging
    Science and Technology
School of Engineering
Tokai University
Hiratsuka, Kanagawa, Japan

1 Basic Theory on Optical Properties of Solids

Moriaki Wakaki and Keiei Kudo (deceased)

CONTENTS

The optical properties of solids are basically described using the dielectric functions of the materials. The function depends on the symmetry of the medium—that is, the crystal structure. The fundamental concept of the geometry of the crystal structure is given in Section 1.1. The propagation of light through the medium or the interaction between medium and light is described by the dispersion characteristics of the light in the medium. The summary of the dispersion of light in the medium is given in Section 1.2. The details of the dielectric properties are also described in Chapter 2 ("Optical Materials for Ultraviolet, Visible, and Infrared"). The detailed behavior of the reflection and transmission of the layers is treated in Chapter 2 and Chapter 6 ("Materials for Optical Thin Films"). The dielectric properties are contributed by many types of elementary excitations. Major excitations contributing to the optical excitations are caused by electrons and phonons. The optical absorptions by electrons and lattice vibrations in solids are reviewed in Sections 1.3 and 1.4, respectively. Light scattering effects like Raman and Brillouin scatterings, which offer the complementary information to the absorption spectra in solids are summarized in Section 1.5. The effects of external fields like magnetic and electric fields on optical properties are also reviewed in Chapter 2. The nonlinear optical properties are also reviewed in Chapter 3 ("Materials for Nonlinear Optics").

1.1   GEOMETRY OF CRYSTAL[1–3]

Materials are classified according to their degrees of aggregation: These are gas, liquid, and solid in the order of increasing densities of molecules or atoms. Recent advanced applications of devices require high-density integration of elements, and the device systems with a solid phase are dominant. Physical properties of solids are the main interest of this text. The solid phase is classified into random arrangements (glasses and amorphous) and regular arrangements (crystals) of molecules or atoms according to their topological forms. Recently, complex materials that are not classified simply to one of the above categories have been actively fabricated and studied for their interesting potentials.[4–7]

Crystalline solids offer the fundamental concepts of the material relating with basic physical properties, and it is still important to learn the basic concept to develop the advanced system of materials, like superlattices, ceramics, and composite materials including nanostructures.[8–10] Recent topics relating to nanophotonics materials are presented in Chapter 7 by K. Asakawa.

The physical properties of crystals are strongly correlated with the symmetry of the lattice that composes the material. In the crystal, the lattice arranges regularly in space and its minimum unit is called a primitive unit cell. The collection of primitive unit cells also arranges regularly to form a crystal. The type of crystal is first classified by the symmetry of the primitive unit cell.

1.1.1 SYMMETRY OPERATIONS

The symmetry is featured by using the symmetry operations not to change the original form by the operations. These operations are described as follows:

1. *Identity (Einhelt)*: The rotation of 360 degrees around an axis does not change the form of the lattice. The operation is usually expressed by the symbol E.
2. *Cyclic axis*: The unit cell does not change its form by some rotation around the proper axis. If the cell returns its original form n times through the rotation of 360 degrees around the axis, the cell has the n-fold rotation axis. These symmetries are expressed as C_n ($n = 1, 2, 3, 4$, and 6) by Schönflies (S) expression and simply as n (1, 2, 3, 4, and 6) by Herman-Mauguin (H-M) expression as shown in Figure 1.1.[11] If there are several cyclic axes, the axis with largest number n is called a *principal axis*.
3. *Symmetry plane (mirror or reflection plane)*: If the unit cell is symmetrical with some plane within the cell, the plane is called the *symmetry plane* as shown in Figure 1.2. The symmetry plane is expressed by the symbol σ (S) or m (H-M).
4. *Rotatory reflection axis*: If the cell does not change form after $2\pi/n$ rotation and the reflection is not perpendicular to the axis, the operation is called the *rotatory reflection axis Sn* (S) or n (H-M) as shown in Figure 1.3.[11]
5. *Inversion*: If the cell does not change form after the operation of the shift of the position of lattice point r to $-r$ referring to the point within the unit cell, the operation is called *inversion i* (S) or 2 (H-M).

The unit cell has several symmetrical elements. For example, the square unit cell C_{4v} (4mm) has the symmetry elements, $E\ 2C_4\ C_4^2\ 2\sigma_v\ 2\sigma_d$ as shown in Figure 1.4(a),

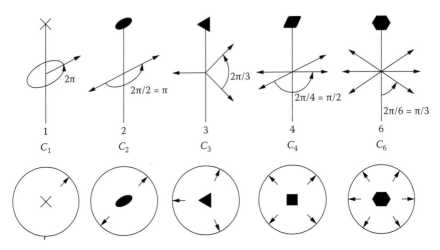

FIGURE 1.1 Rotation around the axis of symmetry. (See also K. Kudo: *Fundamentals of Optical Properties of Solids*, Ohmsha (1977) (in Japanese).) The upper scheme is a stereophonic description, and the lower is a projection description.

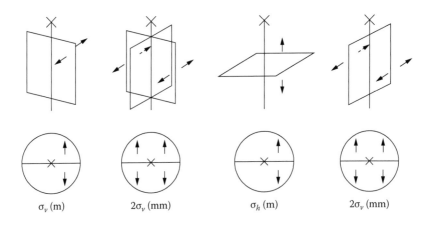

FIGURE 1.2 Reflection through a plane that contains the principal axis σ_v and is perpendicular to the principal axis σ_h. The upper scheme is a stereophonic description, and the lower is a projection description.

and order of 8, which is the total number of the symmetry elements. The cubic unit cell O_h (*m3m*) has the symmetry elements, $E\ 8C_3\ 3C'_2\ 6C_2\ 6C_4\ I\ 8S_6\ 3\sigma_h\ 6\sigma_d\ 6S_4$ and the order of 48.

1.1.2 Point Groups and Crystalline Groups

If the symmetry elements of the unit cell are written as $A_1, A_2, \cdots, A_n$, these elements constitute a group by satisfying the following four conditions:

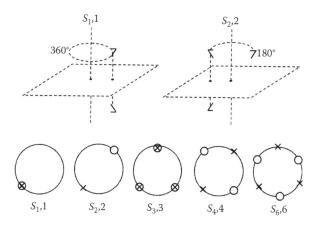

FIGURE 1.3 Rotation about an axis followed by reflection through a plane perpendicular to the axis of rotation. (See also K. Kudo: *Fundamentals of Optical Properties of Solids*, Ohmsha (1977) (in Japanese).) The upper scheme is a stereophonic description, and the lower is a projection description. Cross and circle symbols mean upper and lower position to the plane, respectively.

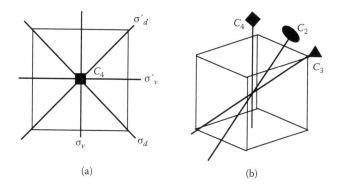

FIGURE 1.4 Symmetry operations for square unit cell C_{4v} (*4mm*) (a), and for cubic unit cell O_h (*m3m*) (b).

1. The state obtained after two successive symmetry operations is expressed by a single operation within the set. That is, the following relation holds, $A_i A_j = A_k$.
2. The identity operation E gives the following relation for any element within the set. That is,

$$E A_j = A_j E = A_j$$

3. The following relation of combination holds—that is, $A_i (A_j A_k) = (A_i A_j) A_k$.
4. If the inverse operation of A_j is written as A_j^{-1}, the following relation holds, $A_j^{-1}A_j = A_j A_j^{-1} = E$, which means A_j^{-1} is also an element of this set (from condition 1).

TABLE 1.1

Multiplication Table for the Elements of C_{4v} (4mm)

	E	C_4	C_4^2	C_4^3	σ_v	$\sigma_{v'}$	σ_d	$\sigma_{d'}$
E	E	C_4	C_4^2	C_4^3	σ_v	$\sigma_{v'}$	σ_d	$\sigma_{d'}$
C_4	C_4	C_4^2	C_4^3	E	$\sigma_{d'}$	σ_d	σ_v	$\sigma_{v'}$
C_4^2	C_4^2	C_4^3	E	C_4	$\sigma_{v'}$	σ_v	$\sigma_{d'}$	σ_d
C_4^3	C_4^3	E	C_4	C_4^2	σ_d	$\sigma_{d'}$	$\sigma_{v'}$	σ_v
σ_v	σ_v	σ_d	$\sigma_{v'}$	$\sigma_{d'}$	E	C_4^2	C_4	C_4^3
$\sigma_{v'}$	$\sigma_{v'}$	$\sigma_{d'}$	σ_v	σ_d	C_4^2	E	C_4^3	C_4
σ_d	σ_d	$\sigma_{v'}$	$\sigma_{d'}$	σ_v	C_4^3	C_4	E	C_4^2
$\sigma_{d'}$	$\sigma_{d'}$	σ_v	σ_d	$\sigma_{v'}$	C_4	C_4^3	C_4^2	E

For example, the operation $A_i\, A_j$ for the square unit cell C_{4v} (4mm) gives the result shown in Table 1.1.

The result satisfies the condition indicated above. These groups of symmetry elements express the lattice points of the crystal and are called the *point group*. Each unit cell is kept invariant for respective symmetry operation in the point group. But each unit cell must also cover the whole crystal by the translational movement. To meet the requirement, the rotation angles in C_n and S_n are limited to the integral multiple of 60° or 90°, which is derived from the properties of the point group. Only 32 point groups are allowed from these results. The 32 point groups give the possible arrangement of the crystal lattice and are often called the *crystal class*.

Two types of symbols are usually used to express these point groups (crystal class): Schönflies and Hermann–Maugin expressions. As shown in the example of a square unit cell, both expressions are shown as C_{4v} (4mm).

1.1.3 Space Groups

The point groups described in Section 1.1.2 are the set of operations not to change the unit cell. The unit cell after the operation cannot be distinguished with the original form. In the case of a simple lattice, the entire crystal lattice can be constructed only by the primitive translations. As the unit cell becomes complex to have ions or ion clusters around or between lattice points, the same arrangement as the original is obtained after a nonprimitive translation (translation by several parts of the lattice constant) and rotation or reflection.

1.1.3.1 Screw Axis

If the cell does not change after a rotation $2\pi/n$ ($n = 1, 2, 3, 4, 6$) around the certain axis and a translation by a distance T along the axis, the lattice has a screw axis.

1.1.3.2 Glide Plane

If the cell does not change after some translation along a plane and a reflection against the plane, the lattice has a glide plane.

The number of the nonprimitive space group is 157, and that of the primitive one is 73. As a result, the total number of the space group is 230 as shown in Table 1.2.[11]

TABLE 1.2

Crystalline System (I, II, III, ..., VII), Crystal Class ((1), (2), (3),...,(32)) and Space Group (1, 2, 3,..., 230)[11]

I. Triclinic

(1) $C_1(1)$, [E]

1. C_1^1 (P1)

(2) $C_i(\bar{1})$, [Ei]

2. C_i^1 (P$\bar{1}$)

II. Monoclinic

(3) $C_2(2)$, [EC$_2$]

3. C_2^1 (P2)

4. C_2^2 (P2$_1$)

5. C_2^3 (B2)

(4) $C_s(m)$, [E σ_h]

6. C_s^1 (Pm)

7. C_s^2 (Pb)

8. C_s^3 (Bm)

9. C_s^4 (Bb)

(5) $C_{2h}(2/m)$, [EC$_2$i σ_h]

10. C_{2h}^1 (P2/m)

11. C_{2h}^2 (P2$_1$/m)

12. C_{2h}^3 (B2/m)

13. C_{2h}^4 (P2/b)

14. C_{2h}^5 (P2$_1$/b)

15. C_{2h}^6 (B2/b)

III. Orthorhombic

(6) $D_2(222)$, [EC$_2$C$_2'$C$_2''$]

16. D_2^1 (P222)

17. D_2^2 (P222$_1$)

18. D_2^3 (P2$_1$2$_1$2)

19. D_2^4 (P2$_1$2$_1$2$_1$)

20. D_2^5 (C222$_1$)

21. D_2^6 (C222)

22. D_2^7 (F222)

23. D_2^8 (I222)

24. D_2^9 (I2$_1$2$_1$2$_1$)

(7) C_{2v} (mm2), [EC$_2\sigma_v\sigma'_v$]

25. C_{2v}^1 (Pmm2)

26. C_{2v}^2 (Pmc2$_1$)

27. C_{2v}^3 (Pcc2)

28. C_{2v}^4 (Pma2)

29. C_{2v}^5 (Pca2$_1$)

30. C_{2v}^6 (Pnc2)

31. C_{2v}^7 (Pmn2$_1$)

32. C_{2v}^8 (Pba2)

33. C_{2v}^9 (Pna2$_1$)

34. C_{2v}^{10} (Pnn2)

35. C_{2v}^{11} (Cmm2)

36. C_{2v}^{12} (Cmc2$_1$)

37. C_{2v}^{13} (Ccc2)

38. C_{2v}^{14} (Amm2)

39. C_{2v}^{15} (Abm2)

40. C_{2v}^{16} (Ama2)

41. C_{2v}^{17} (Aba2)

42. C_{2v}^{18} (Fmm2)

43. C_{2v}^{19} (Fdd2)

44. C_{2v}^{20} (Imm2)

45. C_{2v}^{21} (Iba2)

46. C_{2v}^{22} (Ima2)

(8) D_{2h}(mmm) = V_h, [EC$_2$C$_2'$C$_2''$ i$\sigma_h\sigma'_v\sigma''_v$]

47. D_{2h}^1 (Pmmm)

48. D_{2h}^2 (Pnnn)

49. D_{2h}^3 (Pccm)

50. D_{2h}^4 (Pban)

51. D_{2h}^5 (Pmma)

52. D_{2h}^6 (Pnna)

53. D_{2h}^7 (Pmna)

54. D_{2h}^8 (Pcca)

55. D_{2h}^9 (Pbma)

56. D_{2h}^{10} (Pccn)

57. D_{2h}^{11} (Pbcm)

58. D_{2h}^{12} (Pnnm)

59. D_{2h}^{13} (Pmmn)

60. D_{2h}^{14} (Pbcn)

61. D_{2h}^{15} (Pbca)

62. D_{2h}^{16} (Pnma)

63. D_{2h}^{17} (Cmcm)

64. D_{2h}^{18} (Cmca)

65. D_{2h}^{19} (Cmmm)

66. D_{2h}^{20} (Cccm)

67. D_{2h}^{21} (Cmma)

68. D_{2h}^{22} (Ccca)

69. D_{2h}^{23} (Fmmm)

(Continued)

TABLE 1.2 (Continued)

Crystalline System (I, II, III, …, VII), Crystal Class ((1), (2), (3),…,(32)) and Space Group (1, 2, 3,…, 230)[11]

70. $D_{2h}^{24}(Fddd)$	85. $C_{4h}^3(P4/n)$	102. $C_{4v}^4(P4_2nm)$
71. $D_{2h}^{25}(Immmm)$	86. $C_{4h}^4(P4_2/n)$	103. $C_{4v}^5(P4cc)$
72. $D_{2h}^{26}(Ibam)$	87. $C_{4h}^5(I4/m)$	104. $C_{4v}^6(P4nc)$
73. $D_{2h}^{27}(Ibca)$	88. $C_{4h}^6(I4_1/a)$	105. $C_{4v}^7(P4_2mc)$
		106. $C_{4v}^8(P4_2bc)$
74. $D_{2h}^{28}(Immma)$	(12) $D_4(422), [E2C_4C_4^22C_2'2C_2'']$	107. $C_{4v}^9(I4mm)$
IV. Tetragonal	89. $D_4^1(P422)$	108. $C_{4v}^{10}(I4cm)$
(9) $C_4(4), [EC_4C_4^3C_2]$	90. $D_4^2(P42_12)$	109. $C_{4v}^{11}(I4_1md)$
75. $C_4^1(P4)$	91. $D_4^3(P4_122)$	110. $C_{4v}^{12}(I4_1cd)$
76. $C_4^2(P4_1)$	92. $D_4^4(P4_12_12)$	(14) $D_{2d}=V_d(\bar{4}2m)$ $[E2S_4S_4^22C_2'2\sigma_d]$
77. $C_4^3(P4_2)$	93. $D_4^5(P4_222)$	111. $D_{2d}^1(P\bar{4}2m)$
78. $C_4^4(P4_3)$	94. $D_4^6(P4_22_12)$	112. $D_{2d}^2(P\bar{4}2c)$
79. $C_4^5(I4)$	95. $D_4^7(P4_322)$	113. $D_{2d}^3(P\bar{4}2_1m)$
80. $C_4^6(I4_1)$	96. $D_4^8(P4_32_12)$	114. $D_{2d}^4(P\bar{4}2_1c)$
(10) $S_4(\bar{4}), [E2S_4C_2]$	97. $D_4^9(I422)$	115. $D_{2d}^5(P\bar{4}m2)$
81. $S_4^1(P\bar{4})$	98. $D_4^{10}(I4_122)$	116. $D_{2d}^6(P\bar{4}c2)$
82. $S_4^2(I\bar{4})$	(13) $C_{4v}(4mm), [E2C_4C_4^22\sigma_v2\sigma_d]$	117. $D_{2d}^7(P\bar{4}b2)$
(11) $C_{4h}(4/m), [E2C_4C_2i2S_4\sigma_h]$	99. $C_{4v}^1(P4mm)$	118. $D_{2d}^8(P\bar{4}n2)$
83. $C_{4h}^1(P4/m)$	100. $C_{4v}^2(P4bm)$	119. $D_{2d}^9(I\bar{4}2m2)$
84. $C_{4h}^2(P4_2/m)$	101. $C_{4v}^3(P4_2cm)$	

120. $D_{2d}^{10}(I\bar{4}c2)$
121. $D_{2d}^{11}(I\bar{4}2m)$
122. $D_{2d}^{12}(I\bar{4}2d)$
(15) $D_{4h}(4/mmm), [E2C_4C_4^22C_2'2C_2''\sigma_h 2\sigma_v2\sigma_d2S_4i]$
123. $D_{4h}^1(P4/mmm)$
124. $D_{4h}^2(P4/mcc)$
125. $D_{4h}^3(P4/nbm)$
126. $D_{4h}^4(P4/nnc)$
127. $D_{4h}^5(P4/mbm)$
128. $D_{4h}^6(P4/mnc)$
129. $D_{4h}^7(P4/nmm)$
130. $D_{4h}^8(P4/ncc)$
131. $D_{4h}^9(P4_2/mmc)$
132. $D_{4h}^{10}(P4_2/mcm)$
133. $D_{4h}^{11}(P4_2/nbc)$
134. $D_{4h}^{12}(P4_2/nnm)$
135. $D_{4h}^{13}(P4_2/mbc)$
136. $D_{4h}^{14}(P4_2/mnm)$
137. $D_{4h}^{15}(P4_2/nmc)$

138. $D_{4h}^{16}(P4_2/ncm)$
139. $D_{4h}^{17}(I4/mmm)$
140. $D_{4h}^{18}(I4/mcm)$
141. $D_{4h}^{19}(I4_1/amd)$
142. $D_{4h}^{20}(I4_1/acd)$

V. Trigonal

(16) $C_3(3)$, $[E2C_3]$
143. $C_3^1(P3)$
144. $C_3^2(P3_1)$
145. $C_3^3(P3_2)$
146. $C_3^4(R3)$

(17) $C_{3i} = S_6(\bar{3})$, $[E2C_3 i3CS_6]$
147. $C_{3i}^1(P\bar{3})$
148. $C_{3i}^2(R\bar{3})$

(18) $D_3(32)$, $[E2C_3 3C_2]$
149. $D_3^1(P312)$
150. $D_3^2(P321)$
151. $D_3^3(P3_112)$
152. $D_3^4(P3_121)$
153. $D_3^5(P3_212)$
154. $D_3^6(P3_221)$
155. $D_3^7(R32)$

(19) $C_{3v}(3m)$, $[E2C_3 3\sigma_v]$
156. $C_{3v}^1(P3m1)$
157. $C_{3v}^2(P31m)$
158. $C_{3v}^3(P3c1)$
159. $C_{3v}^4(P31c)$
160. $C_{3v}^5(R3m)$
161. $C_{3v}^6(R3c)$

(20) $D_{3d}(\bar{3}m)$, $[E2S_6 2C_3 i3C_2 3\sigma_d]$
162. $D_{3d}^1(P\bar{3}1m)$
163. $D_{3d}^2(P\bar{3}1c)$
164. $D_{3d}^3(P\bar{3}m1)$
165. $D_{3d}^4(P\bar{3}c1)$
166. $D_{3d}^5(R\bar{3}m)$
167. $D_{3d}^6(R\bar{3}c)$

VI. Hexagonal

(21) $C_6(6)$, $[E2C_6 2C_3 C_2]$
168. $C_6^1(P6)$
169. $C_6^2(P6_1)$
170. $C_6^3(P6_5)$
171. $C_6^4(P6_2)$
172. $C_6^5(P6_4)$
173. $C_6^6(P6_3)$

(22) $C_{3h}(\bar{6})$, $[E2C_3 \sigma_h 2S_3]$
174. $C_{3h}^1(P\bar{6})$

(23) $C_{6h}(6/m)$, $[E2C_6 2C_3 C_2 i2S_b 2S_3 \sigma_{h}]$
175. $C_{6h}^1(P6/m)$
176. $C_{6h}^2(P6_3/m)$

(24) $D_6(622)$, $[E2C_6 2C_3 C_2 3C_2' 3C_2'']$
177. $D_6^1(P622)$
178. $D_6^2(P6_122)$
179. $D_6^3(P6_522)$
180. $D_6^4(P6_222)$
181. $D_6^5(P6_422)$
182. $D_6^6(P6_322)$

(25) $C_{6v}(6mm)$, $[E2C_6 2C_3 C_2 3\sigma_v 3\sigma_d]$
183. $C_{6v}^1(P6mm)$
184. $C_{6v}^2(P6cc)$
185. $C_{6v}^3(P6_3cm)$
186. $C_{6v}^4(P6_3mc)$

(26) $D_{3h}(\bar{6}m2)$, $[E2C_3 3C_2 \sigma_h 2S_3 3\sigma_v]$
187. $D_{3h}^1(P\bar{6}m2)$
188. $D_{3h}^2(P\bar{6}c2)$
189. $D_{3h}^3(P\bar{6}2m)$
190. $D_{3h}^4(P\bar{6}2c)$

(27) $D_{6h}(6/mmm)$, $\left[\begin{array}{c} E2C_6 2C_3 C_2 3C_2' 3C_2'' \\ i2S_6 2S_3 \sigma_h 3\sigma_v 3\sigma_d \end{array}\right]$
191. $D_{6h}^1(P6/mmm)$
192. $D_{6h}^2(P6/mcc)$
193. $D_{6h}^3(P6_3/mcm)$
194. $D_{6h}^4(P6_3/mmc)$

(Continued)

TABLE 1.2 (Continued)

Crystalline System (I, II, III, …, VII), Crystal Class ((1), (2), (3),…,(32)) and Space Group (1, 2, 3,…, 230)[11]

VII.Cubic

(28) T(23), $[E3C_28C_3]$

195. $T^1(P23)$
196. $T^2(F23)$
197. $T^3(I23)$
198. $T^4(P2_13)$
199. $T^5(I2_13)$

(29) $T_h(m3)$, $[E3C_3;8C_3;i3\sigma 8S_6]$

200. $T_h^1(Pm3)$
201. $T_h^2(Pn3)$
202. $T_h^3(Fm3)$
203. $T_h^4(Fd3)$

204. $T_h^5(Im3)$
205. $T_h^6(Pa3)$
206. $T_h^7(Ia3)$

(30) $O(432)$, $[E8C_33C_26C_2''6C_4]$

207. $O^1(P432)$
208. $O^2(P4_232)$
209. $O^3(F432)$
210. $O^4(F4_132)$

211. $O^5(I432)$
212. $O^6(P4_332)$
213. $O^7(P4_132)$
214. $O^8(I4_132)$

(31) $T_d(\overline{4}3m)$, $[E8C_33C_26\sigma_d6S_4]$

215. $T_d^1(P\overline{4}3m)$
216. $T_d^2(F\overline{4}3m)$
217. $T_d^3(I\overline{4}3m)$
218. $T_d^4(P\overline{4}3n)$
219. $T_d^5(F\overline{4}3c)$
220. $T_d^6(I\overline{4}3d)$

(32) $O_h(m3m)$, $\begin{bmatrix} E8C_33C_26C_26C_4 \\ i8S_63\sigma_h6\sigma_d6S_4 \end{bmatrix}$

221. $O_h^1(Pm3m)$
222. $O_h^2(Pn3n)$
223. $O_h^3(Pm3n)$
224. $O_h^4(Pn3m)$

225. $O_h^5(Fm3m)$
226. $O_h^6(Fm3c)$
227. $O_h^7(Fd3m)$
228. $O_h^8(Fd3c)$
229. $O_h^9(Im3m)$
230. $O_h^{10}(Ia3d)$

1.2 PROPAGATION AND DISPERSION OF LIGHT[11–15]

The propagation of light in a medium, typically in a solid, is described using Maxwell's equations within the medium as shown in the following equations:

$$divD = \nabla \cdot D = \varepsilon \nabla \cdot E = 0 \tag{1.1a}$$

$$divB = \nabla \cdot D = \mu \nabla \cdot H = 0 \tag{1.1b}$$

$$rotE = \nabla \times E = -\mu \dot{H} \tag{1.1c}$$

$$rotH = \nabla \times H = \sigma E + \varepsilon \dot{E} \tag{1.1d}$$

where E and H are the electric and magnetic fields of the light, and D and B are the electric and magnetic flux densities, respectively. ε and μ are the dielectric constant and magnetic permeability of the medium, and σ is the conductivity of the material. By differentiating Equation (1.1d) relating with time t and replacing $\dot{H}$ with that in Equation (1.1c), the next wave equation for the electric field E is derived.

$$\nabla^2 E = \mu(\sigma \dot{E} + \varepsilon \ddot{E}). \tag{1.2}$$

The same type of wave equation is obtained for the magnetic field H by operating the rotation to Equation (1.1d) and replacing $\nabla \times E$ with that in Equation (1.1c).

$$\nabla^2 H = \mu(\sigma \ddot{H} + \dot{\varepsilon} \ddot{H}) \tag{1.3}$$

1.2.1 LIGHT PROPAGATION IN HOMOGENEOUS MEDIUM

In a homogeneous medium, ε and σ do not depend on the direction and become scalars. By considering a plane wave for the propagation, E and H of the light are written in the same form as

$$(E, H) = (E_0, H_0)\exp i(\omega t - k \cdot r + \varphi) \tag{1.4}$$

where ( ) means the quantities in the parentheses take the same expression. k and φ are the wavevector and initial phase of the electromagnetic wave, respectively. The magnitude of wavevector k is given using Equation (1.2) as

$$k = |k| = \omega(\mu \tilde{\varepsilon})^{\frac{1}{2}} = \frac{\omega}{\tilde{\upsilon}}, \tag{1.5}$$

where

$$\tilde{\upsilon} = \frac{1}{(\mu \tilde{\varepsilon})^{\frac{1}{2}}} \tag{1.6}$$

$$\tilde{\varepsilon} = \varepsilon_1 - i\varepsilon_2 = \varepsilon_1 - i\frac{\sigma}{\omega} \tag{1.7}$$

We define the complex refractive index $\tilde{n}$ as

$$\tilde{n} = \frac{c}{\upsilon} = (\mu\tilde{\varepsilon})^{\frac{1}{2}} \tag{1.8}$$

The complex refractive index is also written as follows regarding the dielectric constant:

$$\tilde{n} = n - i\kappa = (\mu\tilde{\varepsilon})^{\frac{1}{2}} = \left[\mu(\varepsilon_1 - i\varepsilon_2)\right]^{\frac{1}{2}} = \left[\mu\left(\varepsilon_1 - i\frac{\sigma}{\omega}\right)\right]^{\frac{1}{2}} \tag{1.9}$$

where the real part n and imaginary part κ of the complex refractive index $\tilde{n}$ are called a *refractive index* and an *extinction coefficient*, respectively. The relation between the real and imaginary parts of $\tilde{n}$ and $\tilde{\varepsilon}$ in Equation (1.9) gives the following:

$$\mu\varepsilon_1 = n^2 - \kappa^2 \tag{1.10}$$

$$\mu\varepsilon_2 = 2n\kappa = \mu\frac{\sigma}{\omega} \tag{1.11}$$

On the contrary, the real and imaginary parts of the refractive index are written as follows using the dielectric constants:

$$2n^2 = \mu\left\{\varepsilon_1 + \left[\varepsilon_1^2 + \varepsilon_2^2\right]^{\frac{1}{2}}\right\} = \mu\left\{\varepsilon + \left[\varepsilon^2 + \left(\frac{\sigma}{\omega}\right)^2\right]^{\frac{1}{2}}\right\} \tag{1.12}$$

$$2\kappa^2 = \mu\left\{-\varepsilon_1 + \left[\varepsilon_1^2 + \varepsilon_2^2\right]^{\frac{1}{2}}\right\} = \mu\left\{-\varepsilon + \left[\varepsilon^2 + \left(\frac{\sigma}{\omega}\right)^2\right]^{\frac{1}{2}}\right\} \tag{1.13}$$

Dielectric constants and refractive indices of the medium can be transformed using these relations.

E and H of the electromagnetic wave traveling through an absorbing medium in z direction is written using Equations (1.4) and (1.7):

$$(E,H) = (E_0, H_0)\exp i\left(\omega t - \frac{\omega}{c}\tilde{n}z + \varphi\right) = (E_0, H_0)\exp i\left(\omega t - \frac{\omega}{c}(n - i\kappa)z + \varphi\right)$$

$$= (E_0, H_0)\exp\left(-\frac{\omega}{c}\kappa z\right)\exp i\left(\omega t - \frac{\omega}{c}nz + \varphi\right). \tag{1.14}$$

The amplitude of the electromagnetic wave damps according to $\exp\left(-\frac{\omega}{c}\kappa z\right)$ during travel of the absorbing medium. The energy absorbed by the medium works to excite the atoms, ions, or electrons depending on the photon energy and also to excite elementary excitations or collective excitations like exciton, polariton, plasmon, phonon, magnon, and so forth in solids. Any physical quantities that interact with the electromagnetic wave have correlations with the optical constants of the medium. By analyzing the optical constants obtained by various optical measurements, we can get the static and dynamic physical information of the medium.

In the optical measurements, typically, using a spectrometer, the intensity of the electromagnetic wave is usually observed after some temporal average depending on the measuring system. The flow of the electromagnetic energy is described by the Poynting vector S, which is expressed by the amplitude using the following relations:

$$S = E \times H = |E||H|k_0 = Ik_0 \tag{1.15}$$

$$I = \frac{|\tilde{n}|}{\mu}|E|^2 \tag{1.16}$$

where k_0 is the unit vector with the direction of S, and I is the intensity of the light. When we characterize the optical properties of materials, transmission spectra are often measured. The damping of the light intensity through the absorbing medium is written as

$$I = I_0 \exp(-\alpha z) \tag{1.17}$$

where α is the absorption coefficient, related with the optical constant κ (extinction coefficient) as

$$\alpha = \frac{4\pi}{\lambda}\kappa. \tag{1.18}$$

When we measure the transmission spectra using a spectrometer, the following coefficients are often used:

Transmissivity:

$$T = \frac{I}{I_0} = e^{-\alpha d} \tag{1.19}$$

Absorptivity:

$$k = \frac{(I_0 - I)}{I_0} = 1 - e^{-\alpha d} = 1 - T \tag{1.20}$$

Absorbance A or Optical Density OD:

$$A \text{ or } OD = -\log_{10} T = -\log_{10}\frac{I}{I_0} = \alpha d \log_{10} e \tag{1.21}$$

As a result, the absorption coefficient α that represents the specific absorption of the material is derived from Equations (1.19) and (1.21) as

$$\alpha = -\frac{1}{d}\ln\frac{I}{I_0} = \frac{1}{d}\ln\frac{1}{T} = \frac{A}{d\log_{10}e} = \frac{OD}{d\log_{10}e} \tag{1.22}$$

In these expressions, the reflectance R at the surfaces of the sample is not considered. Then, the α calculated above gives an almost correct value for the sample with small reflectivity. The transmissivity for the practical materials with parallel boundaries between front and back surfaces is given as follows considering multiple reflections:

$$T = \frac{(1-R)^2(1+(\frac{\kappa}{n})^2)\exp(-\alpha d)}{1-R^2\exp(-2\alpha d)} \tag{1.23}$$

If the absorption is relatively large and $n \gg \kappa$, Equation (1.23) is simplified as

$$T = (1-R)^2\exp(-\alpha d) \tag{1.24}$$

The reflectivity at the boundary between the surrounding medium and sample is derived as

$$R = \frac{(n-n_0)^2+\kappa^2}{(n+n_0)^2+\kappa^2} \tag{1.25}$$

where n and n_0 are the refractive indices of the sample and surrounding medium, respectively. When the surrounding medium is air, Equation (1.25) can be written as follows by assuming the small refractive index of the air as 1:

$$R = \frac{(n-1)^2+\kappa^2}{(n+1)^2+\kappa^2} \tag{1.26}$$

If the material is transparent like a glass in the visible region, the equation becomes more simple, like Equation (1.27):

$$R = \frac{(n-1)^2}{(n+1)^2} \tag{1.27}$$

1.2.2 Light Propagation in Inhomogeneous Medium

In inhomogeneous crystals, both the dielectric constant and conductivity are tensors and depend on the direction of the crystal lattice. Optical axis or optic axis is the transmitting direction where two fundamental waves have the same refractive indices. The crystals are classified as isotropic, uniaxial, and biaxial depending on the number of the optical axis.

1.2.2.1 Isotropic Crystal

The refractive index ellipsoid of the crystal belonging to the cubic crystalline system takes the spherical form as shown in Figure 1.5. The number of the optical axis is

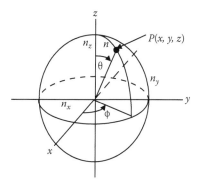

FIGURE 1.5 Refractive index sphere of the crystal belonging to the isotropic crystalline system.

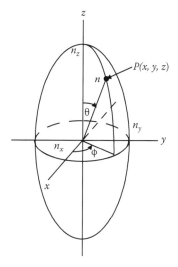

FIGURE 1.6 Refractive index ellipsoid of the crystal belonging to the uniaxial crystalline system.

infinite in the system, and the refractive index does not depend on the transmitting direction of the light. The crystals that have infinite number of axes are called *isotropic* or *anaxial* crystals.

1.2.2.2 Uniaxial Crystal

The refractive index ellipsoid of the crystals belonging to the tetragonal, hexagonal, and trigonal crystalline systems is the rotatory ellipsoid around the axis parallel to the rotatory axis (c axis, z axis). The plane perpendicular to the axis of the ellipsoid is isotropic, and the z axis is the only optical axis. The cross-sectional plane perpendicular to the wave vector k with the angle θ to the optical axis has the short axis along the x axis and long axis in the y-z plane as shown in Figure 1.6. The long axis of the cross-sectional elliptic plane is given in the following equation:

$$\frac{1}{n^2(\theta)} = \frac{\cos^2\theta}{n^2(O)} + \frac{\sin^2\theta}{n^2(E)}. \tag{1.28}$$

Here $n(O)$ and $n(E)$ are the refractive indices $n_x (= n_y)$, and n_z of the refractive index ellipsoid, respectively. The light traveling along k direction is divided into D_1 and D_2 as the fundamental waves. D_1 corresponds to the $n(O)$ and D_2 corresponds to $n(\theta)$. $n(\theta)$ depends on the angle θ. In the fundamental wave D_1, electric displacement D_1 is parallel to the electric field E_1 and has the refractive index $n(O)$ irrespective of the direction of k. This light is the same with the light traveling through the isotropic medium and is called *ordinary ray* (*O*-ray). On the other hand, in the fundamental wave D_2, D_2 is not parallel to the E_2 and $n(\theta)$ depends on the angle θ. Such light behaves as if not to obey the law of refraction and is called *extraordinary ray* (E-ray). The crystal with $n(E) > n(O)$ is called a *positive uniaxial crystal* and the crystal with $n(E) < n(O)$ is called a *negative uniaxial crystal*. The sign (positive or negative) of the crystal is not uniquely determined because the refractive index changes depending on the wavelength. As typical examples, the crystals of quartz and ice are positive and calcite is negative in the visible region.

1.2.2.3 Biaxial Crystal

The refractive index ellipsoid of the crystals belonging to the orthorhombic, monoclinic, and triclinic crystalline systems has three different lengths of axis and two cross-sectional planes of a circle as shown in Figure 1.7.[11, 16] The system has only two optical axes P_1 and P_2 and is called a *biaxial crystal*. The two optical axes are in the plane formed by the longest axis and the shortest axis of the ellipsoid. The angles of the optical axis P_1 and P_2 relative to the z axis are determined using the following equation:

$$\tan\theta_1 = \pm\left[\left(\frac{1}{n_y^2} - \frac{1}{n_x^2}\right) \bigg/ \left(\frac{1}{n_x^2} - \frac{1}{n_z^2}\right)\right]^{\frac{1}{2}}. \tag{1.29}$$

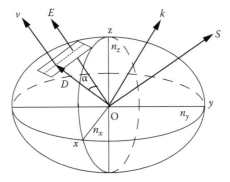

FIGURE 1.7 Refractive index ellipsoid of the crystal belonging to the biaxial crystalline system. (See also K. Kudo: *Fundamentals of Optical Properties of Solids*, Ohmsha (1977) (in Japanese).) E is determined from the tangential plane at the crossing point of D to the ellipsoid. (See also S. Flugge: *Handbuch der Physik*, Band xxv/1, Kristalloptik 1955–1984.)

The equation means optical axes P_1 and P_2 have the same angle and opposite direction to the z axis. The relation between the direction of electric field E and displacement D is determined using the diagram on the ellipsoid for the arbitrary direction as shown in Figure 1.7.

1.2.3　Dispersion of Light

1.2.3.1　Polarization and Dielectric Constant

If the electric field E_0 is applied on the crystal, the local field E' is induced in the crystal. The local field E' induces the polarization P as the result of the displacement of charged particles from the equilibrium positions.

The local electric field induced in the crystal is given as

$$E' = E + \gamma_0 P = E_0 - \gamma P + \gamma_0 P = E_0 + \gamma_1 P, \tag{1.30}$$

$$\gamma_1 = \gamma + \gamma_0, \gamma_0 = 1/3\varepsilon_0 \tag{1.31}$$

$$E = E_0 - \gamma P \tag{1.32}$$

In the equations, E is the macroscopic electric field in the crystal, and γ takes different values depending on the shape of the sample.

1.2.3.2　Dipole Dispersion

1.2.3.2.1　Dispersion by Transverse Wave

The equation of motion for the jth ion or electron with the mass m_j is given as

$$m_j \ddot{u}_j + m_j \Gamma_j \dot{u}_j + \sum_{i=1}^{n} k_{ji} u_i = e_j^* E' = e_j^* (A_t E_0 + A_t \lambda_0 N \sum_{i=1}^{n} e_i^* u_i). \tag{1.33}$$

By solving the equation, the dielectric constant by the dipole ε_d is derived as follows:

$$\tilde{\varepsilon}_d = 4\pi A_t N e^* \frac{u_1 - u_2}{E_0} = \frac{S_1 \omega_1^2}{\omega_1^2 - \omega^2 + i\Gamma_1 \omega}. \tag{1.34}$$

If the k types of oscillators are independent of each other, the dielectric constant for any crystal is written as

$$\tilde{\varepsilon} = \frac{c^2 k^2}{\omega^2} = \varepsilon_1 - i\varepsilon_2 = (n - i\kappa)^2 = 1 + 4\pi A_t N_a \alpha_a + \sum_{j=1}^{k} \frac{S_j \omega_j^2}{\omega_j^2 - \omega^2 + i\Gamma_j \omega} \tag{1.35}$$

where

$$S_j = \frac{4\pi A_t^2 N_j e^{*2}}{\mu_j \omega_j^2}. \tag{1.36}$$

The oscillator strength S_j is described more accurately using quantum mechanics:

$$S_j = 4\pi \left(\frac{A_t^2 N_j e^{*2} f_j}{\mu_j \omega_j^2} \right) = 4\pi \rho_j,$$

(1.37)

where

$$f_j = \left(\frac{2m\hbar\omega}{\hbar^2} \right) |x_{ji}|^2$$

(1.38)

x_{ji} is the transition probability, and $4\pi\rho_j$ is called *oscillator strength*.

The dispersion relation (Equation 1.35) is expressed using various physical variables as

$$\tilde{\varepsilon} = \varepsilon_1 - i\varepsilon_2 = (n - i\kappa)^2 = \varepsilon_\infty + \sum_{j=1}^{k} \frac{S_j \omega_j^2}{\omega_j^2 - \omega^2 + i\Gamma_j\omega} = \varepsilon_\infty + \sum_{j=1}^{k} \frac{S_j(\nu_j)\nu_j^2}{\nu_j^2 - \nu^2 + i\nu_{\gamma j}\nu}$$

$$= \varepsilon_\infty + \sum_{j=1}^{k} \frac{S_j(\lambda_j)}{1 - (\lambda/\lambda_j)^2 + i(\lambda_j^2/\lambda_{\gamma j}\lambda)}$$

(1.39)

where

$$\varepsilon_\infty = 1 + 4\pi A_t N_a \alpha_a \equiv \varepsilon(\omega \to \infty)$$

(1.40)

The real and imaginary parts of the dielectric function are written as

$$\varepsilon_1 = n^2 - \kappa^2 = \varepsilon_\infty + \sum_{j=1}^{k} \frac{S_j \omega_j^2 (\omega_j^2 - \omega^2)}{(\omega_j^2 - \omega^2)^2 + \Gamma_j^2 \omega^2}$$

(1.41)

$$\varepsilon_2 = 2n\kappa = \sum_{j=1}^{k} \frac{S_j \omega_j^2 \Gamma_j \omega}{(\omega_j^2 - \omega^2)^2 + \Gamma_j^2 \omega^2}$$

(1.42)

The typical behaviors of ε_1 and ε_2 are illustrated in Figure 1.8 for three oscillators $k = 3$. The optical constants n and κ are calculated as

$$2n^2 = \varepsilon_1 + (\varepsilon_1^2 + \varepsilon_2^2)^{\frac{1}{2}}$$

(1.43)

$$2\kappa^2 = -\varepsilon_1 + (\varepsilon_1^2 + \varepsilon_2^2)^{\frac{1}{2}}$$

(1.44)

The observed optical spectrums like transmittance and reflectance are reproduced by using the optical constants.

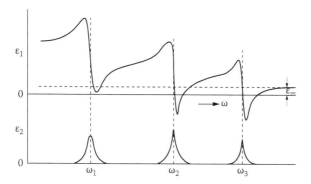

FIGURE 1.8 Dispersion curves of the real ε_1 and imaginary ε_2 parts of the dielectric constant ε. ε_∞ is the real part of the dielectric constant at a higher frequency limit. In this diagram, three independent oscillators are shown.

1.2.3.3 Free Carrier Dispersion

The equation of motion for free electrons under the transverse electric field E of the light is expressed as

$$\ddot{u} + \Gamma \dot{u} = \frac{e^*}{m^*} E \tag{1.45}$$

By considering the response of electrons to the electric field E with the frequency of ω, the following dispersion relations are derived:

$$\tilde{\varepsilon} = \varepsilon_1 - i\varepsilon_2 = (n - i\kappa)^2 = \varepsilon_\infty - \frac{\omega_p^2}{\omega^2 - i\Gamma\omega}, \tag{1.46}$$

where ω_p is a plasma frequency, written as

$$\omega_p^2 = \frac{4\pi N e^{*2}}{m^*}. \tag{1.47}$$

The real and imaginary parts of the dielectric constant are written as

$$\varepsilon_1 = n^2 - \kappa^2 = \varepsilon_\infty - \frac{\omega_p^2}{\omega^2 + \Gamma^2}, \tag{1.48}$$

$$\varepsilon_2 = 2n\kappa = \frac{\omega_p^2 \Gamma}{\omega(\omega^2 + \Gamma^2)}. \tag{1.49}$$

The dispersion due to free carriers is shown in Figure 1.9.[11]

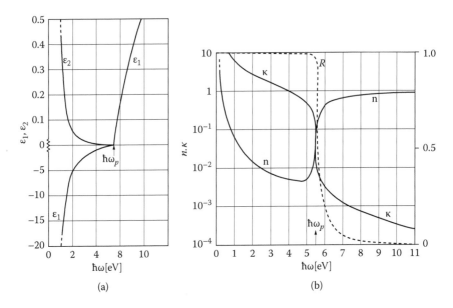

FIGURE 1.9 Dispersion curves of free electrons. (See also K. Kudo: *Fundamentals of Optical Properties of Solids*, Ohmsha (1977) (in Japanese).) (a) Real part ε_1 and imaginary part ε_2 of dielectric constant and (b) real part n and imaginary part κ of refractive index.

1.2.3.4 Coexistence of Dipoles and Free Electrons

In degenerated semiconductors and metals, dispersions of both the lattice vibration and the free carriers coexist. The dielectric constant for such a crystal is written as

$$\varepsilon(v) = \varepsilon_\infty + \sum_j \frac{S_j v_{tj}^2}{v_{tj}^2 - v^2 + i\gamma_j v} - \frac{\varepsilon_\infty \omega_p^2}{v^2 - i\Gamma v} \tag{1.50}$$

The dispersions due to free carriers and lattice vibrations of nondoped In_2O_3 single crystal are shown in Figure 1.10.[17] Far-infrared reflection spectra of nondoped In_2O_3 single crystal is fitted by using the dielectric function expressed by Equation (1.50).

1.3 OPTICAL ABSORPTIONS BY ELECTRONS IN SOLIDS[11,18]

The major optical absorptions are induced by electrons and phonons. Optical absorptions in ultraviolet, visible, and near-infrared regions are mainly caused by electronic transitions. The fundamental theory for the transition and the selection rule are derived using the Hamiltonian considering the interactions of electrons and photons, and the absorption coefficients for such transitions are calculated. Some typical examples for interband transitions are shown for metals, semiconductors, and ionic crystals.

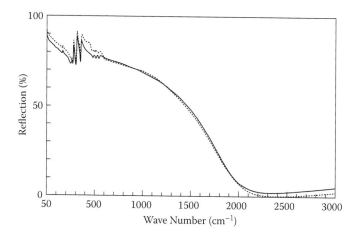

FIGURE 1.10 Infrared reflectivity spectrum of nondoped In_2O_3 single crystal (dotted curve), and reflectivity calculated using classical oscillators and free carrier dispersion (solid curve). (See also M. Wakaki and Y. Kanai: *Jpn. J. Appl. Phys.*, 25, 502–503 (1986).)

1.3.1 TRANSITION PROBABILITY AND TRANSITION RATE

Schrodinger's equation is generally written as follows for the Hamiltonian of the system with the steady state H_0, eigenfunction $\varphi_n(r)$, and eigenvalue E_n:

$$H_0\varphi_n(r)=E_n\varphi_n(r) \tag{1.51}$$

The Hamiltonian $H=H_0+H'(t)$ is considered with an external stimulus $H'(t)$, where $H'(t)$ is smaller enough than H_0 that it is possible to consider as a perturbation. The wavefunction for the case is derived by solving the next time dependent equation:

$$i\hbar\frac{\partial}{\partial t}\Psi(r,t)=(H_0+H')\Psi(r,t) \tag{1.52}$$

We cannot obtain rigorous solutions and try to derive itinerant solutions by using the next relation:

$$\Psi(r,t)=\sum_n a_n(t)\varphi_n(r)e^{-iE_nt/\hbar} \tag{1.53}$$

By substituting this equation in Equation (1.52) and utilizing Equation (1.51), the next relation is given:

$$i\hbar\sum\frac{da_n(t)}{dt}\varphi_n(r)e^{-iE_nt/\hbar}=a_n(t)H'\varphi_n(r)e^{-iE_nt/\hbar} \tag{1.54}$$

By multiplying $\varphi_m^*(r)e^{iE_mt/\hbar}$ from the left-hand side and integrating in space, the next relations are derived utilizing the orthogonality and normalization of $\varphi(r)$:

$$i\hbar\frac{da_m(t)}{dt} = \sum_n a_n(t)H'_{mn}e^{i(E_m-E_n)t/\hbar},$$

$$H_{mn} \equiv \int \varphi_m^*(r)H'\varphi_n(r)dr,$$

(1.55)

where H_{mn} shows the matrix element of the perturbation.

We assume the system occupies the ground state with the energy of E_0 at $t = 0$— that is,

$$a_n = \delta_{n0}(t = 0)$$

(1.56)

Equation (1.55) is changed to the next form within the first-order perturbation:

$$i\hbar\frac{da_m}{dt} = H'_{m0}e^{i(E_m-E_0)t/\hbar} = H'_{m0}e^{i\omega_{m0}t},$$

$$\omega_{m0} = \frac{(E_m-E_0)}{\hbar} = \omega_m - \omega_0.$$

(1.57)

As a result, the transition moment under the first-order perturbation is given as

$$a_m(t) = -\frac{i}{\hbar}\int_{-\infty}^{t} H'_{m0}e^{i\omega_{m0}t}\,dt \quad (m \neq 0).$$

(1.58)

If the perturbation is independent of time, that is,

$$H'_{m0} = 0 \quad at \; t < 0,$$

$$H'_{m0} = const \; at \; t > 0,$$

(1.59)

Equation (1.58) is described as follows for $t > 0$:

$$a_m(t) = -\frac{H'_{m0}}{\hbar}\frac{e^{i\omega_{m0}t}-1}{\omega_{m0}} \quad (m \neq 0)$$

(1.60)

The probability for the system found in the state (m) at t is given as

$$|a_m(t)|^2 = \frac{4|H'_{m0}|^2\sin^2(\frac{1}{2}\omega_{m0}t)}{\hbar^2\omega_{m0}^2} \quad (m \neq 0)$$

(1.61)

If the perturbation changes periodically in time,

$$H' = H'_0\cos(\omega t)$$

(1.62)

Equation (1.58) becomes

$$a_m(t) = \frac{X_{m0}}{\hbar}\left[\frac{1-e^{i(\omega_{m0}+\omega)t}}{\omega_{m0}+\omega} + \frac{1-e^{i(\omega_{m0}-\omega)t}}{\omega_{m0}-\omega}\right] \quad (1.63)$$

where X_{m0} is a transition matrix written as

$$X_{m0} = \frac{1}{2}\int \varphi_m^* H_0' \varphi_0 dr \quad (1.64)$$

The transition from the ground state is given by the next condition:

$$E_m - E_0 \simeq \hbar\omega > 0,$$

by which the first term in Equation (1.63) becomes negligibly small compared with the second term, and the probability found at the state (m) is given as

$$|a_m(t)|^2 = \frac{4|X_{m0}|^2 \sin^2\frac{1}{2}(\omega_{m0}-\omega)t}{\hbar^2(\omega_{m0}-\omega)^2} \quad (m \neq 0) \quad (1.65)$$

For the actual system with the density of state

$$\rho(E_m)dE_m = \rho(E_m)\hbar d\omega_{m0} \quad (1.66)$$

the probability to find at one of these states is written as follows using Equation (1.65):

$$|A_m(t)|^2 = \int |a_m|^2 \rho(E_m)dE_m = \frac{4}{\hbar}\int |X_{m0}|^2 \rho(E_m)\frac{\sin^2\frac{1}{2}(\omega_{m0}-\omega)t}{(\omega_{m0}-\omega)^2}d\omega_{m0} \quad (1.67)$$

The integral over $-\infty\sim+\infty$ for the equation is approximated as follows, because the value around $\omega_{m0}\simeq\omega$ is dominant in the integral, and $|X_{m0}|^2$ and $\rho(E_m)$ are nearly constant for such a narrow frequency region:

$$|A_m(t)|^2 = \frac{2\pi}{\hbar}|X_{m0}|^2 \rho(E_m)t = \frac{2\pi t}{\hbar^2}|X_{m0}|^2 \delta(E_{m0}-\hbar\omega). \quad (1.68)$$

The transition rate from the state (0) to (m) is calculated as

$$W_{mo} = \frac{d}{dt}|A_m(t)|^2 = \frac{2\pi}{\hbar^2}|X_{m0}|^2 \delta(E_{m0}-\hbar\omega) \quad (1.69)$$

This relation is well cited as Fermi's golden rule.

1.3.2 Interband and Intraband Transitions

Typically in semiconductor materials, electrons occupy the valence band, and the upper conduction band is empty. Electrons are excited to upper empty levels by the energy E. Interband transition or intraband transition occurs when the upper level belongs to a different band or the same band.

1.3.2.1 Direct Interband Transition

The transition matrix element given in Equation (1.64) for the electron to transit from the valence band (wavevector k') to conduction band (wavevector k'') induced by the light with the wavevector k_l and frequency ω_l at lth plane wave is given as follows:

$$X_{m0} = X_{cv} = \int \varphi_c^* H' \varphi_v dr$$

$$= \frac{i\hbar e}{2mcN} \int u_c^*(r,k'')\left[\pi_l \cdot \nabla u_v(r,k') + i(\pi_l \cdot k')u_v(r,k')\right]\exp i\left[(k' - k'' + k_l)\cdot r\right]dr \tag{1.70}$$

where the integral is calculated over a unit cell as the amplitudes of the Bloch function u_c and u_v are periodic along a unit cell. The perturbation H' is written as follows for the number of N unit cells using one electron approximation:

$$H' = \frac{e}{2mc}\sum_i^N (p_i \cdot A + A \cdot p_i) = -\frac{e}{mc}A \cdot p = \frac{ie\hbar}{mc}A \cdot \nabla \tag{1.71}$$

Vector potential A is expressed as follows:

$$A = \frac{1}{2}A_0\pi_l\left[\hat{a}^\dagger \exp(ik_l \cdot r) + \hat{a}\exp(-ik_l \cdot r)\right] \tag{1.72}$$

where $\hat{a}^\dagger$ and $\hat{a}$ are the creation and annihilation operators, respectively. π_l is the polarization of the lth plane wave in the volume V, and A_0 is defined as follows:

$$A_0 \equiv \left(\frac{8\pi\hbar c^2}{eV\omega_l}\right)^{\frac{1}{2}}$$

The position vector r is the summation of a lattice vector R_n and a position vector r' within the unit cell.

$$r = R_n + r'$$

Equation (1.70) is calculated as follows:

$$X_{cv} = \frac{i\hbar e}{2mcN} A_0 \int u_c^*(r,k'')\left[\pi_l \cdot \nabla u_v(r,k') + i(\pi_l \cdot k')u_v(r,k')\right]$$

$$\times \exp i\left[(k' - k'' + k_l) \cdot (R_n + r')\right] d(R_n + r')$$

$$= \frac{i\hbar e}{2mcN} A_0 \int u_c^*(r,k'')\left[\pi_l \cdot \nabla u_v(r,k') + i(\pi_l \cdot k')u_v(r,k')\right] dr \qquad (1.73)$$

$$\times \sum_n \exp i\left[(k' - k'' + k_l) \cdot R_n\right]$$

The nonzero condition for the equation is written as

$$k'' - k' = k_l \quad \text{or} \quad \hbar k'' - \hbar k' = \hbar k_l \qquad (1.74)$$

The second relation means the conservation of momentum. The momentum of light is negligibly smaller than that of the electron, and Equation (1.74) is written as

$$k' = k'' = k \qquad (1.75)$$

The transition of an electron by the light occurs vertically in k space. Such a transition is called a direct interband transition.

Equation (1.70) is written as follows using Equation (1.74):

$$X_{m0} = X_{cv} = \frac{i\hbar e}{2mc} A_0 \int u_c^*(r,k'')\left[\pi_l \cdot \nabla u_v(r,k') + i(\pi_l \cdot k')u_v(r,k')\right] dr$$

$$= \frac{i\hbar e}{2mc} A_0 \left[\int u_c^*(r,k'')\pi_l \cdot \nabla u_v(r,k') dr + i \int u_c^*(r,k'')(\pi_l \cdot k')u_v(r,k') dr\right] \qquad (1.76)$$

In the calculation, the Σ term of Equation (1.73) gives the number of cells N within the crystal.

The selection rules are derived depending on the relative amount between the first and second terms of Equation (1.76).

1.3.2.1.1 Direct Allowed Transition

The first term is dominant compared with the second term in this case. By introducing the matrix element of the momentum operator p_{cv}, the transition rate is written as

$$X_{m0} = X_{cv} = -\frac{e}{2mc} A_0 \pi_l \cdot p_{cv} \qquad (1.77)$$

where the matrix element p_{cv} is defined as

$$p_{m0} = p_{cv} = -i\hbar \int_{cell} u_c^*(r,k') \cdot \nabla u_v(r,k') dr \qquad (1.78)$$

The transition rate is obtained by using Equation (1.69).

$$W_{mo} = W_{cv} = \frac{\pi e^2}{2m^2\hbar^2 c^2} A_0^2 |\pi_1 \cdot p_{cv}|^2 \delta(E_{cv} - \hbar\omega). \tag{1.79}$$

The total transition rate is obtained by integrating over allowed k state in the k space considering the density of state $1/4\pi^2$:

$$W_{cv}^0 = \frac{\pi e^2}{2m^2\hbar^2 c^2} A_0^2 \int \frac{d^3 k}{4\pi^2} |\pi_1 \cdot p_{cv}|^2 \delta(E_{cv} - \hbar\omega). \tag{1.80}$$

The absorption coefficient α is the specific quantity derived from the optical measurements. The relation between the absorption coefficient and the transition rate is given as follows:

$$\alpha = \frac{\hbar\omega_{cv} W_{cv}^0}{u} \tag{1.81}$$

where u is the energy density of an incident light given as

$$u = \frac{\varepsilon E^2}{4\pi} = \frac{\varepsilon \omega^2}{8\pi} A_0^2. \tag{1.82}$$

As a result, the absorption coefficient for the direct allowed transition is given as

$$\alpha_d = \frac{8\pi\hbar}{\varepsilon\omega} A_0^{-2} W_{cv}^0 = \frac{e^2}{\pi\varepsilon m^2 c^2 \hbar\omega} \int d^3 k |\pi_1 \cdot p_{cv}|^2 \delta(E_{cv} - \hbar\omega). \tag{1.83}$$

The relation is approximated as follows by considering parabolic bands for valence and conduction bands:

$$\alpha_d = \frac{2e^2(2m_r)^{\frac{3}{2}}}{\varepsilon m^2 c^2 \hbar^4 \omega} |\pi_1 \cdot p_{cv}|^2 (\hbar\omega - E_g)^{\frac{1}{2}}, \tag{1.84}$$

where E_g is the band gap energy at $k = 0$ and m_r is the reduced mass defined as

$$\frac{1}{m_r} = \frac{1}{m_e} + \frac{1}{m_h}$$

If the band has the spherical symmetry, the relation is changed by using a square mean momentum operator defined as

$$<|\pi_1 \cdot p_{cv}|^2> = <p_{cv}^2(k_x)> + <p_{cv}^2(k_y)> + <p_{cv}^2(k_z)> = <3 p_{cv}^2>. \tag{1.85}$$

As a result, the absorption coefficient near the absorption edge is expressed as[2]

$$\alpha_d = \frac{A_1}{\hbar\omega}(\hbar\omega - E_g)^{\frac{1}{2}},\tag{1.86}$$

where A_1 is defined as

$$A_1 \equiv \frac{6e^2(2m_r)^{\frac{3}{2}} <p_{cv}^2>}{\varepsilon m^2 c^2 \hbar^2}\tag{1.87}$$

From Equation (1.86), the next relation is derived:

$$(\alpha_d \hbar\omega)^2 = A_1^2(\hbar\omega - E_g)\tag{1.88}$$

The band gap energy E_g is obtained from the linear relation between $(\alpha_d \hbar\omega)^2$ and $\hbar\omega$ as the crossing point to the $\hbar\omega$ axis. Many materials obey the relation. But the departure from Equation (1.86) is observed for many materials in the small value of $\hbar\omega - E_g$, which originates from the exciton absorption.

1.3.2.1.2 Direct Forbidden Transition

When the first term of Equation (1.76) is negligibly small, the transition occurs through the second term:

$$X_{m0} = X_{cv} = -\frac{\hbar e}{2mc} A_0(\pi_1 \cdot k') \int_{cell} u_c^*(r,k'')u_v(r,k')dr\tag{1.89}$$

In this case, $X_{m0}=0$ is given due to the orthogonality between $u_c^*(r, k'')$ and $u_v(r, k')$, if $k'' = k'$ holds strictly, and the slight difference between k'' and k' is assumed, which gives the next relation:

$$f' = \left| \int_{cell} u_c^*(r,k'')u_v(r,k')dr \right|^2 < 1\tag{1.90}$$

The transition rate is derived by averaging the polarization components over x, y, and z directions:

$$W_{mo} = W_{cv}' = \frac{\pi e^2}{12m^2 c^2} A_0^2 k'^2 f' \delta(E_{cv} - \hbar\omega).\tag{1.91}$$

The total transition rate is obtained for parabolic bands by integrating over the allowed k state in the k space considering the density of state $1/4\pi^2$:

$$W_{cv}'^0 = \frac{\pi e^2 (2m_r)^{\frac{5}{2}} f'}{24\pi m^2 \hbar^2 c^2} A_0^2(\hbar\omega - E_g)^{\frac{3}{2}}.\tag{1.92}$$

The absorption coefficient for the direct forbidden transition is expressed near the absorption edge as

$$\alpha_d' = \frac{A_2}{\hbar\omega}(\hbar\omega - E_g)^{\frac{3}{2}}, \tag{1.93}$$

where A_2 is defined as

$$A_2 \equiv \frac{e^2\omega_{cv}(2m_r)^{\frac{5}{2}}f'}{3\pi m^2\hbar^4c^2\omega^2}. \tag{1.94}$$

From Equation (1.93), the next relation to determine the band gap energy E_g is also derived:

$$(\alpha_d'\hbar\omega)^{\frac{2}{3}} = A_2^{\frac{2}{3}}(\hbar\omega - E_g). \tag{1.95}$$

In Equation (1.94), f' takes the small value and the absorption coefficient is very small compared with that of the direct allowed transition.

1.3.2.2   Indirect Interband Transition

The absorption coefficient in Ge exhibits a weak absorption in the lower energy side than the absorption edge of the direct transition. The absorption cannot be explained by the theory using the direct transition but can be explained by considering the interband transition between different k assisted by the lattice vibration for the band shown in Figure 1.11.[11] The second-order perturbation theory is used to obtain the transition rate between the valence band $k = 0$ and conduction band $k = k_c$.

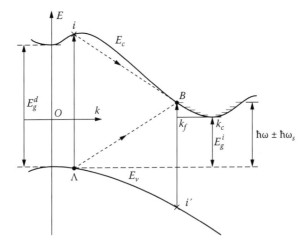

FIGURE 1.11   Band structure to show the indirect transitions, where i and i' indicate the intermediate states of the indirect transitions. (See also K. Kudo: *Fundamentals of Optical Properties of Solids*, Ohmsha (1977) (in Japanese).)

As a perturbation term H', the system composed of photon $\hbar\omega$ and phonon $\hbar\omega_s$ is considered as

$$H' = \frac{1}{2}A_1(r)(e^{i\omega t} + e^{-i\omega t}) + \frac{1}{2}A_2(r)(e^{i\omega_s t} + e^{-i\omega_s t}) \tag{1.96}$$

The next relation is obtained for the transition moment by using the same procedure to derive Equation (1.63).

$$
\begin{aligned}
a_n^{(1)}(t) = & -\frac{H_{n0}^{(1)}}{\hbar}\left[\frac{e^{i(\omega_{n0}+\omega)t}-1}{\omega_{n0}+\omega}+\frac{e^{i(\omega_{n0}-\omega)t}-1}{\omega_{n0}-\omega}\right] \\
& -\frac{H_{n0}^{(2)}}{\hbar}\left[\frac{e^{i(\omega_{n0}+\omega_s)t}-1}{\omega_{n0}+\omega_s}+\frac{e^{i(\omega_{n0}-\omega_s)t}-1}{\omega_{n0}-\omega_s}\right]
\end{aligned}
\tag{1.97}
$$

Similar to Equation (1.69), the next relation is obtained for the transition rate:

$$W_{m0} = i\hbar\frac{da_m^{(2)}(t)}{dt} = \sum_{n\neq0}a_n^{(1)}\begin{cases}H_{mn}^{(1)}\left[e^{i(\omega_{mn}+\omega)t}+e^{i(\omega_{mn}-\omega)t}\right]\\+H_{mn}^{(2)}\left[e^{i(\omega_{mn}+\omega_s)t}+e^{i(\omega_{mn}-\omega_s)t}\right]\end{cases} \tag{1.98}$$

where matrix elements $H_{mn}^{(1)}$ and $H_{mn}^{(2)}$ are defined as

$$H_{mn}^{(1)} = \frac{1}{2}\int\varphi_m^*A_1\varphi_n dr \tag{1.99}$$

$$H_{mn}^{(2)} = \frac{1}{2}\int\varphi_m^*A_2\varphi_n dr \tag{1.100}$$

By putting Equation (1.97) into Equation (1.98), the equation with the denominator having various combinations of ω, ω_{n0}, ω_{mn}, ω_s is obtained. Only the term with the denominator $\omega_{m0}-\omega\pm\omega_s$ is dominant for the energy bands shown in Figure 1.11.

1.3.2.2.1 Indirect Allowed Transition

It is assumed the wavefunction takes $k = k_0$ at the top of the conduction band for the starting state and $k' = k_f$ at the bottom of the conduction band for the final state. Next, matrix elements are selected.

(a) $H_{cc}^{(1)}(k,k') = \frac{1}{2}\int\varphi_c^*(k,r)A_1\varphi_c(k',r)dr$

(b) $H_{cv}^{(1)}(k,k') = \frac{1}{2}\int\varphi_c^*(k,r)A_1\varphi_v(k',r)dr$

(c) $H_{vv}^{(1)}(k,k') = \frac{1}{2}\int\varphi_v^*(k,r)A_1\varphi_v(k',r)dr$

(d) $H_{cc}^{(2)}(\boldsymbol{k},\boldsymbol{k}') = \dfrac{1}{2}\displaystyle\int \varphi_c^*(\boldsymbol{k}',\boldsymbol{r})A_2\varphi_c(\boldsymbol{k},\boldsymbol{r})d\boldsymbol{r}$

(e) $H_{cv}^{(2)}(\boldsymbol{k},\boldsymbol{k}') = \dfrac{1}{2}\displaystyle\int \varphi_c^*(\boldsymbol{k}',\boldsymbol{r})A_2\varphi_v(\boldsymbol{k},\boldsymbol{r})d\boldsymbol{r}$

(f) $H_{vv}^{(2)}(\boldsymbol{k},\boldsymbol{k}') = \dfrac{1}{2}\displaystyle\int \varphi_v^*(\boldsymbol{k}',\boldsymbol{r})A_2\varphi_v(\boldsymbol{k},\boldsymbol{r})d\boldsymbol{r}$ (1.101)

The (a) and (c) become 0 neglecting the momentum of photon, and (e) means the interband transition induced by a phonon that leads also to 0. Parts (d) and (f) are the intraband transitions induced by a phonon within a conduction band and valence band, respectively. The vertical transition with $\boldsymbol{k}=\boldsymbol{k}'$ is allowed for (b). As a result, $H_{cv}^{(1)}(\boldsymbol{k},\boldsymbol{k}')$ and $H_{vv}^{(2)}(\boldsymbol{k},\boldsymbol{k}')$, $H_{cc}^{(2)}(\boldsymbol{k},\boldsymbol{k}')$ are not zero.

Similar to the direct transition, the transition rate is given by

$$|a_f^2|^2 = \dfrac{|H_0^p|^2 |H_c^s|^2}{\hbar^4(\omega_{i0}-\omega)^2}A_c^v + \dfrac{|H_f^p|^2 |H_v^s|^2}{\hbar^4(\omega_{i0}-\omega)^2}A_c^v \qquad (1.102)$$

where $H_{cv}^{(1)}(k_0,k_0) \equiv H_0^p$, $H_{cv}^{(1)}(k_f,k_f) \equiv H_f^p$, $H_{cc}^{(2)}(k_f,k_0) \equiv H_c^s$, $H_{vv}^{(2)}(k_0,k_f) \equiv H_v^s$, and A_c^v are defined as

$$A_c^v \equiv \dfrac{4\sin^2\dfrac{t}{2}(\omega_{f0}-\omega-\omega_s)}{(\omega_{f0}-\omega-\omega_s)^2} + \dfrac{4\sin^2\dfrac{t}{2}(\omega_{f0}-\omega+\omega_s)}{(\omega_{f0}-\omega+\omega_s)^2} \qquad (1.103)$$

The condition that the denominator of the first term becomes 0 corresponds to the transition $A{\rightarrow}i{\rightarrow}B$, and the condition for the second term becomes 0 corresponds to the transition $A \rightarrow i' \rightarrow B$ shown in Figure 1.11. In each term, the term with $\omega_{f0}-\omega-\omega_s$ is the transition accompanying phonon absorption and $\omega_{f0}-\omega+\omega_s$ is the phonon emission.

The transition rate is derived by differentiating Equation (1.102) as follows:

$$W_{f0} = \dfrac{2\pi t |H_0^p|^2 |H_c^s|^2}{\hbar^4(\omega_{ic}-\omega)^2}\left[\delta(E_{f0}-\hbar\omega_s-\hbar\omega)+\delta(E_{f0}+\hbar\omega_s-\hbar\omega)\right]$$

$$+\dfrac{2\pi t |H_f^p|^2 |H_v^s|^2}{\hbar^4(\omega_{i0}-\omega)^2}\left[\delta(E_{f0}-\hbar\omega_s-\hbar\omega)+\delta(E_{f0}+\hbar\omega_s-\hbar\omega)\right]$$

(1.104)

The final state excited by the photon energy $\hbar\omega$ is dominated by the conservation of both energy and momentum as given by the delta function.

The absorption coefficient for indirect transition with phonon absorption is written as

$$\alpha_{ca} \propto \dfrac{|\bar{H}_0^p|^2}{(\omega_{n0}-\omega)^2}\cdot\dfrac{(\hbar\omega+\hbar\omega_s-E_g)^2}{e^{\frac{\hbar\omega_s}{kT}}-1} \qquad (\hbar\omega \geq E_g-\hbar\omega_s) \qquad (1.105)$$

$$\alpha_{ca} = 0 \qquad (\hbar\omega < E_g - \hbar\omega_s)$$

where $\left|\bar{H}_0^p\right|$ means an average over the polarization of photon.

In the equation, the probability of phonon absorption is considered proportional to the phonon density, written as

$$\left|H_c^s\right|^2 \propto N_p = (e^{\frac{\hbar\omega_s}{kT}} - 1)^{-1}$$

The absorption coefficient for indirect transition with phonon emission is written as

$$\alpha_{ce} \propto \frac{\left|\bar{H}_0^p\right|^2}{(\omega_{i0} - \omega)^2} \cdot \frac{(\hbar\omega - \hbar\omega_s - E_g)^2}{1 - e^{\frac{\hbar\omega_s}{kT}}} \qquad (\hbar\omega \geq E_g + \hbar\omega_s) \qquad (1.106)$$

$$\alpha_{ce} = 0 \qquad (\hbar\omega < E_g + \hbar\omega_s).$$

As a result, the absorption coefficient for the indirect transition allowed with vertical transition is summarized as the summation of Equation (1.105) and (1.106).

$$\alpha = A\frac{(\hbar\omega + \hbar\omega_s - E_g)^2}{e^{\frac{\hbar\omega_s}{kT}} - 1} + B\frac{(\hbar\omega - \hbar\omega_s - E_g)^2}{1 - e^{-\frac{\hbar\omega_s}{kT}}}. \qquad (1.107)$$

The square law of the equation is modified to the 3/2 power law by considering the effect of exciton.

1.3.2.2.2 *Indirect Forbidden Transition*

If the direct transition is forbidden, it is necessary to consider the second term of Equation (1.76). The absorption coefficient is given as follows:

$$\alpha' \propto \int_0^{\hbar\omega \pm \hbar\omega_s - E_g} (\hbar\omega \pm \hbar\omega_s - E_g)^{\frac{3}{2}} E^{\frac{1}{2}} dE \propto (\hbar\omega \pm \hbar\omega_s - E_g)^3 \qquad (1.108)$$

The absorption coefficient for the indirect forbidden transition is derived by applying the phonon density function:

$$\alpha' = A'\frac{(\hbar\omega + \hbar\omega_s - E_g)^3}{e^{\frac{\hbar\omega_s}{kT}} - 1} + B'\frac{(\hbar\omega - \hbar\omega_s - E_g)^3}{1 - e^{-\frac{\hbar\omega_s}{kT}}} \qquad (1.109)$$

The third power law of the equation is also modified to the 5/2 power law by considering the effect of exciton.

1.3.3 INTERBAND TRANSITIONS FOR VARIOUS MATERIALS

1.3.3.1 Metals

In metals, the bottom of the conduction band is occupied by the electron, and the intraband transition is dominant, which gives the high reflectance over a wide range

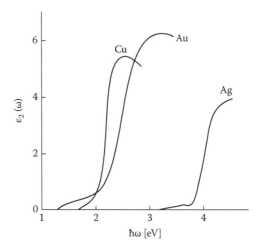

FIGURE 1.12 Imaginary part of dielectric constant ε_2 for Cu, Ag, and Au due to interband transitions. (See also K. Kudo: *Fundamentals of Optical Properties of Solids*, Ohmsha (1977) (in Japanese); M. L. Theye: *Phys. Rev.*, B2, 3060 (1970).)

of spectra from the visible to the far-infrared region. As a result, it is quite difficult to observe the interband transition in these regions due to the quite small penetration depth of the incident light. In alkaline metals like Na, K, and Cs, it is considered to obey to direct transitions because the band structures of these metals are simple and take minimum or maximum at Γ point. But it is difficult to compare the observed optical conductivities with the theoretical ones because these metals are quite weak to humidity.

In noble metals like Cu, Ag, and Au, reflectance becomes lower in the low energy region than the plasma frequency, and it is possible to observe transmission spectra in such a region as shown in Figure 1.12.[11,20]

The absorption edges determined assuming the direct transitions are 2.08eV (Cu), 3.87eV (Ag), and 2.45eV (Au).[21–24] $\varepsilon_2(\hbar\omega)^2$ versus $\hbar\omega$ is shown in Figure 1.13.[11] The observed spectrum is well fitted by Equation (1.86) for the direct allowed transition at higher than the photon energy of 2.5eV. The departure from the theory is considered due to the imperfection of the crystal lattice or the impurities.

The absorption spectrum for the trivalent metal Al was also reported.[25] The spectrum was measured at low temperature 4K to reduce the free carriers. The absorption peaks were observed at about 0.5eV, and 1.5eV, which were assumed due to the direct transitions, but clear assignment has not been done owing to the overlap of several transitions.

1.3.3.2 Semiconductors

In considering the interband transitions in semiconductors, several features must be considered. The interactions between conduction and valence bands are strong due to the small band gap of the semiconductors. The light and heavy valence bands are degenerated at the Γ point ($k = 0$), and a spin split band is located below by the energy Δ. The heavy valence band becomes nonspherical when the minimum of the conduction

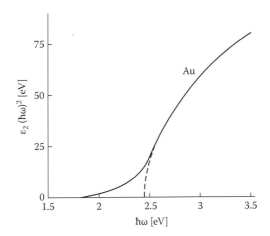

FIGURE 1.13 Photon energy $\hbar\omega$ dependence of $\varepsilon_2(\hbar\omega)^2$ in Au. (See also K. Kudo: *Fundamentals of Optical Properties of Solids*, Ohmsha (1977) (in Japanese).)

band does not locate at the Γ point. As a result, transitions from the three valence bands are necessary to be considered for interband transitions in semiconductors.

Among III-V compound semiconductors, InSb and GaAs are well known as the typical materials to show the direct transition. The absorption spectrum of InSb at 5K is shown in Figure 1.14.[11,26] The spectrum is well fitted by Equation (1.86) for the direct allowed transition at the band gap energy Eg = 0.24eV, which corresponds to the transition at the Γ point.

GaAs has a similar band structure to InSb and shows the direct interband transition at the Γ point. The minimum band gap locates at Γ point, and transition $\Gamma_{15} \rightarrow \Gamma_1$ must show the direct transition at the photon energies 1.55eV (77K) and 1.53eV (300K). The result of the analysis for the absorption spectrum is shown in Figure 1.15,[11,27] which showed agreement only at the higher energy side and could not reproduce the observed value near the absorption edge. The reason for the disagreement is supposed to come from the imperfection of the crystal to cause a p- or n-type degenerated state.

Among III-V compound semiconductors, GaP is the typical material to show indirect transition. $\alpha\hbar\omega$ is plotted for the photon energy in Figure 1.16.[11,28] The observed values are reproduced by the relation Equations (1.106) and (1.107) for the indirect transition.

The detailed absorption spectra at different temperatures are shown in Figure 1.17.[11,29] $\sqrt{\alpha}$ deflects at the energy where the absorption or emission of phonon begins, and phonon energies are derived by analyzing the spectrum as shown in Figure 1.17. At low temperature 1.6K, only the phonon emission is possible, and two-phonon emission (LO + TA) is also observed.

1.3.3.3 Ionic Crystals

The fundamental absorptions in ionic crystals are very strong and the transmission measurements are possible only in quite thin samples. The optical density of

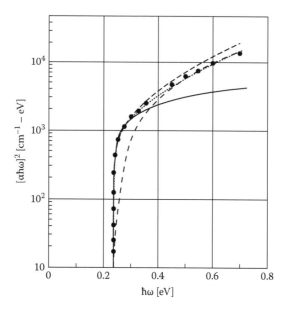

FIGURE 1.14 Photon energy $\hbar\omega$ dependence of the absorption coefficient $\alpha\hbar\omega$ of InSb near the absorption edge. (See also K. Kudo: *Fundamentals of Optical Properties of Solids*, Ohmsha (1977) (in Japanese); E. J. Johnson: *Semiconductors and Semimetals*, 3, p. 170, Academic Press (1967).)

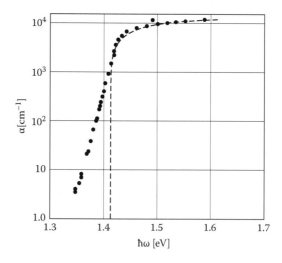

FIGURE 1.15 Photon energy $\hbar\omega$ dependence of the absorption coefficient α of GaAs. (See also K. Kudo: *Fundamentals of Optical Properties of Solids*, Ohmsha (1977) (in Japanese); T. S. Moss and T. D. F. Hawkins: *Infrared Phys.*, 1, 111 (1962).)

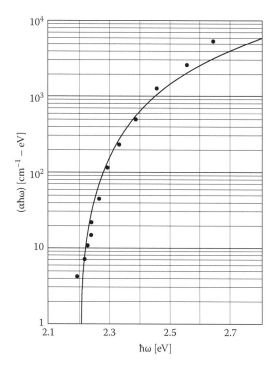

FIGURE 1.16 Photon energy $\hbar\omega$ dependence of absorption coefficient $\alpha\hbar\omega$ in GaP. (See also K. Kudo: *Fundamentals of Optical Properties of Solids*, Ohmsha (1977) (in Japanese); W. G. Spitzer et al.: *J. Phys. Chem. Solids*, 11, 339 (1959).)

a NaCl film deposited on a LiF crystal substrate is shown in Figure 1.18.[11,30] The strong absorption band around 8eV is the exciton absorption typical for the ionic crystals, and other absorptions correspond to interband absorptions. The absorption spectrum increases rapidly with $\hbar\omega$ after subtracting an exciton absorption band from the observed spectrum, and this corresponds to the direct transition with the band gap of 8.4eV. From the band structure of NaCl, the fundamental absorption corresponds to the transition $\Gamma_{15} \rightarrow \Gamma_1$, and the absorption around 10.3eV is assigned to the transition $L_3' \rightarrow L_2'$. Other ionic crystals with a NaCl structure like KCl and KBr have strong exciton absorptions near the absorption edge, and it is difficult to assign that the fundamental absorption is either direct or indirect. But consistent assignment is possible by analyzing exciton absorptions and higher interband transitions.

1.3.4 INTRABAND TRANSITIONS

1.3.4.1 *p*-Type Semiconductors

In many semiconductors, the light valence band (E^l) and heavy valence band (E^h) are degenerated at the Γ point ($k=0$), and a spin split band (E^s) is located below by

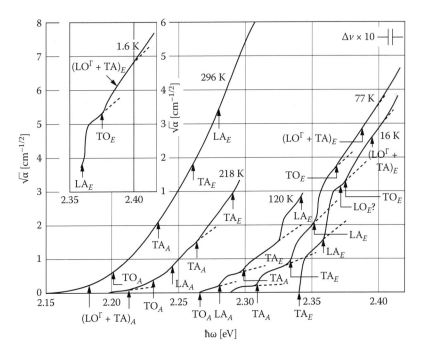

FIGURE 1.17 Photon energy $\hbar\omega$ dependences of absorption coefficient $\sqrt{\alpha}$ in GaP at different temperatures. (See also K. Kudo: *Fundamentals of Optical Properties of Solids*, Ohmsha (1977) (in Japanese); P. J. Dean and D. G. Thomas: *Phys. Rev.*, 150, 690 (1966).)

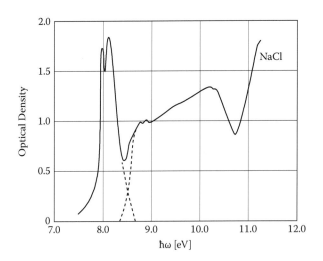

FIGURE 1.18 Photon energy $\hbar\omega$ dependence of optical density of NaCl near the absorption edge. (See also K. Kudo: *Fundamentals of Optical Properties of Solids*, Ohmsha (1977) (in Japanese); K. Teegarden and G. Baldini: *Phys. Rev.*, 155, 896 (1967).)

the energy Δ as shown in Figure 1.19. If the top of the valence bands are occupied with holes and the Fermi level E_F is located within the valence bands as shown in the figure, three kinds of intravalence band transitions are considered.[31,32] In a *p*-type Ge, narrow absorption bands at 0.3 and 0.4eV, and a wide band at a lower energy are observed as shown in Figure 1.20.[11,33] These bands correspond to the transitions shown in Figure 1.19 as transitions a, b, and c, respectively.

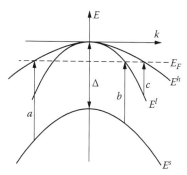

FIGURE 1.19   Schematic band model showing the intraband transitions a, b, and c.

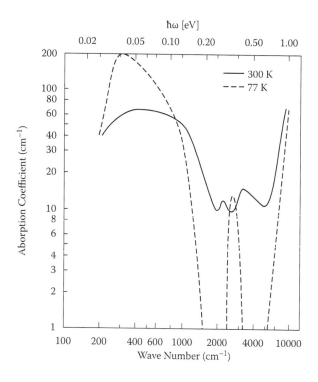

FIGURE 1.20   Absorption spectra of *p*-type Ge due to the intraband transitions. (See also K. Kudo: *Fundamentals of Optical Properties of Solids*, Ohmsha (1977) (in Japanese); W. Kaizer, R. J. Collins, and H. Y. Fan: *Phys. Rev.*, 91, 1380 (1953).)

1.3.4.2   *n*-Type Semiconductors

In *n*-type semiconductors like GaP[28] and AlSb[34], intersubband direct transitions are reported. The absorption spectra of *n*-type GaP are shown in Figure 1.21[11,28] for different temperatures.[28] The transition giving such absorption was assigned as <100> direction $X_1 \rightarrow X_3$ transition.[35] But a careful treatment is required for such absorptions, because the absorption band at 0.25eV in *n*-type GaSb[36] and the weak absorption around fundamental absorption edge in InP[37] were assigned as intervalley transitions from <000> valley to <100> valley.

1.3.4.3   Free Carrier Absorption

In metals, electrons occupy a part of the conduction band, and in intrinsic semiconductors, electrons occupy a conduction band by thermal excitation even at room temperature. *n*-Type semiconductors also have electrons in a conduction band excited from donor levels. In general, metals show strong absorptions from the ultraviolet (UV) to the infrared (IR) region and semiconductors exhibit continuous absorptions that increase with the wavelength. Pure free electrons cannot absorb light. Such absorption is induced by the transition within a conduction band. But such indirect

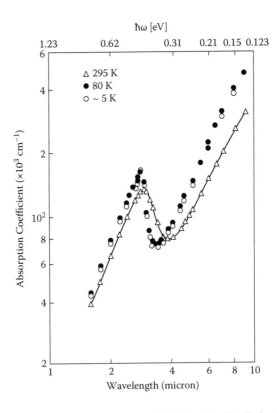

FIGURE 1.21   Absorption spectra of *n*-type GaP. (See also K. Kudo: *Fundamentals of Optical Properties of Solids*, Ohmsha (1977) (in Japanese); W. G. Spitzer et al.: *J. Phys. Chem. Solids*, 11, 339 (1959).)

transition cannot occur unless electrons gain momentum through scattering effects by some media because light has negligibly small momentum compared with that of electrons. Phonons generated by the thermal vibration of the lattice, disorders of crystal potential due to the imperfections of the crystal, ionized impurity centers, and electron-electron scatterings are considered as the main media for the scattering. The feature of the free carrier absorptions is the monotonous wavelength dependence increasing with λ^a ($a = 1.5 \sim 3.5$). These free carrier absorptions are understood by deriving the transition probability using the second-order perturbation with a corresponding appropriate scattering medium. The absorption coefficient for the transverse electric field of the light is given as

$$\alpha_f = \frac{\omega}{cn}\varepsilon_2 = \frac{\omega_p^2 \tau}{cn(\omega^2 \tau^2 + 1)} \tag{1.110}$$

The plasma frequency in metals is $\omega_p \approx 10^{15}\,\text{sec}^{-1}$ and the mean free time is $\tau \approx 10^{-14}\,\text{sec}$, which leads to the following Drude theory around the plasma frequency using $\omega\tau \approx \omega_p\tau \gg 1$:

$$\alpha \simeq \frac{\omega_p^2}{4\pi^2 c^2 n\tau}\lambda^2 = \frac{Ne^{*2}}{\pi c^3 nm^{*}\tau}\lambda^2 \tag{1.111}$$

where N, e^*, and m^* are carrier density, effective charge, and effective mass of the electron, respectively.

But the power a is reported to take different values for compound semiconductors like GaAs ($a = 3$), InAs (3), GaAb (3.5), InSb (2), InP (2.5), and GaP (1.8).[38] The treatments considering the scattering by acoustic phonons give $a = 1.5$,[39] by optical phonons $a = 2.5$,[40] and by ionized impurities $a = 3 \sim 3.5$.[41] In the practical case, these scatterings occur simultaneously, and the absorption coefficient is written as follows depending upon the weight of scattering coefficients A, B, and C for respective scattering:

$$\alpha = A\lambda^{1.5} + B\lambda^{2.5} + C\lambda^{3.5} \tag{1.112}$$

1.3.5   Exciton Absorption

Electrons and holes in the crystal moving around the crystal lattice after interband transition are excited by the light exceeding the band gap energy. These electrons and holes attract each other by the Coulomb force to form electron-hole pairs. Such electon-hole pairs are in a kind of excited state and are called excitons, which can move around in the crystal through the interatomic interaction. The exciton is similar to the hydrogen atom for an excited electron and gives the discreet absorption lines below the band gap photon energy Eg. The attraction force between electron and hole increases the probability to become close and enhance the continuous absorption to a level higher than that calculated by the band model.

As a result, a line spectrum is observed at the photon energy lower than Eg and a continuous absorption at the photon energy higher than Eg.

The absorption intensity by the exciton is derived by using the Hamiltonian including Coulomb interaction and solving the Schrodinger equation.[42,43]

1.3.5.1 Excitons in the Direct Transition

The Hamiltonian for the electron and hole is given as

$$H = -\frac{\hbar^2}{2m_e^*}\nabla_e^2 - \alpha_f\frac{\hbar^2}{2m_h^*}\nabla_h^2 - \frac{e^2}{\varepsilon|r_e - r_h|} \tag{1.113}$$

The energy eigenvalue is given as

$$E_n(k) = E_g - \frac{\mu}{2\hbar^2 n^2}(\frac{e^2}{\varepsilon})^2 + \frac{\hbar^2 k^2}{2M^*} \qquad (n = 1,2,3\cdots) \tag{1.114}$$

where n is a principal quantum number, μ a reduced mass, and M^* a total effective mass of exciton given as $M^* = m_e^* + m_h^*$. The transition probability is calculated like the interband transition, and two terms are given similar to Equation (1.76). In direct allowed transition, hydrogen-like energies are obtained at $k = 0$.

$$E_n(0) = h v_n = E_g - \frac{1}{n^2}E_{ex} \qquad (E_{ex} \equiv \frac{\mu\, e^4}{2\hbar^2\varepsilon^2}) \tag{1.115}$$

The absorption intensity at v_n is given by Elliot[42]:

$$|F(0)|^2 = \frac{1}{\pi a^{*3} n^3} \tag{1.116}$$

where a^* is an effective Bohr radius. The absorption intensity and the separation between the next absorption line becomes smaller as the number n increases, and the photon energy for the line becomes E_g for $n\rightarrow\infty$. The coupling energy of exciton E_{ex}, which corresponds to the ground state ($n = 1$), is written as follows using the ionization energy of hydrogen atom: $\dfrac{me^4}{2m\varepsilon^2} = 13.55eV,$

$$E_{ex} = \frac{13.55\mu}{2m\varepsilon^2} \tag{1.117}$$

This value takes about 0.01eV for typical semiconductors Si and Ge. The Bohr radius is given as

$$a^* = \frac{0.052n^2\varepsilon m}{\mu} = \frac{0.104n^2\varepsilon m}{m^*}[nm] \tag{1.118}$$

The absorption intensity decreases according to n^{-3} and becomes subcontinuous for larger n. Elliot called such subcontinuous spectra an overlapping continuum and calculated the absorption coefficient.

$$\alpha_{d,a}^{(0,c)} = \frac{8\pi\omega|z|^2\,\varepsilon}{n_0 a^* c} \qquad (\hbar\omega < E_g) \qquad\qquad (1.119)$$

where n_0 and $|z|^2$ are the refractive index and the transition probability, respectively.

The real continuous spectrum is given for $\hbar\omega > E_g$, and the absorption coefficient is given as[44]

$$\alpha_{d,a}^{(t,c)} = \frac{2\pi\omega e^2|z|^2\,\gamma\exp(\pi\gamma)(2\mu)^{\frac{3}{2}}}{n_0 c\hbar^3\,\sinh(\pi\gamma)}\,(\hbar\omega - E_g)^{\frac{1}{2}} \qquad\qquad (1.120)$$

where γ is defined as

$$\gamma \equiv \left[\frac{E_{ex}}{(\hbar\omega - E_g)}\right]^{-\frac{1}{2}}$$

Equation (1.120) is approximated as follows in the photon energy $\hbar\omega - E_g \gg E_{ex}$:

$$\alpha_{d,a}^{(t,c)} = \alpha_{d,a}^{(0,c)} + \frac{1}{2}\alpha_{d,a}^{(b,t)} \qquad\qquad (1.121)$$

where $\alpha_{d,a}^{(b,t)}$ is the absorption coefficient for the interband allowed transition without exciton (Equation 1.84).

In the case of forbidden transition where the first term in Equation (1.76) is negligibly small, the absorption intensity for the excitons is calculated as[44]

$$|F(0)|^2 \simeq \frac{n^2-1}{\pi^5\varepsilon^5} \qquad\qquad (1.122)$$

As seen in the equation, the absorption line $n = 1$ is forbidden and the intensities for $n \geq 2$ are extremely weak.

The absorption coefficient for the subcontinuous region is given as

$$\alpha_{d,a}^{(0,c)} = \frac{8\pi\omega\varepsilon|z|^4}{3n_0 ca_{ex}^4}\left(1 + \frac{\hbar\omega - E_g}{E_{ex}}\right) \qquad (\hbar\omega < E_g) \qquad\qquad (1.123)$$

and for the real continuous region

$$\alpha_{d,f}^{(t,c)} = \frac{2\pi\omega e^2|z|^2\,(1+\gamma^3)\gamma\exp(\pi\gamma)(2\mu)^{\frac{5}{2}}}{3n_0 c\hbar^5\,\sinh(\pi\gamma)}\,(\hbar\omega - E_g)^{\frac{3}{2}} \qquad\qquad (1.124)$$

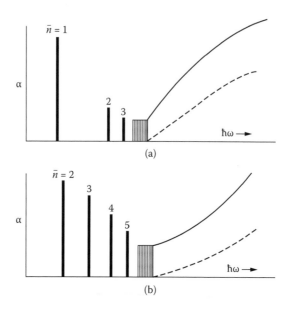

FIGURE 1.22 Absorption spectra of excitons for allowed transition (a) and forbidden transition (b). Dashed lines show the absorption spectra without exciton effect. (See also K. Kudo: *Fundamentals of Optical Properties of Solids*, Ohmsha (1977) (in Japanese).)

The schematic diagrams of the absorption spectra for the allowed and forbidden transitions are shown in Figures 1.22a and 1.22b,[11] respectively. The shape and width of the exciton are discussed in the literature.[45,46]

1.3.5.2 Excitons in the Indirect Transition

The indirect transition by excitons is caused through the absorption and emission of phonons the same as the indirect interband transition. The final state of the transition is the excitonic levels attached to the bottom of the conduction band with $k = k_c$, and the selection rule is relaxed by the effect of phonons, which allows the transitions to any wavevector k as shown in Figure 1.23.[11] As a result, the absorption spectrum becomes not like a line but like a band.

The absorption coefficient near the absorption edge is given as follows for allowed and forbidden transitions.[43,47,48]

1.3.5.2.1 Allowed Transition

The absorption coefficient for the excitonic band is given as

$$\alpha_{i,a}^{(b)} \propto \frac{A}{e^{\frac{\hbar\omega_p}{kT}} - 1}(\hbar\omega - E_-)^{\frac{1}{2}} + \frac{A}{1 - e^{-\frac{\hbar\omega_p}{kT}}}(\hbar\omega - E_+)^{\frac{1}{2}} \quad (1.125)$$

and for the true band is given as

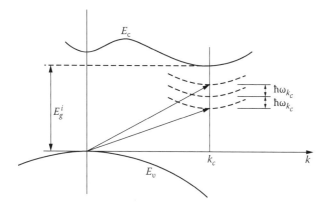

FIGURE 1.23 Indirect exciton transitions. (See also K. Kudo: *Fundamentals of Optical Properties of Solids*, Ohmsha (1977) (in Japanese).)

$$\alpha_{i,a}^{(u)} \propto \frac{A'}{e^{\frac{\hbar\omega_p}{kT}} - 1}(\hbar\omega - E_-)^{\frac{3}{2}} + \frac{A'}{1 - e^{-\frac{\hbar\omega_p}{kT}}}(\hbar\omega - E_+)^{\frac{3}{2}} \qquad (1.126)$$

Absorption occurs at the following energy:

$$E_{\pm} = E_g - E_{ex} \pm \hbar\omega_p \qquad (1.127)$$

where $\hbar\omega_p$ is a phonon energy, and $\pm$ corresponds to the phonon emission and absorption.

1.3.5.2.2 Forbidden Transition
The absorption coefficient for the excitonic band is given as

$$\alpha_{i,f}^{(b)} \propto (\hbar\omega - E_{\pm})^{\frac{3}{2}} \qquad (1.128)$$

and for the true band is given as

$$\alpha_{i,f}^{(u)} \propto (\hbar\omega - E_{\pm})^{\frac{5}{2}} \qquad (1.129)$$

1.3.6 OTHER ELECTRONIC ABSORPTIONS

As the other typical absorption processes, the absorption by impurities is considered. Detailed studies have been carried out for the assignment of the energy levels of the impurities in typical semiconductor materials like Ge, Si, and GaAs,[49–53] which are important for electronic and optical applications.

The effects of the external fields like the magnetic and electric fields on the electronic optical absorptions are summarized in Chapter 2.

1.4   OPTICAL ABSORPTION BY LATTICE VIBRATION[54-62]

1.4.1   Dipole Moments

The crystal is composed with the bonding between valence electrons of constituent atoms, and each atom has some kind of ionic character to generate an electric field around it. When the electromagnetic wave is incident on the crystal, some parts of it interact with the crystal field and are absorbed. The Hamiltonian for the N charges are written as follows:

$$H = H_0 + H' + H'' \tag{1.130}$$

$$H_0 = \sum_{i=1}^{N} \left[\frac{1}{2m} p_i^2 + V_i(r_i) \right] + \frac{1}{2} \sum_{ij}' \frac{e^2}{|r_i - r_j|} + I \tag{1.131}$$

where I is the interaction between ions and the first perturbation part H' is expressed as

$$H' = \frac{e}{2mc} \sum_{i=1}^{N} (p_i \cdot A + A \cdot p_i) + \sum_{i=1}^{N} e\phi_i \tag{1.132}$$

The second perturbation H'' corresponds to the energy of the light written as

$$H'' = \sum_{i=1}^{N} \frac{e^2}{2mc^2} A^2 \tag{1.133}$$

The vector potential of an incident light A is usually considered small, and this term is neglected in the following. The perturbed Hamiltonian for the electrons and ions are expressed as the summation of both contributions and written as

$$H' = -\frac{e}{mc} \sum_{i=1}^{N} p_i \cdot A(r_i) + \frac{G_s}{M_s c} \sum_{i=1}^{N} p(ls) A(x(ls)) \tag{1.134}$$

In the equation, $x(ls) = R(ls) + u(ls)$, where $R(ls)$ shows the equilibrium position of the ion and $u(ls)$ is the displacement of the ion as shown in Figure 1.24.

The operators for the displacement of the phonon are expressed as

$$u^\pm = \sum_{jq} \left(\frac{\hbar}{2NM_s\omega_j} \right)^{\frac{1}{2}} e_{ja}(sq) \tilde{b}_{ja}^\pm e^{\pm iq \cdot R(l)} \tag{1.135}$$

The momentum operators of the phonon are given as

$$p^\pm = \pm i M_s u^\pm \omega_j \tag{1.136}$$

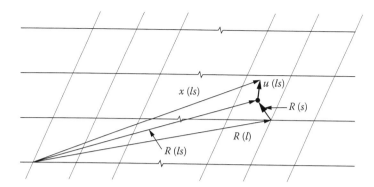

FIGURE 1.24 Position vectors and displacement vector of the atom used to explain the lattice vibration. R(*l*): position vector for the unit cell *l* relative to the original point of the crystal lattice; R(*ls*): position vector for the ion *s* in the unit cell *l* relative to the original point of the crystal lattice; R(*s*): position vector for the ion *s* relative to the original point of the unit cell *l*; u(*ls*): displacement vector of the ion *s* in the unit cell *l*; and x(*ls*): position vector relative to the original point of the crystal lattice for the ion *s* at its displaced position.

The vector potentials are

$$A^{\pm} = \sum_{\tau k} \left(\frac{2\pi\hbar c^2}{\omega_k Q} \right)^{\frac{1}{2}} \pi_{ik} \tilde{a}_{\tau}^{\pm} e^{\pm k \cdot R(l)} \tag{1.137}$$

In these expressions, the superscripts + and − correspond to the creation and annihilation operators of phonon and photon. In the equation, N and Q are the number and the volume of the unit cell, respectively. $\mathbf{e}_{ja}$ is the eigenvector of the vibration, and $\tilde{b}^-$ and $\tilde{b}^+$ are the annihilation and creation operators of phonon, respectively. $\tilde{a}^-$ and $\tilde{a}^+$ are the annihilation and creation operators of photon, and π_{ik} is two unit vectors ($i = 1, 2$) perpendicular to the wavevector $\mathbf{k}$. The phase term for the electron is also approximated as the equilibrium position of the ion. As a result, $\mathbf{u}$ and $\mathbf{p}$ for the electron are expressed by replacing M_s in Equation (1.135) and (1.136) to m. The perturbed Hamiltonian of Equation (1.134) is rewritten to the following equation:

$$H' = i\frac{\omega_j}{c} \left[\sum_i e(u_e^- - u_e^+) - \sum_{ls} G_s(u_l^- - u_l^+) \right] A \tag{1.138}$$

where suffix e of u means for electrons and I for ions. eu_e and $G_s u_l$ correspond to the dipole moments of the valence electron M_e and the ion M_s induced by the vibration of the ion. By adding some constant, Equation (1.138) is transformed to the next equation:

$$H' = -\frac{i\omega_j}{c} \sum_{ls} \left\{ \left[\sum_i er_i^+(ls) - G_s x^+(ls) \right] - \left[\sum_i er_i^-(ls) - G_s x^-(ls) \right] \right\} \cdot A = -\mathbf{MA} \tag{1.139}$$

After H' is obtained, the transition rate W from the initial state $|i\rangle$ to the final state $|f\rangle$ is given as follows:

$$W = \frac{2\pi}{\hbar^2}|\langle f|H'|i\rangle|^2 \, \delta\big[(E_f - E_i) - \omega_k\big] \tag{1.140}$$

By using adiabatic approximation that electrons respond to any movements of ions (Born–Oppenheimer), the next relations are given:

$$|i\rangle = \psi_i(r,x) = \varphi_{am}(x)\varphi_a(r,x) \tag{1.141a}$$

$$|f\rangle = \psi_f(r,x) = \varphi_{bn}(x)\varphi_b(r,x) \tag{1.141b}$$

The dipole moment of the crystal M can be expanded by $u(ls)$ as

$$M = M_0 + M_1 + M_2 + \ldots \tag{1.142a}$$

Each component of the moment is given as

$$M_1 = \sum_{ls} B(ls) \cdot u(ls) \tag{1.142b}$$

$$M_2 = \sum_{ls,l's'} C(ls,l's') \cdot u(ls)u(l's') \tag{1.142c}$$

The transition matrix elements X_{if} between the initial state $|i\rangle$ and the final state $|f\rangle$ are written as follows:

$$X_{if} = \frac{i\omega_j}{c}\int \varphi_f^*(x)MA\varphi_i(x)dx = X_{if}^{(0)} + X_{if}^{(1)} + X_{if}^{(2)} + \cdots \tag{1.143a}$$

where each element is given as

$$X_{if}^{(0)} = \frac{i\omega_j}{c}\langle f|M_0(A^- + A^+)|i\rangle \tag{1.143b}$$

$$X_{if}^{(1)} = \frac{i\omega_j}{c}\langle f|M_1A|i\rangle = \frac{i\omega_j}{c}\left\langle f\left|\sum B(ls)(u^-(ls)+u^+(ls))(A^- + A^+)\right|i\right\rangle \tag{1.143c}$$

$$X_{if}^{(2)} = \frac{i\omega_j}{c}\langle f|M_2A|i\rangle$$

$$= \frac{i\omega_j}{c}\left\langle f\left|\sum C(ls,l's')(u^-(ls)+u^+(ls))(u^-(l's')+u^+(l's'))(A^- + A^+)\right|i\right\rangle \tag{1.143d}$$

The elements other than $X_{if}^{(0)}$ relate with the optical absorptions and if $X_{if}^{(1)}$, $X_{if}^{(2)}$ take nonzero values, these transition processes are called one-phonon process, two-phonon process. The processes over two-phonon processes are called multiphonon processes.

1.4.2 Selection Rule and Absorption Coefficient in Lattice Absorption

1.4.2.1 Selection Rule and Absorption Coefficient for One Phonon

The matrix element for one-phonon process is written as follows considering all phonon branches:

$$X_{if}^{(1)} = i\hbar \langle f | \sum_{ls,jq,\tau k} \left(\frac{\pi \omega}{N\Omega M_s \omega_k} \right)^{\frac{1}{2}} B(s)(e_{ja}(sq) \cdot \pi_{\tau k})(b_{jq}^+ \hat{a}^- e^{i(q-K) \cdot R(l)}$$

$$+ b_{jq}^- \hat{a}^+ e^{-i(q-K) \cdot R(l)}) | i \rangle$$

(1.144)

To have a nonzero value for the equation, the condition for the effective charges $B(s) \neq 0$ and the polarization $e_{ja}(sq) \cdot \pi_{\tau k} \neq 0$ should be satisfied simultaneously. The condition for the effective charges is checked by the charge neutrality of the unit cell given as

$$\sum_s B(ls) = \sum_s B(s) = 0$$

(1.145)

For example, homopolar crystals consisting of one kind of atom are considered. $B(1) = 0$ clearly holds when an atom is included in a unit cell. $B(1) + B(2) = 0$ holds when two atoms are included in a unit cell. But $B(s) = 0$ holds, because $B(1) = B(2)$ due to no difference among the atoms. As a result, in almost all mono-atom crystals like Si and Ge with the diamond structure, typically no transition occurs due to the first-order process.

$B(1)=B(2)\neq0$ holds clearly when two species of ions are included in a unit cell, and it is possible for the optical absorption due to one phonon to occur.

To satisfy $e_{ja}(sq) \cdot \pi_{\tau k} \neq 0$, the electric vector of the incident light $\pi_{\tau k}$ must have the parallel component to the eigenvector of the vibration $e_{ja}(sq)$. In such a condition, the transition rate from $|i\rangle$ to $|f\rangle$ is derived after the time and space integral under the next condition for the space integral.

$$q-k=G$$

(1.146)

where G is the reciprocal lattice vector. When the relation holds, the summation for (ls) gives the number of unit cells N.

As a result, the transition rate for the phonon with q in the branch j induced by the light labeled by (τk) is given as follows:

$$W = 2\pi \left| \sum_s \left(\frac{\pi N \omega B^2(s)}{\Omega M_s \omega_k} \right)^{\frac{1}{2}} (e_{ja}(sq) \cdot \pi_{\tau k}) \right|^2 \delta(\omega_{jq} - \omega_k)$$

(1.147)

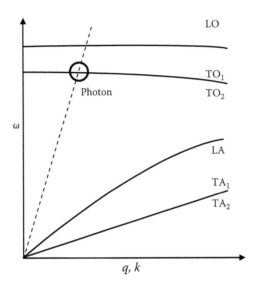

FIGURE 1.25 Dispersion curves for photons and phonons in the crystal with cubic symmetry.

As the nonzero condition for the equation, the energy conservation condition, $\omega_{jq} = \omega_k \equiv \omega$ must be satisfied.

When a unit cell contains two of the same atoms like Si and Ge, there are optical branches in the dispersion curve, but one-phonon absorption does not occur due to the condition for the effective charges, Equation (1.145). In case of the two atoms crystal with cubic symmetry, there are three acoustic and optical branches, and transverse modes are degenerated as shown in Figure 1.25. In such a case, as $B(s) \neq 0$ holds, the selection rule is dominated by $\mathbf{e}_{j\alpha}(s\mathbf{q}) \cdot \boldsymbol{\pi}_{\tau k}$. When $k \| q$ holds, $(\mathbf{e}_{j\alpha}(sq) \cdot \boldsymbol{\pi}_{\tau k})$ is zero for LO phonon and not zero for TO phonon. If we set q not parallel to k, the LO phonon also could be IR active.

In case of $q = 0$, Equation (1.147) is written as follows for the TO modes:

$$W = \frac{2\pi^2 N B^2}{\Omega \, \bar{M}} \delta(\omega_T - \omega) = \frac{2\pi^2 \rho B^2}{M_+ M_-} \delta(\omega_T - \omega) \tag{1.148}$$

where $\omega_{jq} = \omega$, $\omega_k = \omega$, ρ, N, and $\bar{M}$ are the TO phonon frequency, the frequency of light, the density of the crystal, the number of unit cells, and the reduced mass of positive M_+ and negative M_- ions, respectively. The absorption coefficient α can be written using Equation (1.148) as follows:

$$\alpha = \frac{16\pi^3 \hbar \rho B^2}{\varepsilon \omega A_0^2 M_+ M_-} \delta(\omega_T - \omega) \tag{1.149}$$

The value of the equation is generally quite large, and it requires a quite thin crystal to measure transmission spectra.[63] The resonance frequency of ω_T is usually derived

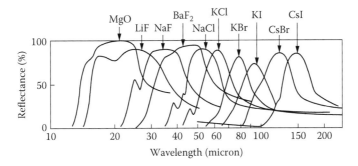

FIGURE 1.26 Infrared reflectance spectra showing reststrahlen bands of typical ionic crystals. (See also M. Wakaki, K. Kudo, and T. Shibuya: *Physical Properties and Data of Optical Materials*, pp. 43, 109, 114, 258, 290, 329, 336, 347, 431, 443 CRC Press (2007).)

from the analysis using the dielectric function on the reflectance spectra of the sample.

The reflectance spectra of typical ionic crystals are shown in Figure 1.26.[64–73] The reflectance peaks around the TO phonon frequencies are at nearly 100% corresponding to the large values of extinction coefficient κ, which is related with α through the relation

$$\alpha = \frac{4\pi\kappa}{\lambda} \qquad (1.150)$$

where λ is the wavelength of the light in vacuum. These strong reflection bands are called *reststrahlen band*, which means "remained" band in German, because the light at the wavelength remains after several reflections on the surface.

The absorption coefficient of NaCl is shown in Figure 1.27.[11,74,75] The bandwidth of the absorption spectrum is unexpectedly seen as rather large from Equation (1.149). To explain the broadness, it is necessary to introduce the damping factor Γ related with the inverse of the lifetime of TO phonon, which gives the width of the absorption peak in the dispersion relation.

1.4.2.2 Selection Rule for Multiphonon

As shown in the previous section, the one-phonon process is allowed in the scheme of the rigid ion model as shown in Figure 1.28a.[11] However, many absorption peaks that cannot be explained by the one-phonon process are observed in the infrared spectra. The weak reflection peaks are seen near the strong reflection bands (Figure 1.26), and also weak absorption peaks are observed around strong peaks as shown in Figure 1.27. In the case of homopolar crystals like Si and Ge in which the lattice absorption is forbidden, weak absorption peaks are observed as shown in Figure 1.29 for Si.[11,76]

In the case of ionic crystals, the origin of the absorptions except one-phonon lattice absorption is explained by considering the third-order term of the potential energy. In the case of homopolar crystals ($M_1 = 0$), the origin is explained by considering the higher terms of dipole moment, M_2 and M_3 as shown in Figures 1.28b and 1.28c.

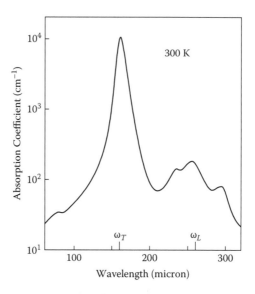

FIGURE 1.27 Absorption spectrum of NaCl in the wavelength region of lattice absorption. (See also K. Kudo: *Fundamentals of Optical Properties of Solids*, Ohmsha (1977) (in Japanese); C. Smart et al.: *Lattice Dynamics*, p. 387 (1964); L. Genzel et al.: *Z. Physik*, 154, 13 (1959).)

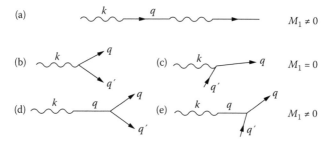

FIGURE 1.28 Feynman diagram for the interaction between photon and phonon. (See also K. Kudo: *Fundamentals of Optical Properties of Solids*, Ohmsha (1977) (in Japanese).)[11]

In the process, the next conservation lows about frequency and momentum must hold. For the summation band,

$$\omega_k = \omega = \omega_j + \omega'_j \tag{1.151a}$$

$$k = q + q' \approx 0 \tag{1.151b}$$

and for the subtraction band,

$$\omega_k = \omega = \omega'_j - \omega_j \tag{1.152a}$$

$$k = q' - q \approx 0 \tag{1.152b}$$

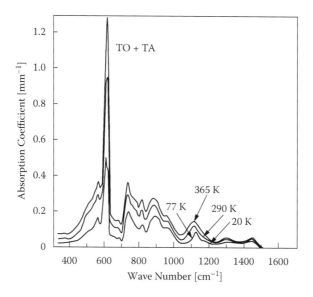

FIGURE 1.29 Infrared absorption spectra of Si at different temperatures. (See also K. Kudo: *Fundamentals of Optical Properties of Solids*, Ohmsha (1977) (in Japanese); F. A. Johnson: *Proc. Phys. Soc. London*, 73, 265 (1959).)

Equations (1.151a) and (1.151b) correspond to the diagram of Figure 4.5b, and two phonons are created by one photon. The absorption band due to such two-phonon absorption is called a *summation band*. Equations (1.152a) and (1.152b) correspond to the diagram of Figure 4.5, and one phonon is created and another phonon is annihilated simultaneously. The absorption band due to such two-phonon absorption is called the *difference band*, and these absorption bands (summation and difference bands) are generally called *combination bands*.

1.4.3 Application of Group Theory for Lattice Absorption[77–83]

1.4.3.1 Symmetry Species and Characters of Normal Vibration Modes

When r numbers of atoms are included in the primitive unit cell, $3r$ branches are formed. It is convenient to use the group theory to know active or inactive for the IR absorption or Raman scattering. The next relation must be satisfied for the phonon to interact with light:

$$|\mathbf{q}|(phonon) = |\mathbf{k}|(phonon) \simeq 0$$

The normal coordinate Q_l is expressed using the following equation, when the unit cell consisting of r atoms vibrates at the frequency of ω_j at $\mathbf{q} = 0$:

$$Q_l = l_j \sum_{\alpha r} \mathbf{e}_{j\alpha}(r)\mathbf{u}_\alpha(r) \tag{1.153}$$

where l_j is a constant.

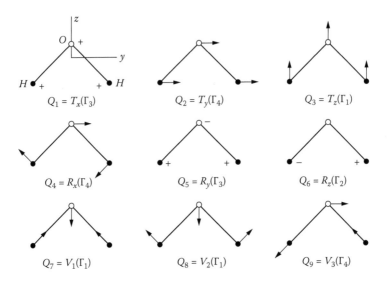

FIGURE 1.30 Normal vibration modes of H_2O molecule. (See also K. Kudo: *Fundamentals of Optical Properties of Solids*, Ohmsha (1977) (in Japanese).)

The displacement of the atom is quite small, and the strain due to the vibration of the unit cell or molecule is negligibly small. A normal vibration mode Q_j is illustrated by drawing arrows to express the displacements proportional to $e_{j\alpha}(r)u_\alpha(r)$ at the unit cell or the molecule at the equilibrium state as shown in Figure 1.30 for a H_2O molecule.[11] Once Q_j is determined, the kinetic energy T and the potential energy φ_2 are given, and these quantities are expressed as follows to satisfy the condition that these are invariant for symmetry operation on the unit cell or molecule:

$$T = \frac{1}{2}\sum_{j=1}^{3r} \dot{Q}_j^2 = \frac{1}{2}\sum_{j=1}^{3r}(R\dot{Q}_j^2) \tag{1.154}$$

$$\varphi_2 = \frac{1}{2}\sum_{j=1}^{3r} \omega_j^2\dot{Q}_j^2 = \frac{1}{2}\sum_{j=1}^{3r}\omega_j^2(R\dot{Q}_j^2) \tag{1.155}$$

If the displacement vector of the normal mode Q takes the same direction after the operation R, such a normal mode is called symmetric to R, and if it takes the reverse direction, it is called asymmetric to R. The symmetric operation is expressed as the character +1, and the asymmetric operation is expressed as −1. The table of such character of the normal mode for the symmetric operations is called a *character table*. The character table for the H_2O molecule is shown in Table 1.3.[11]

1.4.3.2 Factor Group Analysis of Normal Vibration

The character table is generally written as Table 1.4. The following symbols are usually used to classify the normal mode instead of Γ_1, Γ_2, and so on.

TABLE 1.3

Character Table of H_2O

	E	C_2	$\sigma_v(zx)$	$\sigma_v(yz)$	
			H$_2$O (C_{2v})		
$\Gamma_1(A_1)$	1	1	1	1	V_1, V_2, T_z
$\Gamma_2(A_2)$	1	1	−1	−1	R_z
$\Gamma_3(B_1)$	1	−1	1	−1	T_x, R_y
$\Gamma_4(B_2)$	1	−1	−1	1	V_3, T_y, R_x

Source: K. Kudo: Fundamentals of Optical Properties of Solids, Ohmsha (1977) (in Japanese). With permission.

TABLE 1.4

General Form of Character Table

(Class)		R1 R2 ... Rj ... Rk	Number of normal mode
(Mode) (Dimension)		(Character)	
Γ_1	d_1	χ_{1R_1} χ_{1R_2} χ_{1Rj} χ_{1R_k}	n_1
Γ_2	d_2	χ_{2R_1} χ_{2R_2} χ_{2Rj} χ_{2R_k}	n_2
.	.		.
Γ_i	d_i	χ_{iR_1} χ_{iR_2} χ_{iRj} χ_{iR_k}	n_i
.	.		.
Γ_k	d_k	χ_{kR_1} χ_{kR_2} χ_{kRj} χ_{kR_k}	n_k

$A, B, E, F,$ or T modes
  A: Symmetric mode for the operation by the symmetric element C_n of the group D_n
  B: Asymmetric mode for the operation by the symmetric element C_n of the group D_n
  E: Doubly degenerated mode
  F or T: Triply degenerated mode

$A_1, B_1, E_1, T_1 = F_1$ modes: Symmetric modes for the rotation around the two hold rotation axis vertical to the main symmetry axis in the group D_n, or symmetric modes for the mirror reflection σ_v in the group C_{nv}.
$A_2, B_2, E_2, T_2 = F_2$ modes: Asymmetric modes for the rotation around the two hold rotation axis vertical to the main symmetry axis in the group D_n, or asymmetric for the mirror reflection σ_v in the group C_{nv}.
A_g, B_g, A_{1g}, B_{1g} modes: Symmetric modes for the inversion center i.

A_u, B_u, A_{1u}, B_{1u} modes: Asymmetry for the inversion center i.
A', B' modes: Symmetric modes for σ_h when the direct product has the plane
   σ_h as an element.
A'', B'' modes: Asymmetric for σ_h when the direct product has the plane σ_h as
   an element.

By using the character, the number of the vibration species is obtained,[81,83] The normal modes of a unit cell or a molecule are classified into external and internal modes. Pure acoustic mode (translation of a whole unit cell or a molecule: T), translatory vibration in a unit cell or a molecule (translatory: T'), and rotatory vibration within a unit cell or a molecule (rotatory: R') belong to the external vibration, and other vibrations belong to the internal vibration.

To induce the infrared absorption, the component of the dipole moment shown in Equation (1.143c) must not be zero. The first-order dipole moment for kth mode is written as

$$\langle f|H'_k|i\rangle = \langle f|M_{ik}|i\rangle = \sum_{ls\alpha} B(ls)(NM_s)^{-\frac{1}{2}} e_{k\alpha} \langle f|Q_{k\alpha}|i\rangle \qquad (1.156)$$

To have a nonzero value for $\langle f|Q_{k\alpha}|i\rangle$, it is required that the integrand not change the sign by the symmetry operation. For instance, the number of normal modes of a unit cell or a molecule consisting of r atoms is $3r$. The wavefunctions of the ground state φ_0^k and the first excited state φ_1^k are written as

$$\varphi_0^k = A_{0k}e^{-\frac{r_kQ_k^2}{2}} \qquad (1.157a)$$

$$\varphi_1^k = A_{1k}Q_k e^{-\frac{r_kQ_k^2}{2}} \qquad (1.157b)$$

As is seen from the function, φ_0^k is invariant for the symmetry operations, but φ_1^k has the same symmetry of normal coordinate Q_k. As a result, Equation (1.156) is nonzero unless the $B(ls)$ is zero. The components Q_{kx}, Q_{ky}, and Q_{kz} of Q_k have the same dimension as the modes T_x, T_y, and T_z, respectively, which suggests the infrared active modes are the normal modes belonging to the same irreducible representations as T_x, T_y, and T_z. The actual absorption is observed among these modes with $B(ls) \neq 0$. As for the Raman selection rule, it is required that the polarizability tensors are nonzero, which is explained in Section 1.5.

By using the above factor group analysis, we can get the information about the number and the character of the vibration mode, and whether the normal modes of various molecules and crystals are optically active or not. The analysis is correct for the selection rule about the number of bands, but it requires more consideration about the degeneracy of the modes.

1.4.4 TEMPERATURE AND PRESSURE DEPENDENCIES OF LATTICE VIBRATION ABSORPTION INTENSITIES

The probability of transforming photon to phonon is related to the number of phonons. The relation between the number of the phonon and the temperature is given by the Bose–Einstein statistic equation.

$$\langle \upsilon_\mu \rangle = \left[\exp\left(\frac{\hbar \omega_\mu}{kT} \right) - 1 \right]^{-1} \tag{1.158}$$

where $\langle \upsilon_\mu \rangle$ is the average number of phonon $\hbar \omega_\mu$ with the mode μ equilibrium at the temperature T.

The creation and annihilation probabilities of the phonons are proportional to the squares of the matrix element of phonon creation and annihilation operators, like $\left| \langle \upsilon_j + 1 | \hat{b}^+ | \upsilon_j \rangle \right|^2 = \upsilon_j + 1$ or $\left| \langle \upsilon_j - 1 | \hat{b}^- | \upsilon_j \rangle \right|^2 = \upsilon_j$. The rate of absorption of light is proportional to the difference between the probability for creation and annihilation of phonons and is written as follows:

$$(\upsilon_j + 1) - \upsilon_j = 1 \tag{1.159}$$

As a result, the absorption according to a one-phonon process shows no temperature dependence. In this process, the transformations caused by the thermal expansion or the temperature variation are not considered. The transition probabilities for the two-phonon process are proportional to the following relations.

For the harmonic bands, $\omega = 2\omega_j$,

$$(\upsilon_j + 1)^2 - \upsilon_j^2 = 2\upsilon_j + 1 \tag{1.160a}$$

For the summation bands, $\omega = \omega_j + \omega_{j'}$

$$(\upsilon_j + 1)(\upsilon_{j'} + 1) - \upsilon_j \upsilon_{j'} = \upsilon_j + \upsilon_{j'} + 1 \tag{1.160b}$$

and for the subtraction bands, $\omega = \omega_j - \omega_{j'}$,

$$\upsilon_{j'}(\upsilon_j + 1) - (\upsilon_{j'} + 1)\upsilon_j = \upsilon_{j'} - \upsilon_j \tag{1.160c}$$

For a three-phonon process, the following relations are considered:

$$\omega = \omega_j + \omega_{j'} + \omega_{j''} : (1 + \upsilon_j)(1 + \upsilon_{j'})(1 + \upsilon_{j''}) - \upsilon_j \upsilon_{j'} \upsilon_{j''}$$

$$\omega = \omega_j + \omega_{j'} - \omega_{j''} : (1 + \upsilon_j)(1 + \upsilon_{j'})\upsilon_{j''} - \upsilon_j \upsilon_{j'}(1 + \upsilon_{j''})$$

$$\omega = \omega_j + \omega_{j'} - \omega_{j''} : (1 + \upsilon_j)\upsilon_{j'} \upsilon_{j''} - \upsilon_j(1 + \upsilon_{j'})(1 + \upsilon_{j''}) \tag{1.160d}$$

At low temperature, temperature dependences of summation bands become small, because υ_j becomes small for low temperatures. For the subtraction bands, the absorption intensity becomes small due to the cancellation by each other. The assignment of the phonon spectra becomes possible by analyzing such temperature dependence.

By considering a higher term of the potential like φ_3, φ_4, the following eigenvalue is obtained:

$$\varepsilon' = \sum_{jq} \hbar \omega'_{jq} \left(\upsilon_{jq} + \frac{1}{2} \right) \tag{1.161}$$

where quasi-normal-mode frequency ω'_{jq} is slightly different with ω_{jq} and expressed as follows:

$$\Delta\omega_{jq} = \omega'_{jq} - \omega_{jq} = \Delta_{jq}(\varphi_3, \varphi_4 \upsilon_{jq}) \tag{1.162}$$

The frequency of absorption or reflection peak corresponds to ω'_{jq}, which relates with υ_j and consequently depends on the temperature.

One-phonon frequency depends on the anharmonic combination and also on the thermal expansion. The thermal expansion is caused by the anharmonicity, and it affects the quasi-normal frequency through the coupling constant f_{jq}, which depends on the volume change. The volume dependence of ω_{jq} is given by the next equation[84–88]:

$$\frac{\partial \omega_{jq}}{\omega_{jq}} = -\gamma_{jq} \frac{\partial V}{V} \tag{1.163}$$

where γ_{jq} is called the Grüneisen constant for mode j. γ_{jq} is determined experimentally by measuring the pressure dependence of the absorption frequency under the constant temperature to avoid the change of self energy.[86–89] The value of γ is possible to calculate for transverse and longitudinal modes as follows[90]:

$$\gamma_t = \frac{\dfrac{A}{6} B - \dfrac{\pi}{3} C \dfrac{e^2}{r_0^3}}{A - \dfrac{2\pi}{3} C \dfrac{e^2}{r_0^3}}, \tag{1.164a}$$

$$\gamma_l = \frac{\dfrac{A}{6} B + \dfrac{2\pi}{3} C \dfrac{e^2}{r_0^3}}{A + \dfrac{4\pi}{3} C \dfrac{e^2}{r_0^3}} \tag{1.164b}$$

where A, B, and C take different values depending on the kind of ionic model.

Many crystals cause a phase transition at some transition temperature specific to each crystal. The eigenfrequency changes around the transition point, and the dielectric constant also changes. In some crystals, a low-lying TO phonon mode shifts extraordinarily depending on the temperature, and the dielectric constant changes abnormally. These lattice vibrations are called soft phonon and are explained qualitatively as the change of quasi-normal frequency relating to higher terms of the potential. ɼ

1.4.5 Optical Absorption by Impurities[57,58,91–100]

New absorption that is not observed in a pure crystal appears if the crystal has some defects like impurities, vacancies, and dislocations. The difference of the mass or coupling constant of the defect from those of the host crystal causes the change of kinetic energy or potential energy around it to lead to the break of the translational symmetry, which results in the change of the normal mode. Most crystals are not perfect, and some impurities are mixed into the crystal during the growing process or from the starting materials. The additional vibration caused by adding impurities is discussed based on the lattice vibration of a perfect crystal.

1.4.5.1 Rayleigh's Theorem

1. In case of a light impurity $(M' < M, f' > f)$, the frequency of the impurity ω_j' is higher than the normal frequency ω_j and becomes higher as the mass becomes smaller but does not exceed the next higher normal frequency ω_{j+1}.
2. In case of a heavy impurity $(M' > M, f' < f)$, the frequency of the impurity ω_j' is lower than the normal frequency ω_{j+1} and becomes lower as the mass becomes larger but does not exceed the next lower normal frequency ω_j. For the lowest impurity frequency, it is possible to take low frequencies including 0, but it is within the bulk crystal frequency region 0–ω_m.

The bulk mode changes as described above by the introduction of an impurity. The normal mode of the impurity induced within the normal vibration frequency region 0–ω_m is called a *band mode*, and the impurity mode induced at a higher frequency than the maximum bulk frequency ω_m is called a *local mode*. In the same case, the vibration mode within the band mode becomes very strong, which is called a *resonance mode*.

For detailed analyses, the normal vibrations of impurities and the optical absorptions by impurities[95,96,99] and their selection rules[93,100] have been discussed and many references have been issued. There are also many experimental reports on the impurity absorptions typically by semiconductors[100,104–111] and ionic crystals.[93,94,112,113] Please refer to the literature for further understanding.

1.5 LIGHT SCATTERING IN SOLIDS

1.5.1 Light Scattering

It was said that famous painter and multitalented genius Leonardo da Vinci (1452–1519) predicted the blue of the sky is due to the particles floating in the air. This prophecy was proofed by Isaac Newton (1642–1727), and it was found that light scattering occurs due to the particles in the air or liquid. Later, John Tyndall

(1820–1893) found out that the scattered light becomes a tinge of blue when fine particles are included in the air. At the end of the nineteenth century, in 1871 Lord Rayleigh (1842–1919) gave evidence that the origin of the blue of the sky is the molecules of the air, and the scattering intensity is proportionally inverse to the fourth power of the wavelength, which is called Rayleigh scattering.

Between 1919 and 1922, Brillouin proposed a new idea.[114] His theory was that scattered light has a slightly different frequency than the original incident light due to the Doppler effect. This was caused by the sonic wave generated by the incident monochromatic ray, with the frequency shift depending on the direction of observation and the velocity of the sonic wave.

In 1923 Smekel tried the interpretation of quantum theory that two quantized energy levels are required at the scattering.[115] In 1928 Raman experimentally investigated the scattering phenomena in detail.[116] At first, he could not find the scattered light because of its weak intensity. By using the strong light from the mercury lamp emerging at that time, he found that the scattered light contains several new lights other than the incident light, and these lights depend on the scattering materials. These phenomena have been called *Raman scattering*. At almost the same time, similar phenomena were discovered for quartz crystals and called *combinational scattering*. Two or three years later, Gross confirmed the Brillouin theory in solids and liquids, called the *Brillouin scattering*.[117]

Generally, a part of the incident light is scattered inelastically when a light is incident on a transparent material and several lights are contained in the scattered lights. One is the same wavelength to the incident light, which we call an excitation light hereafter, and the others are the newly emerging lights with longer or shorter wavelengths than the excitation light. These effects are generally called Raman scattering. We call it Raman scattering when the scattered light is induced by the molecular vibrations, the optical phonons in the solid, and other elementary excitations such as impurities, plasma, and polariton. On the other hand, when the new scattered lights are caused by the interaction between the acoustic phonon in liquids and solids, we called this Brillouin scattering

We can investigate the elementary excitations within the materials by using the Raman and Brillouin scatterings. We can get the information about the change of the state within the material by observing the change of the scattering relating to the application of heat or pressure that offers very important information about the material.[57,83,118–125] The number of papers published about the Raman effect drastically increased after the realization of lasers in 1960 due to its strong intensity and pure monochromatic property. Before the laser, a high-intensity mercury arc lamp was used, but the intensity and monochromatic properties were not enough and enough results were not expected. On the other hand, laser lights are ideal for the light source of Raman scattering, and its spectrum can be measured with almost the same ease as infrared absorption measurements can be obtained.

1.5.2 CLASSICAL THEORY FOR LIGHT SCATTERING

We consider the simple case of Raman scattering by phonons in the crystal. Excitation light E_i is incident on the crystal, and the scattered light E_s is observed from the

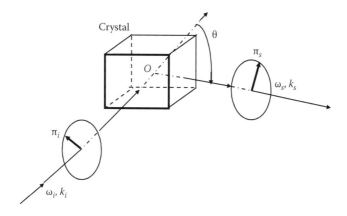

FIGURE 1.31   The scattering process by crystal. ω_i, k_i, and π_i are, respectively, the frequency, wavevector, and polarization of the electric field of incident light, and ω_s, k_s, and π_s are, respectively, those of scattered light.

direction of θ to the incident light as shown in Figure 1.31. We use the notation for the excited and scattered lights as

$$E_i = \pi_i E_{0i} \exp\left[i(\omega t - k_i \cdot r)\right] \tag{1.165a}$$

$$E_s = \pi_s E_{0s} \exp\left[i(\omega t - k_s \cdot r)\right] \tag{1.165b}$$

where ω, k, and π are the frequency, wavevector, and unit vector of polarization, respectively. A kind of dipole moment called electronic polarization P is induced as the excitation light polarizes the atoms in the crystal to induce electric dipoles.

The components of the polarization are written as

$$P = [\alpha] E_i \tag{1.166a}$$

where

$$[\alpha] = \begin{bmatrix} \alpha_{xx} & \alpha_{xx} & \alpha_{xx} \\ \alpha_{xx} & \alpha_{xx} & \alpha_{xx} \\ \alpha_{xx} & \alpha_{xx} & \alpha_{xx} \end{bmatrix} \tag{1.166b}$$

The susceptibility (polarizability) α is modulated according to the lattice vibration. As a result, it is possible to expand α using the displacement u of the atom. If you write the displacement relative with the normal coordinate μ as

$$u_\mu = e_\mu A_\mu \exp[\pm i(\omega_\mu t - q \cdot r)] \tag{1.167}$$

where

$$\mu(=jq) = 1,2,...$$

the components of α are expanded as

$$\alpha_{p\sigma} = \alpha_{p\sigma}^0 + \sum_{\mu} \alpha_{p\sigma,\mu}^0 u_\mu + \frac{1}{2} \sum_{\mu\mu'} \alpha_{p\sigma,\mu\mu'}^0 \cdot u_\mu u_{\mu'} + \cdots \tag{1.168a}$$

where

$$\alpha_{p\sigma,\mu}^0 = \left(\frac{\partial \alpha_{p\sigma}}{\partial u_\mu}\right)_0, \quad \alpha_{p\sigma,\mu\mu'}^0 = \left(\frac{\partial^2 \alpha_{p\sigma}}{\partial u_\mu \partial u_{\mu'}}\right)_0 \tag{1.168b}$$

The notation $(\ )_0$ means the value at the equilibrium position of the atom. The next relation is obtained by substituting Equation (1.168a) into Equation (1.166a):

$$P_p = \sum_{\sigma} \left[\alpha_{p\sigma}^0 \pi_{i\sigma} E_{0i} \exp i(\omega t - k_i r) \right.$$

$$+ \sum_{p} \alpha_{p\sigma,\mu}^0 e_\mu A_\mu \pi_{i\sigma} E_{0i} \exp i\left[(\omega_i \pm \omega_\mu)t - (k_i \pm g_\mu)r\right]$$

$$\left. + \frac{1}{2} \sum_{\mu\mu'} \alpha_{p\sigma,\mu\mu'}^0 \cdot e_\mu e_{\mu'} \cdot A_\mu A_\mu \cdot \pi_{i\sigma} E_{0i} \exp i\left[(\omega_i \pm \omega_\mu + \omega_{\mu'})t - (k_i \pm g_p \pm g_{p'})r\right] + \cdots \right]$$

$$\tag{1.169}$$

As a result, Equation (1.166b) has the tensors like $\left[\alpha_{p\sigma}^0\right], \left[\alpha_{p\sigma,\mu}^0\right], \left[\alpha_{p\sigma,\mu\mu'}^0\right]$. The tensor at the equilibrium position $\left[\alpha_{p\sigma}^0\right]$ does not contribute to the Raman scattering. The other tensors like $\left[\alpha_{p\sigma,\mu}^0\right], \left[\alpha_{p\sigma,\mu\mu'}^0\right]$ are the polarizability tensors induced by the excitation light and related with the scattering. These two tensors are called the first- and second-order Raman tensors, respectively. The intensities of the scattered lights induced by the modulation of the polarization are obtained by using the following Poynting vector:

$$S = \frac{c}{4\pi} \left\{ \frac{1}{Rc^2} \left| \frac{R}{R} \times \left[\frac{R}{R} \times \ddot{P} \right] \right| \right\}^2 . \tag{1.170}$$

The light energy I flowing to the solid angle $d\Omega$ per second is defined by the following relation:

$$I d\Omega = R^2 \overline{S} d\Omega, \tag{1.171}$$

where $\overline{S}$ is the time average over the single cycle of the vibration. The energy I is obtained by using Equation (1.171) as

$$I = \frac{1}{2\pi c^3} \sum_{(i)=(1),(2)} \sum_{p\sigma} n_p^{(1)} n_\sigma^{(2)} \{\omega_i^4 [\alpha_{p\sigma}^0]^2 E_{0i}^2$$

$$+ \sum_{\mu=1}^{3N} (\omega_i \pm \omega_\mu)^4 [\alpha_{p\sigma,\mu}^0]^2 (e_\mu A_\mu \pi_{i0} E_{0i})^2 \tag{1.172}$$

$$+ \frac{1}{2} \sum_{\mu\mu'} (\omega_i \pm \omega_\mu \pm \omega_{\mu'})^4 [\alpha_{p\sigma,\mu\mu'}^0]^2 (e_\mu e_\mu \cdot \pi_{i\sigma})^2 (A_\mu A_\mu \cdot E_{0i})^2 + \cdots\}$$

where $n^{(1)}$ and $n^{(2)}$ are the unit vectors normal to each other and also perpendicular to the vector R. Several frequency terms are included in the scattered lights other than the excitation light as shown in the equation, which means that electronic polarization P_p is composed from the induced dipole oscillators that correspond to the respective frequencies. The lights generated from these oscillators have features described as follows:

The scattered light described by the first term:

$$\omega_s = \omega_i, \quad k_s = k_i \tag{1.173a}$$

The second term of the scattered light:

$$\omega_s = \omega_i \pm \omega_\mu, \quad k_s = k_i \pm q_\mu \tag{1.173b}$$

The third term of the scattered light:

$$\omega_s = \omega_i \pm \omega_\mu \pm \omega_{\mu'}, \quad k_s = k_i \pm q_\mu \pm q_{\mu'} \tag{1.173c}$$

The relation (1.173a) means that the scattered light has the same frequency as the excitation light called oth order scattering or Rayleigh scattering. The relation (1.173b) means that the scattered light has the frequency added or subtracted by a phonon frequency. In the first-order scattering, the light shifted to the longer wavelength—that is $\omega_s = \omega_i - \omega_\mu$—is called the *Stokes scattered light*. On the other hand, the light shifted to shorter wavelength—that is $\omega_s = \omega_i - \omega_\mu$—is called the *anti-Stokes scattered light*. The relation (1.173c) means that the scattered light was generated relating with two phonons and is called second-order scattering.

1.5.3 QUANTUM THEORY FOR LIGHT SCATTERING

Raman scattering is a similar phenomenon to the luminescence in the point that the material absorbs the light at some particular wavelength and emits the light with different wavelengths as shown in Figure 1.32.[11] The atoms in the ground state $|a, \upsilon_j\rangle$ absorb incident light $\hbar\omega$ and are excited to the upper state $|a', \upsilon'_j\rangle$. If some condition

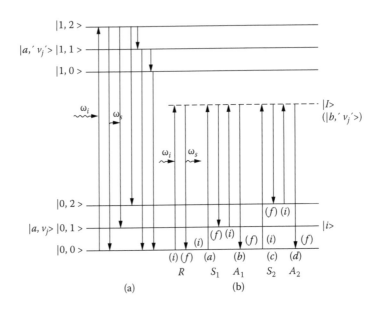

FIGURE 1.32 Transition processes of fluorescence (a) and Raman scattering (b). R, Rayleigh scattering; S_1, first-order Stokes scattering; A_1, first-order anti-Stokes scattering; S_2, second-order Stokes scattering; A_2, second-order anti-Stokes scattering. (See also K. Kudo: *Fundamentals of Optical Properties of Solids*, Ohmsha (1977) (in Japanese).)

is satisfied, the excited state relaxes to the lower state to emit light. The difference between the Raman scattering and the luminescence is that the material does not absorb the incident light in the Raman scattering. Born and Huang explained successfully the Raman scattering by considering the virtual upper level[120] as shown in Figure 1.32. The states of the atom are described as $|i\rangle = |a, n_i, n_s, \upsilon_j\rangle$ for the initial state, $|I\rangle = |b, n'_i, n'_s, \upsilon'_j\rangle$ for the intermediate state, and $|f\rangle = |a, n_i - 1, n_s + 1, \upsilon''_j\rangle$ for the final state.

The Hamiltonian consisting of atoms and the scattering field is described as

$$H = H_0 + H_1 + H_2 \tag{1.174a}$$

where

$$H_1 = -\frac{e}{m} A \cdot P \tag{1.174b}$$

$$H_2 = \frac{e^2}{2m} A^2 \tag{1.174c}$$

H_0 contains the atomic state, p is the momentum of the electron, and A is the vector potential. $H_2 \ll H_1$ is generally held, and H_2 is neglected. The vector potential A is given as

$$A_{i\sigma} = \left(\frac{A_0}{\omega_i}\right)^{\frac{1}{2}} (\hat{a}_{i\sigma}e^{-ik\cdot R} + \hat{a}_{i\sigma}^+ e^{-ik\cdot R})\pi_{i\sigma} \tag{1.175a}$$

where

$$A_0 \equiv \frac{2\pi\hbar c^2}{\varepsilon V} \tag{1.175b}$$

V is the volume of the crystal. The first-order perturbation gives the following relation:

$$\langle f|H_1|i\rangle = 0 \tag{1.176}$$

As a result, second-order perturbation must be considered. The amplitude transition probability $a^{(2)}(t)$ derived from the second-order perturbation theory is given as

$$a^{(2)}(t) = -\frac{1}{\hbar^2}\sum_{I\neq 0}\int_0^t \langle f|H_1|I\rangle e^{i\omega_{fI}t_1}dt_1 \int_0^{t_1}\langle I|H_1|i\rangle e^{i\omega_{Ii}t_2}\,dt_2 \tag{1.177}$$

Atoms are vibrating according to the lattice vibration, and the momentum of the electron p can be expanded using the displacement u of the normal vibration of the atom.

$$p = p_0 + Bu + Cuu' + \cdots \tag{1.178a}$$

where

$$u = \left(\frac{\hbar}{2NM_s\omega_j}\right)^{\frac{1}{2}} (\hat{b}_j e^{-iq\cdot R} + \hat{b}_j^\dagger e^{-iq\cdot R})e_j \tag{1.178b}$$

Equation (1.177) is transformed as follows by using Equations (1.176) and (1.178a):

$$-\hbar^2 a^{(2)}(t) = X_0 + X_1 + X_2 + \cdots \tag{1.179a}$$

where

$$X_0 = \frac{e^2}{m^2}\sum_{I\neq 0}\int_0^t \langle f|A\cdot p_0|I\rangle e^{i\omega_{fI}t_1}dt_1 \int_0^{t_1}\langle I|A'\cdot p_0|i\rangle e^{i\omega_{Ii}t_2}\,dt_2 \tag{1.179b}$$

$$X_1 = \frac{e^2}{m^2}\sum_{I\neq 0}\left\{\int_0^t \langle f|A\cdot Bu|I\rangle e^{i\omega_{fI}t_1}dt_1 \int_0^{t_1}\langle I|A\cdot p'_0|i\rangle e^{i\omega_{Ii}t_2}\,dt_2\right.$$
$$\left. + \int_0^t \langle f|A\cdot p_0|I\rangle e^{i\omega_{fI}t_1}dt_1 \int_0^{t_1}\langle I|A'\cdot Bu|i\rangle e^{i\omega_{Ii}t_2}\,dt_2\right\} \tag{1.179c}$$

$$X_2 = \frac{e^2}{m^2} \sum_{I \neq 0} \left\{ \int_0^t \langle f|A \cdot Cuu'|I\rangle e^{i\omega_{fI}t_1} dt_1 \int_0^{t_1} \langle I|A' \cdot p_0|i\rangle e^{i\omega_{Iit_2}} dt_2 \right.$$

$$+ \int_0^t \langle f|A \cdot Bu|I\rangle e^{i\omega_{fI}t_1} dt_1 \int_0^{t_1} \langle I|A' \cdot Bu|i\rangle e^{i\omega_{Iit_2}} dt_2 \qquad \text{(1.179d)}$$

$$\left. + \int_0^t \langle f|A \cdot p_0|I\rangle e^{i\omega_{fI}t_1} dt_1 \int_0^{t_1} \langle I|A' \cdot Cuu'|i\rangle e^{i\omega_{Iit_2}} dt_2 \right\}$$

X_0, X_1, and X_2 give the amplitude scattering probabilities of the Rayleigh scattering, the first-order scattering, and the second-order scatterings, respectively.

In the case of the first-order scatterings, Stokes $\upsilon'' = \upsilon_j + 1$ and anti-Stokes scatterings $\upsilon'' = \upsilon_j - 1$ are considered. The amplitude scattering probability for the first-order Stokes scattering is calculated as

$$a^{(2)}(t) = \frac{e^2}{m^2} \frac{2\pi c^2}{V} \left(\frac{B_0}{\omega_i \omega_s \omega_j} \right)^{\frac{1}{2}} \left[n_i(n_i + 1)(\upsilon_j + 1) \right]^{\frac{1}{2}}$$

$$\text{(1.180a)}$$

$$\times \sum_{I \neq 0} M_I \frac{2i \sin\frac{t}{2}(\omega_i - \omega_s - \omega_j)}{\omega_{if}} e^{-i\frac{t}{2}\omega_{if}} e^{i(k_i - k_s - q) \cdot R}$$

where

$$M_I \equiv \frac{\left[\pi_f, \langle f|Be_j|I\rangle \right]\left[\pi_i, \langle I|p_0|i\rangle \right]}{\varepsilon_b - \varepsilon_a - \hbar\omega_i} + \frac{\left[\pi_i, \langle f|Be_j|I\rangle \right]\left[\pi_f, \langle I|p_0|i\rangle \right]}{\varepsilon_b - \varepsilon_a + \hbar\omega_i} \qquad \text{(1.180b)}$$

The condition $a^{(2)}(t) \neq 0$ on the summation over the crystal of Equation (1.180a) gives the next momentum condition:

$$k_i - k_s - q = 0 \text{ or } k_i = k_s + q \qquad \text{(1.180c)}$$

The possibility of finding the atom and the scattering field at the final state at t is given as

$$\left|A^{(2)}(t)\right|^2 = \int_0^\infty \frac{2V}{(2\pi c)^3} \omega_s^2 d\omega_s \left|a^{(2)}(t)\right|^2 d\Omega$$

$$\text{(1.181a)}$$

$$= \frac{c\hbar e^4}{2\pi M_s m^4} \frac{n_i(n_s + 1)(\upsilon_j + 1)}{\omega_i \omega_j} \int |M|^2 \omega_s d\omega_s \frac{4\sin^2\frac{t}{2}(\omega_i - \omega_s - \omega_j)}{\omega_{if}^2} d\Omega$$

$$|M|^2 \equiv \sum_l M_l \tag{1.181b}$$

The possibility per second is obtained by differentiating the formula for t as

$$W = \frac{c\hbar e^4}{M_s m^4} \frac{\omega_s}{\omega_i \omega_j} n_i (n_s + 1)(\upsilon_j + 1)|M|^2 \delta(\omega_i - \omega_s - \omega_j)d\Omega \tag{1.182}$$

By the δ function, the following frequency condition is given for the Stokes scattering:

$$\omega_i - \omega_s - \omega_j = 0 \text{ or } \omega_i = \omega_s + \omega_j \tag{1.183}$$

In many cases, the derivative scattering cross section $d\sigma$ or scattering efficiency S is used instead of scattering probability for Raman scattering. $d\sigma$ is defined as the quantity in which W is divided by the number of photons incident on the unit area per second.

$$d\sigma = W \frac{n_i c}{V} \tag{1.184}$$

On the other hand, the scattering efficiency S is defined as

$$S = W \frac{L\varepsilon^{\frac{1}{2}}}{V} d\sigma = \frac{L\varepsilon^{\frac{1}{2}}}{V} d\sigma \tag{1.185}$$

In the formula, L is the crystal length along the direction of the incident light k_i, and $\varepsilon^{\frac{1}{2}}$ is the refractive index of the crystal. By using Equation (1.182), S is given as

$$S = \frac{V\hbar e^4}{M_s m^4} d\Omega \frac{\omega_s}{\omega_i \omega_j} (n_s + 1)(v_j + 1)|M|^2 \delta(\omega_i - \omega_s - \omega_j) \tag{1.186a}$$

$$= \frac{V\hbar e^4}{M_s m^4} d\Omega \frac{\omega_s}{\omega_i \omega_j} (n_s + 1)(v_j + 1)|\pi_s \cdot R \cdot \pi_i|^2$$

In the formula, R is the Raman tensor defined as

$$R = \frac{\langle f|Be_j|I\rangle\langle I|p_0|i\rangle}{\varepsilon_b - \varepsilon_a - \hbar\omega_i} + \frac{\langle i|p_0|I\rangle^*\langle I|Be_j|f\rangle^*}{\varepsilon_b - \varepsilon_a + \hbar\omega_s} \tag{1.186b}$$

For the anti-Stokes scattering, where the final state is written as

$$|f\rangle = |a, n_i - 1, n_s + 1, \upsilon_j - 1\rangle$$

each quantity is derived by replacing $\upsilon_j + 1$ with υ_j in the formula, and the next condition for the frequency and momentum are obtained instead of Equations (1.183) and (1.180c).

$$\omega_i - \omega_s + \omega_j = 0 \quad \text{or} \quad \omega_i = \omega_s - \omega_j \tag{1.187a}$$

$$k_i - k_s + q = 0 \quad \text{or} \quad k_i = k_s - q \tag{1.187b}$$

1.5.4 First-Order Raman Scattering and the Symmetry of the Crystal

The condition for the normal vibration of the crystal to be infrared active is $\langle f|M_1|i \rangle \neq 0$, where M_1 is the first-order electric dipole moment. On the other hand, the condition for Raman active is $\langle f|\alpha_1|i \rangle \neq 0$, where α_1 is the first-order susceptibility as shown in Equation (1.168a). It is required to satisfy the quantum condition and also not to change the polarity of the integral for the symmetry operation by the symmetry element of the unit cell to meet the nonzero condition.

The selection rule that a phonon is related with the Raman scattering depends on the Raman tensor not being zero. The crystal system where all the Raman tensors are not zero is the tricrinic crystal system. On the other hand, some tensor components are zero in other crystal systems.[126,127] Loudon summarized the Raman tensors R or α in 32 crystal groups as shown in Table 1.5.[11,121–123]

To know which component is related for the vibration mode, it is required to use polarized lights for both incident and scattered lights and to calculate the scattering efficiencies for the combination of the polarization directions.

The phonons of NaCl are infrared and Raman active, and those of diamond are infrared inactive and Raman active. Both materials have an inversion symmetry, but the directions of the displacement vectors of the vibration are symmetric (gerade) or antisymmetric (ungerade) for the inversion operation. These properties of the displacement vectors of the vibration for the inversion center are called the parity of the optical phonon.

As the simplest case, we consider the scattering efficiencies of the Raman active phonons A_{1g}, E_g, and F_{2g} of the cubic crystalline group O_h. As seen from Table 1.5, the Raman tensors contain only one component for any vibration mode. If you use a nonpolarized incident light in the geometry shown in Figure 1.33a,[11] the polarized components of the scattered light become $(0, -\pi_{s\parallel}\sin\phi, \pi_{s\parallel}\cos\phi)$ for $S_\parallel$ and $(\pi_{s\perp}, 0, 0)$ for $S_\perp$. As a result, the following scattering efficiencies are given:

$$S_\parallel(A_{1g}) = Aa^2 \pi_{s\parallel}^2 (\pi_i^y \sin\phi - \pi_i^z \cos\phi)^2 \tag{1.188a}$$

$$S_\perp(A_{1g}) = Aa^2 (\pi_i^x \pi_{s\perp})^2 \tag{1.188b}$$

$$S_\parallel(E) = 2Ab^2 \pi_{s\parallel}^2 \left[(\pi_i^y \sin\phi - 2\pi_i^z \cos\phi)^2 + 3(\pi_i^y)^2 \sin^2\phi \right] \tag{1.188c}$$

TABLE 1.5

Raman Tensors and Raman Active Irreducible Representations for Different Crystalline Systems

Crystalline systems	Crystal classes	Raman tensors and Raman active irreducible representations
Triclinic	1 C_1	$A(x,y,z)$ $\begin{bmatrix} a & d & e \\ d & b & f \\ e & f & c \end{bmatrix}$
	$\bar{1}$ C_1	A_g
Monoclinic	2 C_2	$A(y)$ $B(x,y)$
	m C_3	$A'(x,z)$ $A''(y)$ $\begin{bmatrix} a & & d \\ & b & \\ d & & c \end{bmatrix}\begin{bmatrix} & d & \\ d & & e & f \\ & e & f \end{bmatrix}$
	$2/m$ C_{2h}	A_g B_g
Orthorhombic	222 D_2	A $B_1(z)$ $B_2(y)$ $B_3(x)$
	$mm2$ C_{2v}	$A_1(z)$ A_2 $B_1(x)$ $B_2(y)$ $\begin{bmatrix} a & & \\ & b & \\ & & c \end{bmatrix}\begin{bmatrix} & d & \\ d & & \\ & & \end{bmatrix}\begin{bmatrix} & & e \\ & & \\ e & & \end{bmatrix}\begin{bmatrix} & & \\ & & f \\ & f & \end{bmatrix}$
	mmm D_{2h}	A_g B_{1g} B_{2g} B_{3g}

(Continued)

TABLE 1.5 (*Continued*)
Raman Tensors and Raman Active Irreducible Representations for Different Crystalline Systems

Crystalline systems	Crystal classes	Raman tensors and Raman active irreducible representations
Trigonal, Rhombohedral	3 C_3	$A(z)\begin{bmatrix} a & & \\ & a & \\ & & b \end{bmatrix}$ $E(x)\begin{bmatrix} c & d & e \\ d & -c & f \\ e & f & \end{bmatrix}$ $E(y)\begin{bmatrix} d & -c & -f \\ -c & -d & e \\ -f & e & \end{bmatrix}$
	$\bar 3$ C_{3i}	$A_g\begin{bmatrix} a & & \\ & a & \\ & & b \end{bmatrix}$ E_g E_g
	32 D_3	$A_1\begin{bmatrix} a & & \\ & a & \\ & & b \end{bmatrix}$ $E(x)\begin{bmatrix} c & & \\ & -c & d \\ & d & \end{bmatrix}$ $E(y)\begin{bmatrix} & -c & -d \\ -c & & \\ -d & & \end{bmatrix}$
	$3m$ C_{3v}	$A_1(z)\begin{bmatrix} a & & \\ & a & \\ & & b \end{bmatrix}$ $E(y)$ $E(-x)$
	$\bar 3$ D_{3d}	$A_{1g}\begin{bmatrix} a & & \\ & a & \\ & & b \end{bmatrix}$ E_g E_g
Tetragonal	4 C_4	$A(z)\begin{bmatrix} a & & \\ & a & \\ & & b \end{bmatrix}$ $B\begin{bmatrix} c & d & \\ d & -c & \\ & & \end{bmatrix}$ $E(x)\begin{bmatrix} & & d \\ & & c \\ d & c & \end{bmatrix}$ $E(y)\begin{bmatrix} & & e \\ & & f \\ e & f & \end{bmatrix}$
	$\bar 4$ S_4	$A\begin{bmatrix} a & & \\ & a & \\ & & b \end{bmatrix}$ $B(z)\begin{bmatrix} c & d & \\ d & -c & \\ & & \end{bmatrix}$ $E(x)$ $E(x)\begin{bmatrix} & & e \\ & & f \\ -f & e & \end{bmatrix}$
	$4/m$ C_{4h}	$A_g\begin{bmatrix} a & & \\ & a & \\ & & b \end{bmatrix}$ $B_g\begin{bmatrix} c & d & \\ d & -c & \\ & & \end{bmatrix}$ E_g E_g
	$4mm$ C_{4v}	$A_1(z)$ B_1 B_2 $E(x)\begin{bmatrix} & & e \\ & & \\ e & & \end{bmatrix}$ $E(y)\begin{bmatrix} & & \\ & & e \\ & e & \end{bmatrix}$
	422 D_4	A_1 B_1 B_2 $E(-y)$ $E(x)$

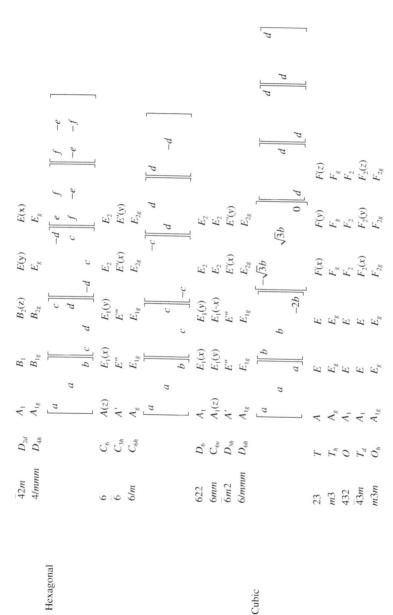

Sources: K. Kudo: *Fundamentals of Optical Properties of Solids*, Ohmsha (1977) (in Japanese). R. Loudon: *Adv. Phys.*, 13, 423 (1964); R. Loudon: *Proc. Roy. Soc.*, A275, 218 (1963); and R. Loudon: *Roy. Phys. Soc.*, 82, 393 (1963). With permission.

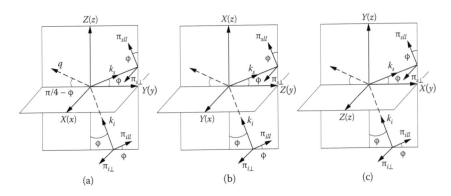

FIGURE 1.33 Arrangements to measure the Raman scattering for three incident planes of YZ (a), ZX (b), and XY (c) in biaxial crystal. X, Y, and Z are the principal axes of the crystal. (See also K. Kudo: *Fundamentals of Optical Properties of Solids*, Ohmsha (1977) (in Japanese).)

$$S_\perp(E) = 4Ab^2(\pi_i^x \pi_{s\perp})^2 \tag{1.188d}$$

$$S_\parallel(F_{2g}) = Ad^2\pi_{s\parallel}^2\left[(\pi_i^y \cos\phi - \pi_i^z \sin\phi)^2 + (\pi_i^x)^2\right] \tag{1.188e}$$

$$S_\perp(F_{2g}) = Ad^2\pi_{s\perp}^2\left[(\pi_i^x)^2 + (\pi_i^y)^2\right] \tag{1.188f}$$

In the case of the anisotropical crystal with centro-symmetry, the tensors of vibration mode have several different components. As a result, it is necessary to perform various scattering measurements for three planes of YZ, ZX, and XY among three principal axes of the crystal as shown in Figures 5.3a, 5.3b, and 5.3c.

1.5.5 SECOND-ORDER RAMAN SCATTERING

The energy and momentum conservation relations are written as

$$\omega_s = \omega_i \pm \omega_{jq} \pm \omega_{j'q'} \tag{1.189a}$$

$$k_s = k_i \pm q \pm q' \tag{1.189b}$$

Double frequency, summing frequency, and subtraction frequency are generated depending on the condition that the frequency shift occurs among the same branch or different branches. From the momentum conservation,

$$\pm q \pm q' \simeq 0 \tag{1.190a}$$

where

$$k_s \approx k_i \simeq 0 \tag{1.190b}$$

The momentums q and q' in either optical or acoustic branch can be allowed. Then we cannot separate the Raman scattering and the Brillouin scattering in a narrow sense in the second-order Raman. The allowed condition for second-order Raman is obtained by analyzing the crystal symmetry by group theory. The selection rule for NaCl was given by Burstein et al.[128] Krauzman performed the detailed assignment of the second-order Raman of NaCl and KCl by using a laser as a light source. The selection rule for the second-order Raman in NaCl-type crystals has almost no limitation. As a result, the profile of the Raman spectra depends on the phonon density of state in the crystal. Karo et al.[129] derived the joint density of state for CsCl and CsBr using ω versus q curve calculated theoretically and showed the good fit to the experimental result.[130]

1.5.6 VARIOUS RAMAN SCATTERINGS

In the previous sections, we mainly discussed Raman scattering by phonons. Almost all elementary excitations in the crystal contribute to Raman scattering. Raman scatterings of the polariton were investigated for various crystals, like GaP,[131] ZnO,[132] SiO$_2$,[133] LiF, and BaTiO$_3$,[134,135] by changing the scattering angle to obtain the dispersion curves of the polariton. The introduction of impurities into the crystal lattice gives the extra vibration modes like band, gap, and resonance modes, which contribute to Raman scattering. Raman scattering is an important technique used to obtain the information about impurities by accompanying infrared measurements. In the magnetic materials like ferromagnetic, antiferromagnetic, and ferrimagnetic materials, collective excitations by the spin wave called magnon are excited at a low temperature. Such magnons also contribute to the Raman scattering like phonons. The Plasmon, Landou splitting and spin flip transition under the magnetic field, and various elementary excitations caused by the strain in the crystal induced by the electric field or the pressure can be investigated by Raman scattering.

1.5.7 BRILLOUIN SCATTERING

The atoms within the unit cell oscillate in phase at the acoustic mode. As a result, the parity rule cannot be applied to the acoustic mode. The first-order Brillouin scattering is induced in any crystals, and three frequency shifts correspond to one LA and two TA modes for the light in the direction of the wavevector q. By considering the small values of q and frequency shifts, it is plausible to obtain the following relation:

$$\frac{\omega_i \sim \omega_s}{\omega_i} = \frac{\omega_j}{\omega_i} \simeq \pm \frac{2n\upsilon_B}{c} \sin\frac{\theta}{2} \qquad (1.191)$$

where + and − correspond to Stokes and anti-Stokes scatterings, respectively. n is the refractive index of the crystal for the incident light at ω_i. The Raman shifts are usually less than 1 cm^{-1} because $\upsilon_B/c \leq 10^{-5}$ is held. The values are quite small and a high-resolution spectrometer is required to separate the scattered light from the Rayleigh scattering. Then interferometers like a Fabry-Perot-type interferometer are typically used. In gas and liquid, only the longitudinal acoustic waves are induced, and Brillouin scattering lines are observed at both higher- and lower-frequency sides of the Rayleigh line. The sound velocities in the materials are in the order gas < liquid < solid, and the shifts of the Brillouin lines are the largest in solids followed by liquids and gas.[136] In the case of solids, there are three acoustic modes (two TA and one LA), and three pairs of scattering lines are observed if the sound velocity related with respective oscillation direction is different. The shift of the Brillouin scattering is quite small, and the intensities of Stokes and anti-Stokes lines are almost the same as explained in the Raman scattering theory. The ratio between the total intensity of the Brillouin line I_B and the Rayleigh line I_R is given as the following Landau-Placzek equation:

$$\frac{I_B}{I_R} = \frac{C_V}{C_V - C_P} \tag{1.192}$$

where C_P and C_V are the specific heat at constant pressure and volume, respectively.

In the case of water, $C_V \approx C_P$ is hold and $I_B \gg I_R$ is estimated. But the Rayleigh scattering in water is enhanced by the introduction of dusts, and the comparable intensities for the Rayleigh and the Brillouin scatterings are observed by purifying the water.

The scattering efficiency of Brillouin scattering is calculated using a similar method as Raman scattering and is given as

$$S^j = \frac{kT\omega_s^4 L}{32\pi^2 c^4 \rho \upsilon_{Bj}^2} \left| \pi_i T^j \pi_s \right| \tag{1.193}$$

where L is the crystal length along the excitation light, T^j is the tensor for the branch j, and $\rho \upsilon_{Bj}^2$ is calculated from the secular equation of the equation of motion for the longitudinal oscillation. The tensor components for various crystals are summarized by Born and Huang[9] and Cummins et al.[137] The tensors T^j and $\rho \upsilon_{Bj}^2$ are related with Pockels coefficients and elastic constants, respectively, and Brillouin scattering is often used to verify these constants obtained by other methods.

REFERENCES

1. R. W. G. Wyckoff: *Crystal Structures*, 2nd ed., Interscience (1961).
2. H. D. Megaw: *Crystal Structures: A Working Approach*, W. B. Saunders (1973).
3. *International Tables for X-ray Crystallography*, Vol. I, Kynoch Press, Birmingham, for the International Union of Crystallography (1952).
4. S. A. Ramakrishna and T. M. Grzegorczyk: *Physics and Applications of Negative Refractive Index Materials*, CRC Press (2008).
5. M. Yan, W. Yan, and M. Qiu: Invisibility cloaking by coordinate transformation, *Prog. Opt.*, 52, 261–301 (2009).
6. T. J. Cui, D. R. Smith, and R. Liu, Eds., *Metamaterials: Theory, Design, and Applications*, Springer, New York (2010).
7. T. G. Mackay and A. Lakhtakia: *Electromagnetic Anisotropy and Bianisotropy*, World Scientific (2010).
8. N. Lane: The grand challenges of nanotechnology, *J. Nanopart. Res.*, 3, 95–103 (2001).
9. R. F. Service: Is Nanotechnology Dangerous?, *Science*, 290, 1526–1527 (2000).

10. L. Esaki and R. Tsu: Superlattice and Negative Differential Conductivity in Semiconductors, *IBM J. Res. Dev.*, 14, 61–65 (1970).

11. K. Kudo: *Fundamentals of Optical Properties of Solids*, Ohmsha (1977) (in Japanese).

12. M. Born and E. Wolf: *Principles of Optics*, Pergamon Press (1975).

13. J. M. Stone: *Radiation and Optics*, McGraw-Hill (1963).

14. B. Rossi: *Optics*, Addison-Wesley (1957).

15. M. Fox: *Optical Properties of Solids*, Oxford University Press (2001).

16. S. Flugge: *Handbuch der Physik*, Band xxv/1, Kristalloptik.

17. M. Wakaki and Y. Kanai: Far-Infrared Reflectivity Spectra of Non-Doped and Mg Doped In2O3 Single Crystals, *Jpn. J. Appl. Phys.*, 25, 502–503 (1986).

18. M. L. Cohen and J. R. Chelikowsky: *Electronic Structure and Optical Properties of Semiconductors*, Springer-Verlag (1988).

19. G. G. Macfarlane, T. P. McLean, J. E. Quarrington, and V. Roberts: Direct Optical Transitions and Further Exciton Effects in Germanium, *Proc. Phys. Soc.*, 71, 863 (1958).

20. M. L. Theye: Investigation of the Optical Properties of Au by Means of Thin Semitransparent Films, *Phys. Rev.*, B2, 3060 (1970).

21. B. R. Cooper and H. Ehrenreich: Optical Properties of Noble Metals. II., *Phys. Rev.*, A138, 494 (1965).

22. G. B. Irani, T. Huer, and F. Wooten: Optical Properties of Ag and α-Phase Ag Al Alloys, *Phys. Rev.* B3, 2385 (1971).

23. F. Wooten: *Optical Properties of Solids*, p. 127, Academic Press (1972).

24. F. Abeles: *Optical Properties of Solids*, p. 127, North-Holland (1972).

25. L. W. Bos and D. W. Lynch: Low-Energy Optical Absorption Peak in Alminum and Al Mg Alloys, *Phys. Rev. Lett.*, 25, 156 (1970).

26. E. J. Johnson: *Semiconductors and Semimetals*, 3, p. 170, Academic Press (1967).

27. T. S. Moss and T. D. F. Hawkins: Infrared absorption in gallium arsenide, *Infrared Phys.*, 1, 111 (1962).

28. W. G. Spitzerm, M. Gershenzon, C. J. Frosch, D. F. Gibbs: Optical absorption in n-type gallium phosphide, *J. Phys. Chem. Solids*, 11, 339 (1959).

29. P. J. Dean and D. G. Thomas: Intrinsic Absorption-Edge Spectrum of Gallium Phosphide, *Phys. Rev.*, 150, 690 (1966).

30. K. Teegarden and G. Baldini: Optical Absorption Spectra of Alkali Halides at 10K, *Phys. Rev.*, 155, 896 (1967).

31. H. R. Briggs and R. C. Fletcher: New Infrared Absorption Bands in *p*-Type Germanium, *Phys. Rev.*, 87, 1130 (1952).

32. H. R. Briggs and R. C. Fletcher: Absorption of Infrared Light by Free Carriers in Germanium, *Phys. Rev.*, 91, 1342 (1952).

33. W. Kaizer, R. J. Collins and H. Y. Fan: Infrared Absorption in *P*-Type Germsanium, *Phys. Rev.*, 91, 1380 (1953).

34. W. J. Turner and W. E. Reese: Infrared Absorption in *n*-Type Aluminum Antimonide, *Phys. Rev.*, 117, 1003 (1960).

35. W. Paul: Band Structure of the Intermetallic Semiconductors from Pressure Experiments, *J. Appl. Phys.*, 32, 2082 (1961).

36. W. M. Becker, A. K. Randas, and H. Y. Fan: Energy Band Structure of Gallium Antimonide, *J. Appl. Phys.*, 32, 2094 (1961).

37. M. R. Lorentz W. Revter, W. P. Dumke, R. J. Chicotke, G. D. Pettit, and J. M. Woodell: Band Structure and Direct Transition Electroluminscence in the $In_{1-x}Ga_xP$ Alloys, *Appl. Phys. Letters*, 13, 421 (1968).

38. H. Y. Fan: *Semiconductors and Semimetals*, Vol. 3, p. 409, Academic Press (1967).

39. H. Y. Fan and M. Becker: *Semiconducting Materials*, p. 132. Butterworth (1951).

40. S. Visvanathan: Free Carrier Absorption Due to Polar Modes in the III-V Compound Semiconductors, *Phys. Rev.*, 120, 376 (1960).

41. H. Y. Fan, W. Spitzer, and R. J. Collins: Infrared Absorption in *n*-Type Germanium, *Phys. Rev.*, 101, 566 (1956).
42. R. J. Elliott: Intensity of Optical Absorption by Excitons, *Phys. Rev.*, 108, 1384 (1957).
43. T. P. Mclean: *Progress in Semiconductors* (ed. A. F. Gibson), Vol. 5, Heywood (1960).
44. R. K. Willardson and A. C. Beer: *Semiconductors and Semimetals,* Vol. 9, p. 366, Academic Press (1972).
45. Y. Toyozawa: Theory of Line-Shapes of Exciton Absorption Bands, *Prog. Theor. Phys.*, 20, 53 (1959).
46. Y. Toyozawa: Interband effect of lattice vibration in the exciton absorption spectra, *J. Phys. Chem. Solids*, 25, 59 (1964).
47. I. Balslev: Influence of Uniaxial Stress on the Indirect Absorption Edge in Silicon and Germanium, *Phys. Rev.*, 143, 636 (1966).
48. W. E. Engler, M. Garfinkel, and J. J. Tiemann: Piezotransmission Measurements of PhonorrAssisted Transmission in Semiconductors. I. Germanium, *Phys. Rev.*, 155, 693 (1967).
49. M. M. Cohen: *Introduction to the Quantum Theory of Semiconductors*, p. 162, Gordon and Breach Science (1972).
50. P. W. Kruse et al.: *Elements of Infrared Technology*, p. 203, Wiley (1962).
51. K. M. Van Vliet: Noise Limitations in Solid State Photodetectors, *Appl. Optics*, 6, 1145 (1967).
52. M. I. Schultz and G. A. Morton: *Proc. Inst. Radio Eng.*, 43, 1819 (1955).
53. R. H. Bube: *Photoconductivity of Solids*, p. 136, Wiley (1960).
54. W. Heitler: *Quantum Theory of Radiation*, Oxford at the Clarendon Press (1954).
55. L. I. Schiff: *Quantum Mechanics*, McGraw-Hill (1968).
56. R. H. Dicke and J. P. Wittke: *Introduction to Quantum Mechanics*, Addison-Wesley (1960).
57. B. Donovan and J. E. Angress: *Lattice Vibrations*, Chapman and Hall (1971).
58. M. Balkanski: *Elementary Excitation in Solids* (eds. A. A. Maradudin and G. F. Nardelli), p. 113, Plenum Press (1969).
59. M. Balkanski: *Optical Properties of Solids* (ed. F. Abeles), p. 529, North-Holland (1972).
60. J. E. Parrott: *Solid State Theory, Methods and Applications* (ed. P. T. Landsberg), Wiley-Interscience (1969).
61. M. Lax and E. Burstein: Infrared Lattice Absorption in Ionic and Homopolar Crystals, *Phys. Rev.*, 97, 39 (1955).
62. B. DI Bartolo: *Optical Interactions in Solids*, Wiley (1968).
63. R. B. Barnes and M. Czerny: Messungen am NaCl und KC1 im Sprktrabereich ihrer ultraroten Eigenschwingungen, *Z. Physik*, 72, 447 (1931).
64. M. Wakaki, K. Kudo, and T. Shibuya: *Physical Properties and Data of Optical Materials*, p. 290, CRC Press (2007).
65. M. Wakaki, K. Kudo, and T. Shibuya: *Physical Properties and Data of Optical Materials*, p. 258, CRC Press (2007).
66. M. Wakaki, K. Kudo, and T. Shibuya: *Physical Properties and Data of Optical Materials*, p. 443, CRC Press (2007).
67. M. Wakaki, K. Kudo, and T. Shibuya: *Physical Properties and Data of Optical Materials*, p. 431, CRC Press (2007).
68. M. Wakaki, K. Kudo, and T. Shibuya: *Physical Properties and Data of Optical Materials*, p. 336, CRC Press (2007).
69. M. Wakaki, K. Kudo, and T. Shibuya: *Physical Properties and Data of Optical Materials*, p. 347, CRC Press (2007).
70. M. Wakaki, K. Kudo, and T. Shibuya: *Physical Properties and Data of Optical Materials*, p. 329, CRC Press (2007).

71. M. Wakaki, K. Kudo, and T. Shibuya: *Physical Properties and Data of Optical Materials*, p. 114, CRC Press (2007).

72. M. Wakaki, K. Kudo, and T. Shibuya: *Physical Properties and Data of Optical Materials*, p. 109, CRC Press (2007).

73. M. Wakaki, K. Kudo, and T. Shibuya: *Physical Properties and Data of Optical Materials*, p. 43, CRC Press (2007).

74. C. Smart, G. R. Wilkinson, A. M. Karo, and J. R. Hardy: *Lattice Dynamics*, (ed. R. F. Wallis) p. 387 Pergaman, Oxford (1965).

75. L. Genzel, H. Happ, and R. Weber: Dispersionmessungen am NaCl, KC1 und KBr zwischen 0.3 und 3mm. Wellen-Lange, *Z. Physik*, 154, 13 (1959).

76. F. A. Johnson: Lattice Absorption Bands in Silicon, *Proc. Phys. Soc. London*, 73, 265 (1959).

77. S. Bhagavantam and T. Venkatarayudu: *Theory of Groups and Its Application to Physical Problems*, Academic Press (1969).

78. R. S. Halford: Motions of Molecules in Condensed Systems: I. Selection Rules, Relative Intensities, and Orientation Effects for Raman and Infra-Red Spectra, *J. Chem. Phys.* 14, 8 (1946).

79. G. Hertzberg: *Molecular Spectra and Molecular Structure, II. Infrared and Raman Spectra of Polyatomic Molecules*, D. Van Nostrand (1951).

80. A. A. Marddudin, E. W. Montroll, and G. H. Weiss: *Solid State Physics*, Supplement 3, Academic Press (1963).

81. E. B. Wilson, J. C. Decious, and P. C. Cross: *Molecular Vibrations*, McGraw-Hill (1955).

82. G. M. Barrow: *Molecular Spectroscopy*, McGraw-Hill (1962).

83. G. Turrel: *Infrared and Raman Spectra of Crystals*, Academic Press (1972).

84. S. Ganesan, and R. Srinivasan: Temperature Variation of the Effective Gruneisen Parameter in Caesium Chloride Structures, *Proc. Roy. Soc.*, 271, 154 (1962).

85. A. Shauer: Thermal Expansion of Solids and the Temperature Dependence on Lattice Vibration Frequencies, *Can. J. Phys.*, 42, 1857 (1964).

86. S. S. Mitra, C. Postmus and J. R. Ferraro: Pressure Dependence of Long-Wavelength Optical Phonons in Ionic Crystals, *Phys. Rev. Lett.*, 18, 455 (1967).

87. S. S. Mitra, O. Bratman, W. B. Daniels and R. K. Crawford: Pressure-Induced Phonon Frequency Shifts Measured by Raman Scattering, *Phys. Rev.*, 186, 942 (1969).

88. J. R. Ferraro, S. S. Mitra and A. Quattroch: Pressure Dependence of Infrared Eigenfrequencies of KI, Rbl, and Their Mixed Crystals, *J. Appl. Phys.*, 42, 3677 (1971).

89. J. F. Vetelino, L. G. Roy, S. S. Mitra: Lattice dynamical calculation of the second Gruneisen constant of CsBr, *J. Phys. Chem. Solids*, 35, 47 (1974).

90. S. S. Mitra: *Optical Properties of Solids* (ed. S. Nudelman and S. S. Mitra), p. 383, Plenum Press (1969).

91. A. A. Maradudin: *Solid State Physics* (ed. F. Seitz and D. Turnbull), pp. 18, 273 (1966).

92. A. A. Maradudin: *Solid State Physics* (eds. F. Seitz and D. Turnbull), pp. 19, 1 (1966).

93. L. Genzel: *Optical Properties of Solids* (ed. Sol. Nudelman and S. S. Mitra), p. 453, Plenum Press (1969).

94. A. A. Maradudin, E. W. Montroll, G. H. Weiss, and I. P. Ipatova: *Theory of Lattice Dynamics in the Harmonic Approximation, Solid State Physics* (eds. F. Seitz and D. Turnbull), Supplement 3, Academic Press (1971).

95. P. G. Dawber and R. J. Elliot: The Vibrations of an Atom of Different Mass in a Cubic Crystal, *Proc. Roy. Soc.*, A273, 222 (1963).

96. P. G. Dawber and R. J. Elliot: Theory of Optical Absorption by Vibrations of Defects in Silicon, *Proc. Phys. Soc.*, 81, 453 (1963).

97. Y. Mitani and S. Takeno: Frequencies of Localized Vibrations of a Diatomic Simple Cubic Lattice with a Point Imperfection, *Prog. Theor. Phys.*, 33, 779 (1965).

98. S. Takeno: Theory of Impurity-Induced Infrared Absorption in Alkali Halide Crystals, *Prog. Theor. Phys.*, 38, 995 (1967).

99. R. E. Wallis and A. A. Maradudin: Impurity Induced Infrared Lattice Absorption, *Prog. Theor. Phys.*, 24, 1055 (1960).

100. R. C. Newman: Infra-red absorption due to localized modes of vibration of impurity complexes in ionic and semiconductor crystals, *Adv. Phys.*, 18, 545 (1969).

101. A. A. Maradudin, E. W. Montroll, G. H. Weiss, and I. P. Ipatova: *Theory of Lattice Dynamics in the Harmonic Approximation, Solid State Physics* (eds. F. Seitz and D. Turnbull), Supplement 3, p. 453, Academic Press (1971).

102. A. J. Sievers and S. Takeno: Isotope Shift of a Low-Lying Lattice Resonant Mode, *Phys. Rev.*, A140, 1030 (1965).

103. S. Takeno and A. J. Sieverse: Characteristic Temperature Dependence for Low-Lying Lattice Resonant Modes, *Phys. Rev. Lett.*, 15, 1020 (1965).

104. J. F. Angress, A. R. Goodwin and S. D. Simth: A Study of the Vibrations of Boron and Phosphorus in Silicon by Infra-Red Absorption, *Proc. Roy. Soc.*, A287, 64 (1965).

105. R. C. Newman and R. S. Smith: Local mode absorption from boron arsenic and boron phosphorus pairs in silicon, *Solid State Commun.*, 5, 723 (1967).

106. T. Arai: *J. Phys. Soc. Japan*, Suppl. 2, 18, 43 (1963).

107. P. Pajot: Structure fine de l'absorption a 9 u des groupements Si20 dans le Silicium a basse temperature, *J. Phys. Chem. Solids*, 28, 73 (1967).

108. P. Pajot and J. P. Deltour: Etats vibrationnels associes au groupement SiO2 dans le silicium, *Infrared Phys.*, 7, 195 (1967).

109. W. L. Bond and W. Kaiser: Interstitial versus substitutional oxygen in silicon, *J. Phys. Chem. Solids*, 16, 44 (1960).

110. R. M. Chrenko, R. S. McDonald, and E. M. Pell: Vibrational Spectra of Lithium-Oxygen and Lithium-Boron Complexes in Silicon, *Phys. Rev.*, A138, 1775 (1965).

111. W. R. Thorson and I. Nakagawa: Dynamics of the Quasi-Linear Molecule, *J. Chem. Phys.*, 33, 994 (1960).

112. G. Schofer: Das ultrarote spektrum des U-Zentrums, *J. Phys. Chem. Solids*, 12, 233 (1960).

113. R. J. Elliott, W. Hayes, G. D. Jones, H. F. Mac Donald and C. T. Sennett: Localized Vibration of H- and D- Ions in the Alkaline Earth Fluorides, *Proc. Roy. Soc.*, A289, 1 (1965).

114. L. Brilloun: Diffusion de la Lumiere et des Rayon x par un Corps Transparent Homogene, *Ann. Phys.* (Paris), 17, 88 (1922).

115. A. Smekel: Zur quantentheorie der dispersion, *Naturwiss*, 11, 873 (1923).

116. C. V. Raman: A new radiation, *Indian J. Phys.*, 2, 387 (1928).

117. E. Gross: Change of Wave-Length of Light due to Elastic Heat Waves at Scattering in Liquids, *Nature* 126, 201 (1930).

118. E. Burstein: *Elementary Excitation in Solids* (eds. A. A. Maradudin and G. F. Nardelli), p. 367, Plenum Press (1969).

119. R. K. Chang, J. M. Ralson, and D. E. Keating: *Light Scattering Spectra of Solids* (ed. G. B. Wright), Springer-Verlag (1969).

120. H. Born and K. Huang: *Dynamical Theory of Crystal Lattices*, Clarendon Press (1954).

121. R. Loudon: The Raman effect in crystals, *Adv. Phys.*, 13, 423 (1964).

122. R. Loudon: Theory of the First-Order Raman Effect in Crystals, *Proc. Roy. Soc.*, A275, 218 (1963).

123. R. Loudon: Theory of Stimulated Raman Scattering from Lattice Vibrations, *Proc. Phys. Soc.*, 82, 393 (1963).

124. A. K. Gangly and J. L. Birman: Theory of Lattice Raman Scattering in Insulator, *Phys. Rev.*, 162, 806 (1967).

125. G. B. Benedek and K. Fritch: Brillouin Scattering in Cubic Crystals, *Phys. Rev.*, 149, 647 (1966).

126. L. N. Ovander: The Form of the Raman Tensor, *Optics and Spectroscopy*, 9, 302 (1960).
127. L. N. Ovander: Raman Scattering in Piezoelectric Crystals, *Optics and Spectroscopy*, 12, 405 (1962).
128. E. Burstein, F. A. Johnson, and R. Loudon: Selection Rules for Second-Order Infrared and Raman Processes in the Rocksalt Structure and Interpretation of the Raman Spectra of NaCl, KBr, and NaI, *Phys. Rev.*, 139, 1239 (1965).
129. A. M. Karo, and A.P. Korolkov J. R. Hardy and I. Morrison: Theoretical interpretation of thesecond-order Raman spectra of caesium halides, *J. de Phys.*, 26, 668 (1965).
130. A. I. Stekhanov and A. P. Korolkov: *Sov. Phys. Solid State*, 4, 945 (1962).
131. C. H. Henry and J. J. Hopfield: Raman Scattering by Polaritons, *Phys. Rev. Lett.*, 15, 964 (1965).
132. T. C. Damen, S. P. S. Porto, and B. Tell: Raman Effect in Zinc Oxide, *Phys. Rev.*, 142, 570 (1966).
133. J. F. Scott, L. E. Cheesman, and S. P. S. Porto-* Polariton Spectrum of α-Quartz, *Phys. Rev.*, 162, 834 (1967).
134. K. Hisano, Y. Okamoto, and O. Matsumura: Spectral Emission by Polaritons in LiF, *J. Phys. Soc. Japan*, 28, 425 (1970).
135. N. Ohama, Y. Okamoto, and O. Matsumura: Polariton Dispersion Relation in Cubic BaTiO$_3$, *J. Phys. Soc. Japan*, 29, 1648 (1970).
136. S. M. Shapiro, R. W. Gammon and H. Cummins: Brillouin Scatteringi Spectra of Crystalline Quartz, Fused Quartz and Glass, *Appl. Phys. Lett.*, 9, 157 (1966).
137. H. Z. Cummins and P. E. Schoen: *Physical Applications of Linear Scattering from Thermal Fluctuations*, Laser Handbook 2 (ed. F. T. Arecchi and E. O. Schultz-Dubois) North-Holland, Amsterdam 1029 (1972), p. 1045.

2 Optical Materials for Ultraviolet, Visible, and Infrared

Toshihiro Arai (deceased) and Moriaki Wakaki

CONTENTS

2.1 INTRODUCTION[1]

Up to the nineteenth century, materials for optics were limited to passive elements in the visible region such as spectacles, lenses, deflectors, beam splitters, and mirrors. Therefore, the main materials for optics were either transparent materials with suitable refractive indices like glasses or opaque materials with the high reflectivity like metals.

Years ago, the aging effect of an optical glass surface was discovered. The effect corresponds to the change in the refractive index of a lens surface caused by the humidity in air. This phenomenon was used for forming an antireflection layer on a glass surface. After time, the multilayer systems began to be used for making high reflection and antireflection coatings, and band pass filters. Such multilayer systems are mainly fabricated by the vacuum evaporation technique. The dielectric materials that fit for the evaporation technique became optical materials. In addition, the wavelength region treated in optics was expanded to both infrared and ultraviolet regions. The sorts of optical materials have increased to meet these needs. For example, some kinds of metals such as tungsten and tantalum have been used in the vacuum ultraviolet region, and some kinds of ionic crystals such as NaCl and KCl, and some kinds of semiconductors like Si and Ge have been used in the near- and middle-infrared regions.

Electromagnetic energy transforms into some kinds of energy by the absorption process. For example, the energy of an electromagnetic wave should be changed into other kinds of energy such as the electric current and voltage for the quantitative detection of the light intensity. Some optically active materials are necessary for these quantitative changes in energy. As is well known, optical absorption occurs by the resonance between the frequencies of the incident radiation and the oscillation in the material. The frequency of oscillation in the classical theory corresponds to the excitation energy (in the quantum mechanical theory) of the elementary particles, such as electrons and phonons. The excitation energies in the electron systems corresponding to the transitions from the valence band to the conduction band or from the impurity level to the band in semiconductors and ionic crystals are in the photon energy region between ultraviolet and near infrared. In this chapter, the materials for active elements in optics are not discussed.

2.2 BASIC MATTERS[2–5]

2.2.1 Refractive Index, Extinction Coefficient, and Dielectric Constant

As is well known, light is one of the electromagnetic waves vibrating periodically in time and space. In the isotropic space, the electric field and the magnetic field vibrate in perpendicular planes of each other. The direction of the propagation wave is also perpendicular to both fields in the isotropic space. So, the light is the transverse wave characterized by the periodicity in space (wavelength), the periodicity in time (frequency), the phase, the direction of the vibrating electric field (polarization), and the amplitude. These characters vary with the change of light-propagating medium, because the electric field induces the electric dipole moment in the medium through the interaction with the electric charges in the medium such as electrons, nuclear charges, molecules, and ions. The induced dipole moments should be calculated by the quantum mechanical method. When the distribution of the dipole moments in the medium is larger compared to the wavelength, the localized approximation cannot be applied for the calculation. In this case, the Maxwell equations and the formula of the transition matrix can be used. However, the localized approximation usually holds. Therefore, the results deduced from the Maxwell equations will be used here. According to the Maxwell equations, the optical property of material can be treated by the electric susceptibility, χ, which is a proportional constant in the relation between the induced polarizability P and the electric field of incident wave E.

$$P = \varepsilon_0 \chi(\omega) E \tag{2.1}$$

Here, ε_0 is the dielectric constant of vacuum. $\chi(\omega)$ relates to the refractive index n and the dielectric constant ε of the material.

$$1+\chi=\varepsilon=n^2 \tag{2.2}$$

χ and n are the first or the second rank tensor when the material has optical anisotropy.

The dielectric constant can be calculated by the mechanical spring model, called the *Lorentz model*. If a particle with a mass m and an electric charge e is moved by the external force $\boldsymbol{E} = \boldsymbol{E}_0 exp(i\omega t)$, the equation of motion is given by

$$m\frac{d^2\boldsymbol{r}}{dt^2}+m\gamma\frac{d\boldsymbol{r}}{dt}+m\omega_0\boldsymbol{r}=e\boldsymbol{E} \tag{2.3}$$

In this equation, the third term on the left-hand side is the restoring force that is proportional to the displacement from the equilibrium position, ω_0 is the angular resonance frequency of this spring system, and γ is the damping rate. The complex dielectric constant $\tilde{\varepsilon}$ induced by the interaction of this harmonic oscillator with the light is given by the following equation:

$$\tilde{\varepsilon}=1+\frac{Ne^2}{m\varepsilon_0}\frac{f}{\omega_0^2-\omega^2+i\gamma\omega} \tag{2.4}$$

Here, N is the number of oscillators per unit volume, and f is the coupling strength between the light and the spring. f is usually called the oscillator strength or f value. The $\tilde{\varepsilon}$ is divided into a real part ε_1 and an imaginary part ε_2.

$$\tilde{\varepsilon}=(\mathrm{Re}\,\varepsilon)+i(\mathrm{Im}\,\varepsilon)=\varepsilon_1+i\varepsilon_2 \tag{2.5}$$

The $\tilde{\varepsilon}$ is related with the complex refractive index $\tilde{n}$ as

$$\tilde{\varepsilon}=\tilde{n}^2=(n+i\kappa)^2 \tag{2.6}$$

where n and κ are the refractive index and the extinction coefficient, respectively. From Equation (2.5) and Equation (2.6), ε_1 and ε_2 are related with n and κ as follows:

$$\varepsilon_1=n^2-\kappa^2, \varepsilon_2=2n\kappa \tag{2.7}$$

Typical behaviors of the dielectric functions ε_1 and ε_2 around the resonance frequency ω_0 are shown in Figure 2.1 as the function of the angular frequency ω. The half width of ε_2 band is almost equal to γ. The difference of ε_1 values at the high-frequency side and at the low-frequency side of the band is due to the contribution of this oscillator.

The dispersion relation in optics means the frequency dependence of the ε value. The dispersion spectrum is determined by three main causes. One is the electron excitation from the valence band to the conduction band in semiconductors and ionic crystals. This transition in these materials is in the wavelength range between near ultraviolet and near infrared. The dispersion relation in the visible and infrared regions is mainly determined by the band-to-band transition. The dispersion relation

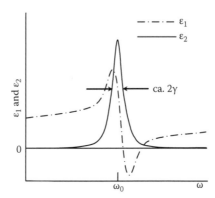

FIGURE 2.1   Dispersion curves of the real part ε_1 and imaginary part ε_2 of dielectric constant ε around resonance frequency ω_0. FWHM (full width at half maximum) of the ε_2 band is almost 2γ.

at the infrared region in ionic crystals is determined by the absorption due to the optical phonon. In addition, the other type of dispersion is caused by the excitation in the same band, called plasma excitation in metals. The real part of the dielectric constant, ε_1, becomes more negative in the range of the higher-frequency side than the resonance frequency of the oscillator as is shown in Figure 2.1. In this frequency region, the reflectivity is almost 100%. The optical ray in this range is termed a *residual ray*.

2.2.2   REFLECTANCE, TRANSMITTANCE, AND ABSORPTION CONSTANT

When a light beam falls on the boundary of the materials that have different dielectric constants, the light beam is reflected and refracted. The *p*-polarization wave where the electric field and displacement are vibrating in the incident plane and the *s*-polarization wave where the electric field and displacement are vibrating in the perpendicular plane to the incident plane have different boundary conditions, if the light beam falls on the boundary obliquely. Therefore, the reflectances and the transmittances of *p*- and *s*-polarized waves differ. When the propagation directions of *p*- and *s*-polarized waves are the same on the surface of isotropic materials, both waves refract according to Snell's law. When a light beam is incident on the surface of a transparent material with the incident angle θ_i, Snell's equation is given by

$$n_i \sin\theta_i = n_t \sin\theta_t \tag{2.8}$$

where n_i and n_t are the refractive indices of the incident side and transmitted side materials, respectively, and θ_t is the refraction angle.

The amplitude transmittance of both polarization waves is given by the following equations:

$$\tilde{t}_p = \frac{E_{t,p}}{E_{i,p}} = \frac{2\sin\theta_t \cos\theta_i}{\sin(\theta_i + \theta_t)\cos(\theta_i - \theta_t)} = \left|\tilde{t}_p\right| e^{i\varphi_p} \tag{2.9a}$$

$$\tilde{t}_s = \frac{E_{t,s}}{E_{i,s}} = \frac{2\sin\theta_t\cos\theta_i}{\sin(\theta_i + \theta_t)} = |\tilde{t}_s|e^{i\varphi_s} \tag{2.9b}$$

The tilde on each symbol means the complex value.

The amplitude reflectances of both waves are given by the following equations:

$$\tilde{r}_p = \frac{E_{r,p}}{E_{i,p}} = \frac{\tan(\theta_i - \theta_t)}{\tan(\theta_i + \theta_t)} = |\tilde{r}_p|e^{i\varphi_p} = 1 - \tilde{t}_p \tag{2.10a}$$

$$\tilde{r}_s = \frac{E_{r,s}}{E_{i,s}} = \frac{\sin(\theta_i - \theta_t)}{\tan(\theta_i + \theta_t)} = |\tilde{r}_s|e^{i\varphi_s} = 1 - \tilde{t}_s \tag{2.10b}$$

The intensity of a light is given by the square of the amplitude. Therefore, the intensity reflectances are given by the square of each value. But the intensity transmittances are multiplied by the factor $n_t\cos\theta_t/n_i\cos\theta_i$ for the square of each amplitude transmittance due to the change of velocity of light at both media.

The absorption and scattering losses are given by the ratio of the intensity at the observing point, I_0, to the incident intensity, I_i.

$$I_0/I_i = e^{-\alpha d} \tag{2.11}$$

Here, d is the propagation distance, and α is the intensity loss per unit length and is called the absorption coefficient when the loss is caused by the absorption. The absorption coefficient relates with the extinction coefficient κ through the following relation:

$$\alpha = \frac{4\pi\kappa}{\lambda} \tag{2.12}$$

If there are multiple causes for the absorption, the principle of superposition holds for α. When the multiple interference effect is taken into consideration, the intensity transmittance T_d of the optically polished thin parallel plate in the air is given as follows for normal incidence:

$$T_d = \frac{I_t}{I_i} = \frac{\left[(1-R)^2 + 4R\sin\varphi\right]e^{-\alpha d}}{(1 - Re^{-\alpha d})^2 + 4Re^{-\alpha d}\sin^2(\varphi + \beta)} \tag{2.13}$$

Here, R is the intensity reflectance on the surface and is given by

$$R = \frac{(n_t - 1)^2 + \kappa_t^2}{(n_t + 1)^2 + \kappa_t^2} \tag{2.14}$$

where φ is the change in the phase at the reflection. If κ_t is almost 0, φ is π. β is the optical thickness of the plate expressed in the phase unit.

$$\beta = \frac{2\pi n_t d}{\lambda} \tag{2.15}$$

Here, λ is the wavelength in vacuum. When $n_t^2 \gg \kappa_t^2$, one can use the approximation of $\kappa_t^2/n_t^2 \sim 0$ and $\varphi = \pi$. In these approximations, T_d becomes

$$T_d = \frac{(1-R)^2 e^{-\alpha d}}{(1-Re^{-\alpha d})^2 + 4Re^{-\alpha d}\sin^2\beta} \tag{2.16}$$

When the plate is very thick or very thin, the interference effect can be neglected, and T_d is given by

$$T_d = \frac{(1-R)^2 e^{-\alpha d}}{1-R^2 e^{-2\alpha d}} \tag{2.17}$$

In addition, R becomes small when κ_t and n_t are small. In that case, T_d can be calculated by the following simple equation:

$$T_d = (1-R)^2 e^{-\alpha d} \tag{2.18}$$

Next, the normal reflectance R_d from the optically polished parallel plate with a thickness d is treated. Corresponding to Equation (2.13), R_d is given by the following equation:

$$R_d = \frac{I_R}{I_i} = \frac{\left[(1-e^{-\alpha d})^2 + 4e^{-\alpha d}\sin^2\beta\right]R}{(1-Re^{-\alpha d})^2 + 4Re^{-\alpha d}\sin^2(\varphi+\beta)} \tag{2.19}$$

In the case corresponding to Equation (2.16), R_d is represented by the following equation:

$$R_d = \frac{R(1-e^{-\alpha d})^2}{(1-Re^{-\alpha d})^2 + 4Re^{-\alpha d}\sin^2\beta} \tag{2.20}$$

In the case corresponding to Equation (2.17), R_d can be written as follows:

$$R_d = R\left[1 + \frac{(1-R)^2 e^{-2\alpha d}}{1-R^2 e^{-\alpha d}}\right] = R(1+T_d e^{-\alpha d}) \tag{2.21}$$

The amplitude reflectance r on a surface of transparent material is treated. The amplitude reflectance of the s-polarized light r_s is always negative at any incident angle, and the amplitude reflectance of the p-polarized light r_p is negative only when the summation of the incident angle θ_i and the refraction angle θ_r, $(\theta_i + \theta_r)$, is larger than $\pi/2$. The phase of reflected light amplitude differs by π from that of the incident light at any incident angle for the s-polarized light. However, for p-polarized light, the phase between incident and reflected lights differs by π, only when the summation of angles of incidence and of refraction is larger than $\pi/2$. The reflectivity of

p-polarized light becomes zero at the turning point in the phase of p-polarized light that is at $(\theta_i + \theta_t) = \pi/2$. Therefore, the reflected light becomes a linear polarized light that has only s-component. This incident angle is called a *Brewster angle*, θ_B. The θ_B is connected with the refractive index of the material n_t by the following equation:

$$\tan \theta_B = \frac{n_i}{n_t} = \frac{1}{n_t} \tag{2.22}$$

When the material is opaque, the ratio of the s-polarized light to the p-polarized light at the reflection is written as follows:

$$\frac{\tilde{r}_s}{\tilde{r}_p} = -\frac{\cos(\theta_i - \theta_t)}{\cos(\theta_i + \theta_t)} \cdot \frac{E_{is}}{E_{ip}} = \tan \psi e^{i\Delta} \frac{E_{is}}{E_{ip}} \tag{2.23}$$

Here, $\tilde{r}_s = |\tilde{r}_s| e^{i\varphi_s}$, $\tilde{r}_p = |\tilde{r}_p| e^{i\varphi_p}$, and $\Delta = \varphi_s - \varphi_p$. When the incident angle is satisfied the condition $\Delta = \pi/2$, one of the principal axes of the elliptically polarized reflection wave is in the plane of incidence. This incident angle is called a *principal angle*.

When the light enters from the optically higher-density media to the lower-density media, θ_i is less than θ_t. Therefore, the refraction angle becomes $\pi/2$ at some incident angle. This incident angle is called a *critical angle* θ_c and is given by

$$\sin \theta_c = \frac{n_t}{n_i} \tag{2.24}$$

When the incident angle becomes larger than θ_c, the light is reflected perfectly. This phenomenon is called a *total reflection* and the light propagates along the boundary face. The electric field normal to the boundary decreases exponentially with the distance from the boundary. The depth d_p where the strength of the electric field attenuates down to $1/e$ is given by

$$d_p = \frac{\lambda}{2\pi \left[\sin^2 \theta_i - (\frac{n_t}{n_i})^2 \right]^{\frac{1}{2}} \cdot n_i} \tag{2.25}$$

This phenomenon is utilized for the analysis of the surface layer and also can be used for the optical fiber.

2.3 MATERIALS FOR POSSIBLE ELEMENTS

Figure 2.2 shows an example of an optical path system. As seen in the figure, various optical elements are used. The main materials used to construct these elements are optically transparent materials such as oxide glasses, chalcogenide glasses, and organic glasses. Some optically anisotropic crystals are used for polarizers, beam splitters, and wavelength plates. The optical thin films deposited by

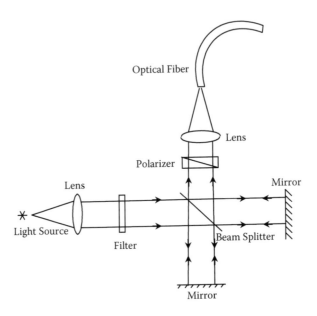

FIGURE 2.2   Model of optical path system.

vacuum evaporation, sputtering, and other methods are also important for filters, lossless high reflectance mirrors, and so forth. Some typical optical materials will be described in the following sections.

2.3.1   OXIDE SILICATE GLASS

Oxide silicate glasses are used widely for lens materials, because the refractive indices of the oxide silicate glasses can be varied almost continuously by changing the composition, and the transmittances of the oxide silicate glasses are very high compared with those of other materials. There are prisms, diffraction gratings, photonic crystals, and so forth, for light dispersion tools, and corner cube prisms and beam splitters can be used for the deflection of an optical beam. Furthermore, there are optical fibers and pipes for the light transmission line and image formation.

The control of the refractive index of the glass is very important to compose the no aberration lens system. Before the nineteenth century, optical glasses were only Crown and Flint glasses. In the first part of the nineteenth century, boric acid glasses doped with rare earth elements were discovered. These glasses had large refractive index ranging around 1.7 ~ 1.9. In the first part of the twentieth century, fluoride glasses doped with La, Y, Hf, and so forth, have been discovered. The characteristic properties of these glasses have large dispersion and small refractive index. Now, over 300 kinds of glasses are available. The characteristic properties of optical glasses are expressed by using the graph of the refractive index n_d at the d-line of He atom, 587.56 nm, versus Abbe's number, $v_d = (n_d - 1/n_f - n_c)$. Here, n_f and n_c are the refractive indices at the F-line, 486.13 nm, and the C-line, 656.27 nm, of the hydrogen atom, respectively. Figures 2.3a and 2.3b show the n_d-v_d curves.[6–8] To built up

(a)

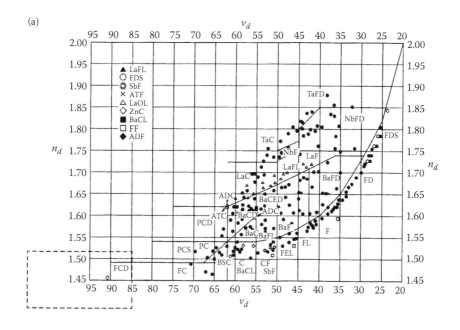

(b)

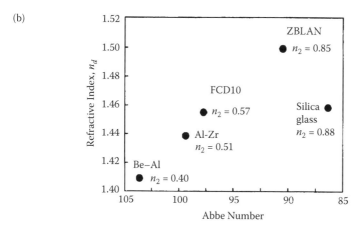

FIGURE 2.3 (a) Refractive index at D emission line (589 nm) of sodium, n_d versus Abbe number of optical glasses, ν_d. (See also K. Asahara: *Fundamentals and Applications of Optical Elements*, M. Ohi (ed.), p. 14, Japan Scientific Societies Press (1996) (in Japanese); *Catalogue of Optical Glasses*, HOYA Corporation (in Japanese).) The square region enclosed by a dashed line is enlarged in (b). (b) n_d versus ν_d of fluoride phosphoric acid and fluoride glasses. (See also *Catalogue of Optical Glasses*, HOYA Corporation (in Japanese); K. Asahara: *Fundamentals and Applications of Optical Elements*, M. Ohi (ed.), p. 15, Japan Scientific Societies Press (1996) (in Japanese).)

the lens system without chromatic aberration at the *F*-line and the *C*-line, the condition $\Sigma(1/f_i\nu_i) = 0$ must be satisfied in the composition of lenses. Here, f is the focal length of the lens. To eliminate the spherical aberration of the Seidel's five aberrations in the system, the glasses that have the same n_d but have large differences in Abbe's numbers must be combined. Figures 2.3a and 2.3b are convenient in the selection of the combination of lens glasses.

The absorption edges of fluoride and fluorine-phosphate glasses are in the shorter wavelength compared with those of typical oxide glasses. As a result, Abbe's numbers are large, and the dispersions of refractive indices in the visible region are very small. Both glasses are also used as transmitting materials in the ultraviolet and near-infrared regions. The second-order nonlinear refractive indices of both glasses are about 5×10^{-14} esu, and these values are much smaller than the value 1.4×10^{-13} esu of the usual silicate and boric acid glasses. Both glasses have disadvantages as materials for nonlinear optics. On the other hand, nonlinear effects do not appear when both glasses are used under intensive light irradiation.

The refractive index is determined by the intensity of induced dipole moments as described before. The values of the molecular refractive indices in usual silicate glasses are determined by the monobonding oxygen ions, for example, Si-O-Ba. The addition theorem holds in the molecular refractive index. The ion refractive index of the positive ion in glass, for example, Si ion in SiO_2 glass or B ion in B_2O_3 glass, is constant, but the ion refractive index of oxygen ion is determined by the ion combined to the bond of the other side of the Si-O or B-O bonds. The molecular refractive index in usual silicate glasses is determined by x in an Si-O-x bond.

2.3.2 CRYSTAL AND AMORPHOUS (FUSED) SILICATE

Crystal silica or quartz and fused silica are transparent in near-ultraviolet, visible, and near-infrared regions. Therefore, both materials are used as various elements in these regions, such as lenses, prisms, windows, and optical guides (optical fibers).

There are natural quartz and synthetic quartz. The thermal expansion coefficient of fused silica is very small. Therefore, a fused silica is used as the substrate for precise elements such as an etalon plate and an interference thin film. The transmittance spectra in ultraviolet and near-infrared regions of fused silica and synthetic quartz plates of 10 mm in thickness are shown in Figures 2.4a and 2.4b.[9-11] Recently, fused silica includes very few H_2O components. As a result, the absorption band at 2.7 µm due to OH bond vibration shown in Figure 2.4b is not observed for such a sample. No difference in quality is found between fused silica and synthetic quartz, except for a crystallography property.

As described in a later section, quartz belongs to the hexagonal system and the optical anisotropic material. Quartz also has the polarizing power, optical rotation power, and piezo power. So, it can be used for the materials of polarizing plates, azimuth polarimeters, oscillators, and so on. However, these properties are not so convenient to make conventional simple optical systems. Fused silica is better than quartz to apply for conventional optical systems.

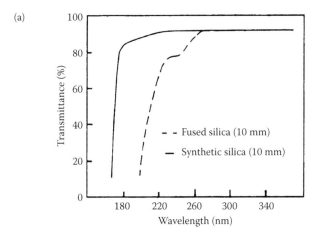

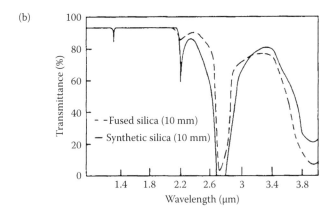

FIGURE 2.4 (a) Transmittance spectra of fused and synthetic silica with the thickness 10 mm in the ultraviolet region. (b) Transmittance spectra of fused and synthetic silica with the thickness 10 mm in near- and middle-infrared regions. (See also T. Ogawa: *Fundamentals and Applications of Optical Elements*, M. Ohi (ed.), p. 16, Japan Scientific Societies Press (1996) (in Japanese); M. Wakaki, K. Kudo, and T. Shibuya: *Physical Properties and Data of Optical Materials*, pp. 357, 370 CRC Press (2007).)

2.3.3 MATERIALS FOR ULTRAVIOLET REGION

Optical materials other than fused silica in the wavelength region between 200 nm and 300 nm are Vycor glass (borosilicate glass) and sapphire. The transmittance spectra of both materials are shown in Figure 2.5.[10–13] Optical material used in the near vacuum ultraviolet region, 100 ~ 250 nm, is only fluoride, such as lithium fluoride (LiF), calcium fluoride (CaF_2), and magnesium fluoride (MgF_2). LiF can be used down to the shortest wavelength as is seen in Figure 2.6.[12,14–16] However, it melts slightly in water and is damaged by longtime ultraviolet light exposure. Recently, zone plates made by nonoxidized hard metals such as Ta and W have been used as an imaging tool.

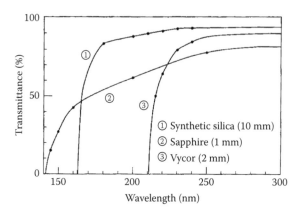

FIGURE 2.5 Transmittance spectra of near-ultraviolet optical materials. (See also M. Wakaki, K. Kudo, and T. Shibuya: *Physical Properties and Data of Optical Materials*, pp. 357, 370, 385, CRC Press (2007); T. Aoki: *Fundamentals and Applications of Optical Elements*, M. Ohi (ed.), p. 19, Japan Scientific Societies Press (1996) (in Japanese).) Numbers within the parentheses are the thicknesses of the samples.

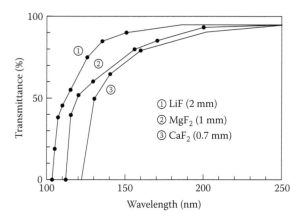

FIGURE 2.6 Transmittance spectra of vacuum ultraviolet optical materials. (See also T. Aoki: *Fundamentals and Applications of Optical Elements*, M. Ohi (ed.), p. 19, Japan Scientific Societies Press (1996) (in Japanese); M. Wakaki, K. Kudo, and T. Shibuya: *Physical Properties and Data of Optical Materials*, pp. 98, 258, 277, CRC Press (2007).) Numbers within the parentheses are the thicknesses of the samples.

2.3.4 Materials for Infrared Region (800 nm ~ 3500 nm)

2.3.4.1 Alkali Halide Crystals

The most popular materials used in the infrared region are alkali halides like NaCl, KBr, and CsI crystals. Large crystals of these materials can be made by the Czochralski method. The transmittance spectra are shown in Figure 2.7.[17-20] These

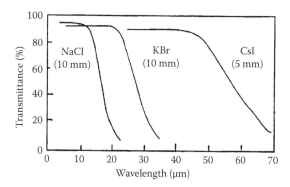

FIGURE 2.7 Infrared transmittance spectra of alkali halide crystals. (See also T. Ogawa: *Fundamentals and Applications of Optical Elements*, M. Ohi (ed.), p. 20, Japan Scientific Societies Press (1996) (in Japanese); M. Wakaki, K. Kudo, and T. Shibuya: *Physical Properties and Data of Optical Materials*, pp. 114, 329, 431, CRC Press (2007).) Numbers within the parentheses are the thicknesses of the samples.

materials have the property of deliquescence. Therefore, NaCl, KBr, and CsI must be kept in humidity lower than 50%, 40%, and 30%, respectively. The KBr crystal has a cleavage surface along (100). These crystals can be polished easily to an optical grade surface.

2.3.4.2 Other Materials

Silver chloride (AgCl) and silver bromide (AgBr) are also used. However, both materials are very soft. It is very difficult to polish these materials optically due to the softness. So, thin films with the thickness of 30 μm can be made by the rolling method. Fluorides such as LiF, MgF_2, CaF_2, and BaF_2 also can be used. The transmission spectra of these materials are shown in Figure 2.8.[14–16,21,22] The reflection losses of these materials are small according to their small refractive indices in the transparent wavelength region. These materials are suitable for optical windows but are weak against heat shock.

The mixed materials of TlBr and TlI, KRS-5, are transparent up to 50 μm. The transmittance spectra are shown in Figure 2.9 together with that of KRS-6.[23–25] KRS-5 has red-orange color and the refractive index in the transparent region is ca. 2.4. As a result, the reflectance of the KRS-5 window is about 25%. KRS-5 and -6 are harmful materials. These materials can be used as the optical fibers transmitting light up to 30 μm.

Some kinds of semiconductors can also be used for the optical materials in the infrared region. For example, chalcogenide materials like ZnS and ZnSe, and arsenic materials like As_2S_3 and As_2Se_3 glasses are transparent in the near-infrared region. The transmission spectra are shown in Figures 2.10a[23,26,27] and 2.10b.[28–30] Quartz and sapphire, Al_2O_3, can also be used in the near-infrared region. Especially, fused silica and a mixture glass of SiO_2 and GeO_2 are used as the core and the cladding materials of the optical fiber for optical communication, respectively.

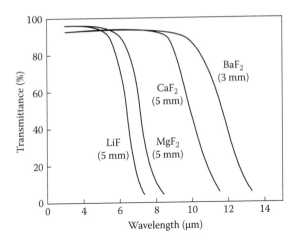

FIGURE 2.8 Infrared transmittance spectra of fluoride crystals. (See also M. Wakaki, K. Kudo, and T. Shibuya: *Physical Properties and Data of Optical Materials*, pp. 43, 98, 258, 277, CRC Press (2007); T. Ogawa: *Fundamentals and Applications of Optical Elements*, M. Ohi (ed.), p. 21, Japan Scientific Societies Press (1996) (in Japanese).) Numbers within the parentheses are the thicknesses of the samples.

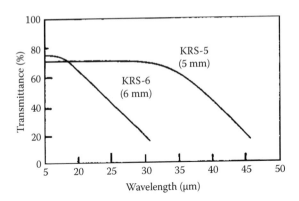

FIGURE 2.9 Infrared transmittance spectra of KRS-5 and -6 mixed crystals. (See also T. Ogawa: *Fundamentals and Applications of Optical Elements*, M. Ohi (ed.), p. 22, Japan Scientific Societies Press (1996) (in Japanese); M. Wakaki, K. Kudo, and T. Shibuya: *Physical Properties and Data of Optical Materials*, pp. 488, 492, CRC Press (2007).) Numbers within the parentheses are the thicknesses of the samples.

2.4   OPTICAL ANISOTROPIC MATERIALS

2.4.1   GENERAL DESCRIPTION

Crystals are classified into seven crystal systems by their structural symmetry. These are triclinic, monoclinic, orthorhombic, tetragonal, cubic, trigonal, and hexagonal systems. The dielectric constants of crystals are expressed by the symmetric tensor

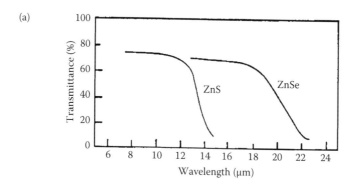

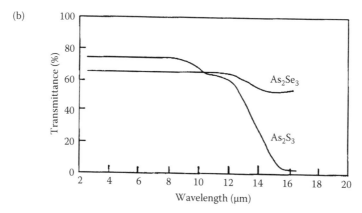

FIGURE 2.10 (a) Infrared transmittance spectra of ZnS and ZnSe crystals. (See also T. Ogawa: *Fundamentals and Applications of Optical Elements*, M. Ohi (ed.), p. 22, Japan Scientific Societies Press (1996) (in Japanese); M. Wakaki, K. Kudo, and T. Shibuya: *Physical Properties and Data of Optical Materials*, pp. 488, 492, 523, 529, CRC Press (2007).) (b) Infrared transmittance spectra of amorphous As_2S_3 and As_2Se_3. (See also T. Ogawa: *Fundamentals and Applications of Optical Elements*, M. Ohi (ed.), p. 23, Japan Scientific Societies Press (1996) (in Japanese); M. Wakaki, K. Kudo, and T. Shibuya: *Physical Properties and Data of Optical Materials*, pp. 31, 34, CRC Press (2007).)

due to the symmetry of crystals. The dielectric tensor can become a diagonal tensor by the suitable rotation of the coordinates. It is expressed as

$$\varepsilon = \begin{pmatrix} \varepsilon_1 & 0 & 0 \\ 0 & \varepsilon_2 & 0 \\ 0 & 0 & \varepsilon_3 \end{pmatrix} \quad (2.26)$$

The ε_1, ε_2, and ε_3 are called principal dielectric constants along the optical axis. When $\varepsilon_1 = \varepsilon_2 = \varepsilon_3$, the crystal is optically isotropic. Only the crystals belonging to the cubic system are optically isotropic. The case of $\varepsilon_1 = \varepsilon_2 \neq \varepsilon_3$ is referred to as optically uniaxial. The crystals belonging to tetragonal, trigonal, and hexagonal systems are optically uniaxial. The case of $\varepsilon_1 \neq \varepsilon_2 \neq \varepsilon_3$ is referred to as optically biaxial.

The crystals belonging to triclinic, monoclinic, and orthorhombic systems are optically biaxial.

Two kinds of light velocity, $c_o = c / \sqrt{\varepsilon_1}$ and $c_e = c / \sqrt{\varepsilon_3}$ exist in the uniaxial crystals. The light propagating with the velocity c_o is termed an *ordinary ray* and its direction is called an *optical axis*. The electric vector of the ordinary ray vibrates in the principal plane, which is the plane including the direction of normal to the wave vector a and the optical axis. The light propagating with the velocity c_e is termed an *extraordinary ray*. The electric vector of the extraordinary ray vibrates in the principal section of the refractive index ellipsoid, which is given at Equations (2.27) and (2.27').

In general, the dielectric constants can be written with following equations:

$$\frac{x^2}{\varepsilon_1} + \frac{y^2}{\varepsilon_2} + \frac{z^2}{\varepsilon_3} = 1 \tag{2.27}$$

$$\frac{x^2}{n_1^2} + \frac{y^2}{n_2^2} + \frac{z^2}{n_3^2} = 1 \tag{2.27'}$$

The ellipsoid given by this equation is called the refractive index ellipsoid. In the optically uniaxial crystal ($\varepsilon_1 = \varepsilon_2$), the index ellipsoid is formed from a sphere and an ellipse, and it has a rotating symmetry axis as is shown in Figure 2.11. Figures 2.12a and 2.12b show the cross section cut by the plane including the optical axis. As is known from this figure, the light velocity of an extraordinary ray varies depending on the propagating direction of the light.

The ray of light incident upon a parallel plate made by an optically uniaxial crystal separates to the ordinary ray and the extraordinary ray. These rays propagate to different directions. This phenomenon is called *birefringence* or *double refraction*.

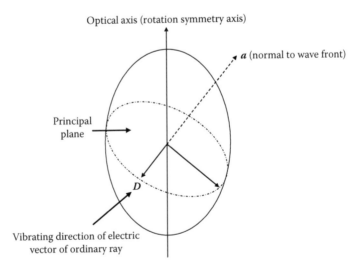

FIGURE 2.11 Refractive index ellipsoid.

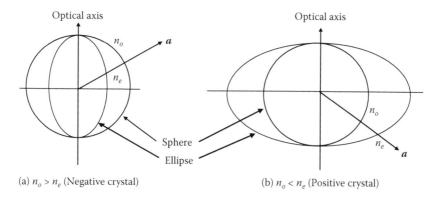

(a) $n_o > n_e$ (Negative crystal)   (b) $n_o < n_e$ (Positive crystal)

FIGURE 2.12   The cross sections of refractive index ellipsoid for positive (a) and negative (b) crystals.

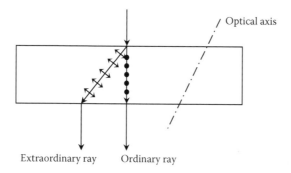

Extraordinary ray   Ordinary ray

FIGURE 2.13   Propagations of ordinary and extraordinary rays in optically uniaxial crystal.

For simplicity, we will treat this phenomenon assuming that the optical axis is parallel to the plane. As is seen in Figure 2.13, the ordinary ray propagates straight, but the extraordinary ray refracts at each boundary surface. The small arrows and the dots in the figure show the vibrating directions of the electric fields of each ray.

In the biaxial crystal, the relation $\varepsilon_1 \neq \varepsilon_2 \neq \varepsilon_3$ holds. Now for simplicity, we assume $\varepsilon_1 < \varepsilon_2 < \varepsilon_3$—that is, $n_z < n_y < n_x$. Here, n_x, n_y, and n_z are the refractive indices in each direction, respectively. In this case, the three cross sections normal to each axis of the refractive index ellipsoid are all ellipses as shown in Figure 2.14. As is seen in Figure 2.14, there are two cross sections to give a circle in the ellipsoid. As a result, there are two optical axes that are perpendicular to the circular cross sections. When the light propagates along the optical axis, the dielectric constant is one kind. However, when the light propagates along another direction, the dielectric constant differs depending on the polarization direction. Light can be divided into two linear polarized lights crossed at a right angle. Therefore, the birefringence phenomenon appears when light propagates along different directions from the optical axis. The birefringence phenomenon is utilized to make polarizing prisms, polarizing beam splitters and phase plates, and so on. The most popular materials used to make that

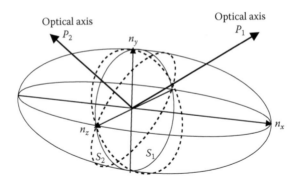

FIGURE 2.14   Refractive index ellipsoid for biaxial crystal. S_1 and S_2 are two cross sections of the ellipsoid to give circular shapes. P_1 and P_2 are the directions of optical axes that are normal to S_1 and S_2, respectively.

equipment in ultraviolet, visible, and near-infrared regions are crystal quartz and calcite, which belong to the trigonal system.

2.4.2   POLARIZERS AND POLARIZING BEAM SPLITTERS

A polarizing film utilizes the birefringence phenomenon due to the ordering in the orientation of the molecular arrangement. In general, polarizing films are made by the following process. As the first step, a polyvinyl-alcohol film in which iodine is absorbed is expanded to one direction by stretching to make the orientation order of molecules. After the process, the film is sandwiched between acetyl-butyl-cellulose sheets. This polarizing film can be used in the temperature range –54 ~ 80°C. These polarizing films are called Polaroid. Polaroid is the trade name of the Polaroid Corporation, Cambridge, Massachusetts. Many kinds of Polaroid sheets are sold by the Polaroid Corporation. The transmission spectra of some of them are shown in Figures 2.15a, 2.15b, 2.15c, and 2.15d.[31] The HN 7 is used in the wavelength range 0.7 ~ 0.9 μm and HN 38 in the visible wavelength range. HR is used for the infrared range (0.5 ~ 2.2 μm), and HNP′B is for the ultraviolet range (260 nm ~ 420 nm).

Several kinds of typical polarizing prisms are shown in Figures 2.16a, 2.16b, 2.16c, 2.16d, 2.16e, and 2.16f. Prisms (a) and (b) are made by two pieces of calcite stuck together with Canada balsam. The refractive index of Canada balsam is 1.55, which has an intermediate value between the refractive indices of ordinary ray, 1.66, and extraordinary ray, 1.47, in calcite. Therefore, ordinary ray reflects totally upon the Canada balsam layer between two pieces of calcite, and extraordinary ray passes through the Canada balsam layer. Prisms (c) and (d) have similar structures to (a) and (b), but they have an air gap instead of a Canada balsam layer. Therefore, these prisms can be used in a wider wavelength region. Prisms (e) and (f) also have similar structures, but the arrangement of the optical axes of two pieces of calcite differs from the above prisms. The optical axes of two pieces of Calcite cross perpendicularly to each other. Furthermore, the two pieces of uniaxial crystals are stuck by vacuum compression. A calcite prism can be used in the range between 300 nm and 2.2 μm, a quartz prism can be used in the range of 190 to 400 nm, and a MgF₂ prism

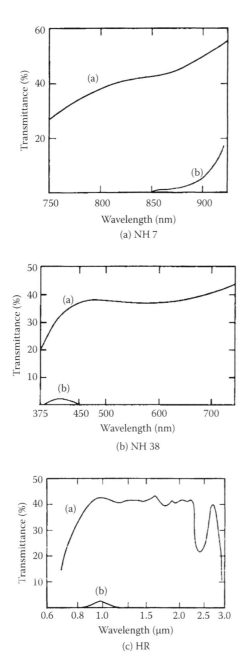

FIGURE 2.15   (a) Transmittance of Polaroid sheet HN 7 for parallel (a) and crossed (b) arrangements of polarizer sheets. (b) Transmittance of Polaroid sheet HN 38 for parallel (a) and crossed (b) arrangements of polarizer sheets. (c) Transmittance of Polaroid sheet HR for parallel (a) and crossed (b) arrangements of polarizer sheets.

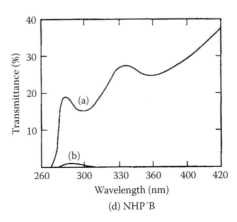

(d) NHP′B

FIGURE 2.15 (*Continued*) (d) Transmittance of Polaroid sheet HNP′B for parallel (a) and crossed (b) arrangements of polarizer sheets. (See also S. Onari: *Fundamentals and Applications of Optical Elements*, M. Ohi (ed.), p. 115, Japan Scientific Societies Press (1996) (in Japanese).)

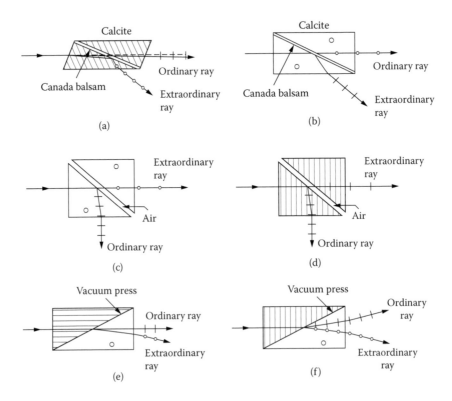

FIGURE 2.16 Typical polarizing prisms: (a) Nicol prism, (b) Glan-Thompson prism, (c) Glan- Foucault prism, (d) Gran-Taylor prism, (e) Rochon prism, and (f) Wollaston prism.

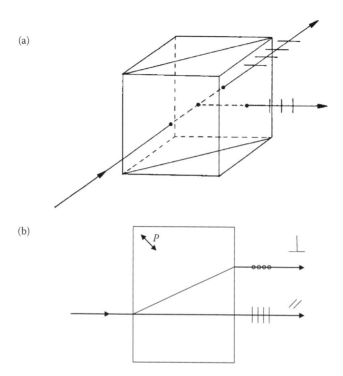

FIGURE 2.17 (a) Function of polarized beam splitter (PBS). (b) Optical paths of Savart plate. The direction of P shows the optical axis of the plate.

can be used within a 130 to 200 nm range. Furthermore, there are two different types of polarizing beam splitters. The structures and the directions of separated beams are shown in Figures 2.17a and 2.17b. As shown in Figure 2.17a, two orthogonal prisms are stuck with sloping surfaces that are coated multiply with dielectric materials. The optical beam enters on the boundary of the multilayer with nearly Brewster angle. Therefore, p-polarized light passes the boundary without reflection, but s-polarized light reflects many times by many boundaries of the multilayer, and the reflectivity reaches over 98%. Figure 2.17b is called a Savart plate. The light beam enters along the diagonal direction to the optical axis. Therefore, one polarized beam propagates along the incident beam direction, and other one refracts on the incident plane and propagates along an oblique direction to the incident beam direction after coming out from the other surface.

2.4.3 OPTICAL ANISOTROPY BY EXTERNAL DISTURBANCE

When the external disturbance such as mechanical force, electric field, and magnetic field is applied to an isotropic material, the material comes to the anisotropic symmetry. Therefore, those materials have optical anisotropic properties.

The effects appearing in the magnetic field are called magneto-optical effects. If atoms, ions, or molecules that form a solid have intrinsic magnetic dipoles, the dipoles are arranged by the magnetic field. Therefore, the material shows double refraction. When the light passes through the material along a perpendicular direction to the applied magnetic field, the light shows the optical rotation.

Similar optical anisotropy occurs on usual solids in the magnetic field. The cause of this effect has a strong relation with the Zeeman effect in atoms. The orbitals of electrons are quantized in the magnetic field. As a result, the energy of an electron rotating clockwise around the axis of a magnetic field differs from that of the electron rotating anticlockwise. The optical selection rule on the magnetic quantum numbers differs in left-handed and right-handed circularly polarized light. Therefore, the double refracting phenomenon appears in the material.

When the linear polarized light passes through the material parallel to the applied magnetic field, the polarized plane rotates to the direction of the induced electric current due to the magnetic field by the angle $\theta = VHl$. Here, V, H, and l are the Verdet constant, the strength of the applied magnetic field, and the optical path length, respectively. This phenomenon is called the Faraday rotation.

When the linear polarized light passes through the material perpendicular to the magnetic field, the polarized plane also rotates, and the phenomenon in this arrangement is called the Voigt effect.

When the strong magnetic field is applied, the effects that are proportional to the second and higher powers of the strength of the magnetic field appear. The double refraction phenomenon that is proportional to the second power of the magnetic field in the Voigt arrangement is called the Cotton-Mouton effect. The second-order Voigt effect, Cotton-Mouton effect, is expressed as follows:

$$\Delta = (n_{\parallel} - n_{\perp})l = clH^2 \tag{2.28}$$

Here, Δ is the phase difference between parallel and perpendicular polarized lights to the magnetic field, and l is the optical path length.

When the linear polarized light is reflected on the surface of the magnetic material, the polarized plane rotates. This phenomenon is called the *magnetic Kerr effect*.

In some magnetic materials, the absorption differs on the left-handed and the right-handed circularly polarized lights. When the cause of this phenomenon is due to the magnetic effect, this phenomenon is called the *magnetic circular dichroism*.

The magneto-optical effects are often used for the elements of optoelectronic devices. Useful magnetic materials for magneto-optical elements are magnetic semiconductors such as chalcopyrite or pyrite-type compounds like MgTe and CdTe doped with Mn, Eu-chalcogenide, and spinel-type compounds. Especially, Eu-chalcogenide and spinel-type compounds show large magneto-optical effects. EuX (X = O, S, Se, Te) has the NaCl-type structure and typical materials of magnetic semiconductors. Spinel-type compounds such as $CdCr_2X_4$ (X = S, Se) are also typical magnetic semiconductors. The Faraday rotation of $CdCr_2Se_4$ is 10^4 deg/cm at the wavelength 1 μm and 80 K. The Kerr rotation angles of $CdCr_2Se_4$ and $CoCr_2Se_4$ are 1° at $\lambda = 600$ nm and 4° at $\lambda = 800$ nm, respectively. The band gap energies of the materials described above are shown in Table 2.1.[32,33]

TABLE 2.1

Band Gap Energies of Several Magnetic Semiconductors

Chalcopyrite (eV)		Pyrite (eV)	
$CuAlS_2$	3.49	FeS_2	0.75
$CuGaS_2$	2.43	NiS_2	0.37
$CuInS_2$	1.53	ZnS_2	2.5
$CuFeS_2$	0.65		

Eu-chalcogenide (eV)		Spinel (eV)	
EuO	1.4	$CdCr_2S_4$	0.8~1.6
EuS	1.7	$CoCr_2S_4$	~1.5
EuSe	1.8		
EuTe	2.0		

Sources: S. Endo: *Phys. Monthly*, 8(8), 441 (1987) (in Japanese); K. Wakakmura, M. Wakaki, and T. Arai: *Phys. Monthly*, 8(8), 480 (1987) (in Japanese). With permission.

TABLE 2.2

Characteristic Properties of Mixed Crystals of $(Y_6T_6Bi)_3Fe_5O_{12}$ and $(GdBi)_3(FeAlGa)_5O_{12}$

Parameters (Unit)	$(Y_6T_6Bi)_3Fe_5O_{12}$		$(GdBi)_3(FeAlGa)_5O_{12}$
Wavelength (µm)	1.31	1.55	1.31
Faraday coefficients (deg/cm)	−1550	−1200	1300
Temperature coefficients of Faraday coefficient (deg/K)	0.04	0.06	−0.10
Magnetic fields for saturation (Oe)	750	1000	200

Source: Catalogue of Products, Sumitomo Metal Industries, (in Japanese). With permission.

The Faraday effects of mixed crystals of $(CaBi)_3(FeAlGa)_5O_{12}$ and $(YbTbBi)_3Fe_5O_{12}$ are used for the optical isolators at λ = 1.31 µm and 1.55 µm in the optical communication. Characteristics of both materials are shown in Table 2.2.[34]

When the electric field is applied to isotropic materials, the materials become uniaxial materials like a hexagonal crystal. The optical axis agrees with the direction of the electric field. Therefore, the refractive index parallel to the electric field, n_o, differs from that perpendicular to the electric field, n_e. This effect is called the *electro-optical effect* or the *electric birefringence effect*. The linear effect to the intensity of the applied field is called the *Pockels effect*. The crystals that have the center of symmetry do not show the Pockels effect from the structure of the crystal symmetry. The effect proportional to the second power of the field intensity is called the *Kerr effect*. The Kerr effect is a similar phenomenon to the Stark effect in atoms. The crystals having Pockels effect certainly show the piezoelectric effect and the electrostriction. Electricity is induced by the mechanical force in the piezo-electric effect. Inversely, strain is induced by the electric field in the electrostriction. If the inverse tensor of the dielectric tensor is expressed with B_{ij}, the refractive index ellipsoid can be written as follows:

$$\sum_{ij} B_{ij} x_i y_j = 1 \ (1 \le i, j \le 3) \tag{2.29}$$

The electro-optical effect is the change of B_{ij} by the electric field. The tensor B_{ij} is written by the following equation if B_{ij} is expanded related to the applied electric field E_k:

$$B_{ij} = B_{ij}^0 + \sum_k Z_{ijk} E_k + \sum_{k,l} g_{ijkl} E_k E_l + \cdots \tag{2.30}$$

The first term on the right-hand side is the value without electric field. The second and the third terms correspond to the Pockels effect and the Kerr effect, respectively. If the strain tensor S_{ij} under electric field is expanded with the applied electric field E_l, S_{ij} is written as follows:

$$S_{ij} = S_{ij}^0 + \sum_l d_{lij} E_l + \sum_{k,l} g_{klij} E_k E_l + \cdots \tag{2.31}$$

S_{ij}^0 is the value without electric field. The second and the third terms on the right-hand side correspond to the piezo and the electrostriction effects. Comparing Equation (2.30) with (2.31), one can understand that the materials showing the Pockels effect should certainly exhibit the piezoelectric effect.

As mentioned before, the Pockels effect occurs in the materials that have no center of symmetry. So, the crystals belonging to 21-point symmetry groups that have no center of symmetry in all 32-point symmetry groups will have Pockels effect. The crystals belonging to 10 groups that have a spontaneous polarization in the 21-point symmetry groups show a large Pockels effect. The crystals having ferroelectricity in the 10 groups show especially large Pockels and piezo effects. The ferroelectricity is shown in the crystals that have plural stable spontaneous-electric phases, and the phase transition occurs among those stable phases when the electric field is applied. The materials that have ferroelectricity are called ferroelectrics. There are over 200 kinds of ferroelectrics, and almost all of those are transparent in the visible region. Therefore, ferroelectrics are suitable for optoelectronics elements. The ferroelectricity disappears at the certain temperature when the temperature rises. The disappearing temperature of ferroelectricity is called the *Curie temperature*. The Curie temperatures, the piezo constants, q_{ijkl}, and the Pockels constants, g_{ijkl}, of several ferroelectrics are shown in Tables 2.3,[35] 2.4,[35] and 2.5,[35] respectively. The time

TABLE 2.3
Curie Temperatures of Typical Ferroelectric Crystals

Crystals	Curie Temperature (°C)
$LiNbO_3$	~1260
$LiTaO_3$	~665
$BaTiO_3$	120, 5, −80
$KNbO_3$	435, 225, −10
$PbTiO_3$	487
SbSI	19
$Pb_5Ge_3O_{11}$	177
$Cd_2(MoO_4)_3$	160
TGS	49
$NaNO_3$	165
KNO_4	163
HCl	−175
Rochelle Salt	23, −18
KH_2PO_4 (KDP)	−150
KD_2PO_4	−160
PVDF	120

Source: S. Shionoya: *Handbook of Optical Properties of Solid,* pp. 322–323, Asakura (1984) (in Japanese). With permission.

TABLE 2.4
Piezo Constants of Typical Ferroelectric Crystals

Crystals	Piezo Constants
α-quartz	$d_{11} = 6.9$, $d_{14} = -2.01$
$LiNbO_3$	$d_{13} = 204$, $d_{22} = 63$, $d_{31} = -3$, $d_{33} = 18$
$LiTaO_3$	$d_{13} = 78$, $d_{22} = 21$, $d_{31} = -6$, $d_{33} = 24$
ZnO	$d_{13} = -41.7$, $d_{31} = -15.6$, $d_{33} = 31.8$
KDP	$d_{14} = 4.2$, $d_{36} = 69.6$
ADP($NH_4H_2PO_4$)	$d_{14} = 5.3$, $d_{36} = 145$

Source: S. Shionoya: *Handbook of Optical Properties of Solid,* pp. 322–323, Asakura (1984) (in Japanese). With permission.

TABLE 2.5

Pockels Constants of Typical Ferroelectric Crystals

Crystals	Pockels Constants
α-quartz	$z_{41} = 0.5$, $z_{11} = 0.2$
LiNbO$_3$	$z_{22} = 21$, $z_{33} = 30.8$, $z_{41} = z_{32} = 28$, $z_{63} = 8.6$
ZnO	$z_{13} = 1.4$, $z_{33} = 3.6$
KDP	$z_{41} = 8.8$, $z_{61} = -10.3$
ADP	$z_{41} = 24.5$, $z_{63} = -8.5$
BaTiO$_3$	$z_{33} - z_{13} = 108$, $z_{42} = 1640$

Source: S. Shionoya: *Handbook of Optical Properties of Solid*, pp. 322–323, Asakura (1984) (in Japanese). With permission.

constant of the Pockels effect is very short. For example, the time constants of ADP, KDP, LiTaO$_3$, and LiNbO$_3$ are ~10^{-8} sec. Therefore, the Pockels effect is used for various kinds of optoelectronic devices such as optical modulators in the intensity and the frequency. The Pockels effect is also used for Q-switching and short pulse generation of lasers.

When the ultrasonic wave is applied to a material, the standing compression wave of the density in the material is built up due to the acoustic effect. Therefore, the standing compression wave of the refractive index is built up. It can be used for optical gratings. The intensity of the diffracted light varies as a function of the intensity of the ultrasonic wave. The diffraction angle varies as a function of the frequency of the ultrasonic wave, and the frequency of the diffracted light shifts by the frequency of the ultrasonic wave. These properties can be used as optical deflectors, optical frequency shifters, and intensity modulators. For these devices, PbMoO$_4$, TiO$_2$, Te doped SiO$_2$ glass, and so on, are used. The modulated frequency width and the frequency shift value are several tens of MHz. The diffraction efficiency is about 85% with the power of 1 W.

The band edge absorption of semiconductor in the electric field shows the Frantz–Keldish effect by the electron tunneling from the valence band to the conduction band. The electron tunneling caused by the incline of the energy bands due to the electric field induces at the same time the optical absorption of electrons. Therefore, optical absorption occurs at a smaller energy level than the absorption edge energy, and the absorption spectrum becomes a stair-step-like shape.

REFERENCES

1. E. Hecht: *Optics*, Addison-Wesley (2002), pp. 1–9.
2. M. Fox: *Optical Properties of Solids*, Oxford University Press (2001).
3. K. Kudo and M. Wakaki: *Fundamentals of Quantum Optics*, Tokai University Press (2009) (in Japanese).

4. K. Kudo: *Fundamentals of Optical Properties of Solids*, Ohmsha (1996) (in Japanese).
5. S. Onari: *Spectroscopy of Solids*, SHOKABO (1994) (in Japanese).
6. K. Asahara: *Fundamentals and Applications of Optical Elements*, M. Ohi (ed.), p. 14, Japan Scientific Societies Press (1996) (in Japanese).
7. *Catalogue of Optical Glasses*, HOYA Corporation (in Japanese).
8. K. Asahara: *Fundamentals and Applications of Optical Elements*, M. Ohi (ed.), p. 15, Japan Scientific Societies Press (1996) (in Japanese).
9. T. Ogawa: *Fundamentals and Applications of Optical Elements*, M. Ohi (ed.), p. 16, Japan Scientific Societies Press (1996) (in Japanese).
10. M. Wakaki, K. Kudo, and T. Shibuya: *Physical Properties and Data of Optical Materials*, p. 357, CRC Press (2007).
11. M. Wakaki, K. Kudo, and T. Shibuya: *Physical Properties and Data of Optical Materials*, p. 370, CRC Press (2007).
12. T. Aoki: *Fundamentals and Applications of Optical Elements*, M. Ohi (ed.), p. 19, Japan Scientific Societies Press (1996) (in Japanese).
13. M. Wakaki, K. Kudo, and T. Shibuya: *Physical Properties and Data of Optical Materials*, p. 385, CRC Press (2007).
14. M. Wakaki, K. Kudo, and T. Shibuya: *Physical Properties and Data of Optical Materials*, p. 98, CRC Press (2007).
15. M. Wakaki, K. Kudo, and T. Shibuya: *Physical Properties and Data of Optical Materials*, p. 258, CRC Press (2007).
16. M. Wakaki, K. Kudo, and T. Shibuya: *Physical Properties and Data of Optical Materials*, p. 277, CRC Press (2007).
17. T. Ogawa: *Fundamentals and Applications of Optical Elements*, M. Ohi (ed.), p. 20, Japan Scientific Societies Press (1996) (in Japanese).
18. M. Wakaki, K. Kudo, and T. Shibuya: *Physical Properties and Data of Optical Materials*, p. 114, CRC Press (2007).
19. M. Wakaki, K. Kudo, and T. Shibuya: *Physical Properties and Data of Optical Materials*, p. 329, CRC Press (2007).
20. M. Wakaki, K. Kudo, and T. Shibuya: *Physical Properties and Data of Optical Materials*, p. 431, CRC Press (2007).
21. T. Ogawa: *Fundamentals and Applications of Optical Elements*, M. Ohi (ed.), p. 21, Japan Scientific Societies Press (1996) (in Japanese).
22. M. Wakaki, K. Kudo, and T. Shibuya: *Physical Properties and Data of Optical Materials*, p. 43, CRC Press (2007).
23. T. Ogawa: *Fundamentals and Applications of Optical Elements*, M. Ohi (ed.), p. 22, Japan Scientific Societies Press (1996) (in Japanese).
24. M. Wakaki, K. Kudo, and T. Shibuya: *Physical Properties and Data of Optical Materials*, p. 488, CRC Press (2007).
25. M. Wakaki, K. Kudo, and T. Shibuya: *Physical Properties and Data of Optical Materials*, p. 492, CRC Press (2007).
26. M. Wakaki, K. Kudo, and T. Shibuya: *Physical Properties and Data of Optical Materials*, p. 523, CRC Press (2007).
27. M. Wakaki, K. Kudo, and T. Shibuya: *Physical Properties and Data of Optical Materials*, CRC Press, p. 529 (2007).
28. T. Ogawa: *Fundamentals and Applications of Optical Elements*, M. Ohi (ed.), p. 23, Japan Scientific Societies Press (1996) (in Japanese).
29. M. Wakaki, K. Kudo, and T. Shibuya: *Physical Properties and Data of Optical Materials*, p. 31, CRC Press (2007).
30. M. Wakaki, K. Kudo, and T. Shibuya: *Physical Properties and Data of Optical Materials*, p. 34, CRC Press (2007).

31. S. Onari: *Fundamentals and Applications of Optical Elements*, M. Ohi (ed.), p. 115, Japan Scientific Societies Press (1996) (in Japanese).
32. S. Endo: *Physical properties and applications of ternary chalcopyrite compounds,* 8(8), 441 (1987) (in Japanese).
33. K. Wakakmura, M. Wakaki, and T. Arai: *Optical properties of spinel compounds,* 8(8), 480 (1987) (in Japanese).
34. *Catalogue of Products*, Sumitomo Metal Industries, (in Japanese).
35. S. Shionoya: *Handbook of Optical Properties of Solid*, pp. 322–323, Asakura (1984) (in Japanese).

3 Materials for Nonlinear Optics

Toshihiro Arai (deceased) and Moriaki Wakaki

CONTENTS

3.1 BASIC MATTERS[1–7]

Materials show nonlinear responses under an intense light irradiation. These responses are used widely in electro-optical devices as especially higher harmonic, sum, and difference frequencies generations. In these cases, the electric displacement D can be written as follows:

$$D = \varepsilon E = \varepsilon_0 E + P = \varepsilon_0 E + P^{(1)}(\omega_s) + P^{(2)}(\omega_s) + P^{(3)}(\omega_s) + \cdots$$

$$= \varepsilon_0 [\chi^{(1)} E + \chi^{(2)} E^2 + \chi^{(3)} E^3 + \cdots]$$

(3.1)

where $P^{(1)}(\omega_s)$ is a linear polarization, and $P^{(n)}(\omega_s)$ are higher-order polarizations. The linear polarization is proportional to the electric field and the nth-order polarization is proportional to the nth power of the electric field at the frequency ω_s generated in the material, respectively. Each $\chi^{(n)}$ is the electric susceptibility corresponding to each $P^{(n)}$.

The even-power terms do not appear in the materials that have the center of symmetry in the structure, but the odd-power terms appear in all materials. The second term in Equation (3.1), the $\chi^{(2)}$ term, shows that the refractive index can be controlled with the external electric field. This property is well known as the electro-optic effect described in the materials for linear optics. The property is important for many applications such as the second harmonic generation, optical rectification, down conversion, sum frequency mixing, difference frequency mixing, and electro-optic modulation.

The third term in Equation (3.1), the $\chi^{(3)}$ term, can be expected to play a key role in all-optical switching devices, because their optical properties can be controlled by light. They are employed in solid-state lasers, absorbers in high-power lasers, and for pulse forming. Inorganic, organic, and low-dimensional materials are being investigated for $\chi^{(3)}$.

As will be described later, the phase-matching condition should be satisfied to get the nonlinear response with high efficiency.

Lasers are suitable optical sources for the purpose. Laser light has the features of high power, good monochromaticity, good directivity, good polarization, large coherence length, and further, ultra-short pulses can be realized. There are several kinds of lasers such as gas lasers, liquid lasers, solid-state lasers, and junction-type semiconductor lasers. The oscillation frequencies of these lasers cover typically from 0.10µm to 10µm in wavelength. Some of them, for example dye lasers, can change oscillation frequency continuously. Dielectric materials, such as Al_2O_3, MgO, CaF_2, and YAl_5O_{13}, doped with transition or rare earth elements such as Cr^{3+}, Nd^{3+}, and Li^{3+}, are used for solid-state lasers. The qualities of emitted lights of solid-state and gas lasers are better than those of semiconductor and liquid lasers. However, gas lasers are large, and the handling of them is not so easy. Therefore, solid-state lasers are mainly used for the generation of second higher harmonics. Semiconductor lasers are small and can be easily excited by electric current. So these lasers are used for electro-optic devices, such as in optical communication.

Now we consider the vibration of the electronic charge e with the mass m in a nonharmonic potential. For simplicity, the problem is treated in one-dimensional x. The equation of motion is expressed as follows:

$$\ddot{x} + \gamma\dot{x} + (\omega_0^2 + \alpha_1 x + \alpha_2 x^2 + \cdots)x = \frac{e}{m}(E + c.c.) \tag{3.2}$$

where $c.c.$ is a complex conjugate term.

Here, ω_0 is a natural resonant frequency, γ is a damping rate, and $\alpha_1, \alpha_2, \alpha_3 \cdots$ are shape factors of the potential. E is the electric field of the incident light. The general solution of Equation (3.2) can be written as

$$x(t) = x_1 e^{-i\omega_1 t} + x_2 e^{-i2\omega_1 t} + x_3 e^{-i3\omega_1 t} + \cdots + c.c. \tag{3.3}$$

The first-order polarization $P^{(1)}$ in Equation (3.1) is given from the first term of Equation (3.3):

$$P^{(1)}(\omega_1) = Nex_1 e^{-i\omega_1 t} = \varepsilon_0 \chi^{(1)}(\omega_1)E(\omega_1) \tag{3.4}$$

where N is the number of electrons per unit volume. The second- and third-order polarizations are given from the second and third terms of Equation (3.3), respectively.

$$P^{(2)}(2\omega_1) = \varepsilon_0 \chi^{(2)}(2\omega_1)E(\omega_1)^2, \tag{3.5}$$

$$P^{(2)}(0) = \varepsilon_0 \chi^{(2)}(0)E(\omega_1)^2, \tag{3.6}$$

$$P^{(3)}(3\omega_1) = \varepsilon_0 \chi^{(3)}(3\omega_1)E(\omega_1)^3, \tag{3.7}$$

$$P^{(3)}(\omega_1) = \varepsilon_0 \chi^{(3)}(\omega_1)E(\omega_1)^3. \tag{3.8}$$

Therefore, polarization at the frequency ω_1 is given by the following equation:

$$P(\omega_1) = P^{(1)}(\omega_1) + P^{(3)}(\omega_1) + higher\ odd\ terms \tag{3.9}$$

When two light beams of frequencies ω_1 and ω_2 come into the medium at the same time, the first- and second-order polarizations are given as follows:

$$\begin{aligned} P^{(1)}(\omega_1 + \omega_2) &= P^{(1)}(\omega_1) + P^{(1)}(\omega_2) \\ &= \varepsilon_0 \chi^{(1)}(\omega_1)E_1(\omega_1) + \varepsilon_0 \chi^{(1)}(\omega_2)E_2(\omega_2) \\ &= \varepsilon_0 \sum_{p=1}^{2} \chi^{(1)}(\omega_s, \omega_p)E(\omega_p), \end{aligned} \tag{3.10}$$

where ω_s is a signal frequency induced by ω_p.

$$P^{(2)}(\omega_1 + \omega_2) = \varepsilon_0 \chi^{(2)}(\omega_s, \omega_1 + \omega_2)E_1(\omega_1)E_2(\omega_2) \tag{3.11}$$

$$P^{(2)}(\omega_1 - \omega_2) = \varepsilon_0 \chi^{(2)}(\omega_s, \omega_1 - \omega_2)E_1(\omega_1)E_2^*(\omega_2) \tag{3.12}$$

$$\bar{P}^{(2)}(\omega_s, \omega_1 + \omega_1) = \varepsilon_0 \chi^{(2)}(\omega_s, 2\omega_1)E_1(\omega_1)^2 \tag{3.13}$$

$$P^{(2)}(\omega_s, \omega_2 + \omega_2) = \varepsilon_0 \chi^{(2)}(\omega_s, 2\omega_2)E_2(\omega_2)^2 \tag{3.14}$$

$$P^{(2)}(\omega_s, \omega_1 - \omega_1) = \varepsilon_0 \chi^{(2)}(\omega_s, 0)E_1(\omega_1)E_1^*(\omega_1) \tag{3.15}$$

$$P^{(2)}(\omega_s, \omega_2 - \omega_2) = \varepsilon_0 \chi^{(2)}(\omega_s, 0)E_2(\omega_2)E_2^*(\omega_2) \tag{3.16}$$

where $E^*(\omega) = E(-\omega)$ for no absorption.

The polarizations for higher-order polarization waves are given as similar forms.

3.2 SECOND HARMONIC GENERATION (SHG), SUM FREQUENCY GENERATION (SFG), AND DIFFERENCE FREQUENCY GENERATION (DFG)

The generation of 2ω wave is called second harmonic generation (SHG), and the generations of $\omega_1 + \omega_2$ and $\omega_1 - \omega_2$ are called sum frequency generation (SFG) and difference frequency generation (DFG), respectively. The schematic arrangements of SHG, SFG, and DFG are shown in Figure 3.1. In this figure, we assume $\omega_1 < \omega_2 < \omega_3$ for simplification. Exciting waves ω_1 and ω_2 are called *pumping waves*, and the weaker one in both pumping waves is called the *idler wave*. The condition $\omega_3 = \omega_1 \pm \omega_2$ is the energy conservation low in SHG, SFG, and DFG.

The electromagnetic wave propagating to the z direction is expressed by $E_i(\omega_i; zt) = A_i(\omega_i, z)e^{i(K_i z - \omega_i t)}$, where $K_i = |\mathbf{K}_i| = (\omega_i / c)n_i$ is the absolute value of the wave vector, and c and n_i are the light velocity in vacuum and the refractive index, respectively. From the wave equations for three waves and the assumption that the variation of the amplitude of the wave in the medium is slow so as to satisfy $d^2A/dz^2 \ll |KdA/dz|$, we obtain the next relation:

$$\Delta K = K_3 - (K_1 + K_2) \qquad (3.17)$$

This equation shows the momentum conservation low in the interaction among ω_1, ω_2, and ω_3 frequencies. As shown in Figure 3.1, a nonlinear optical material is

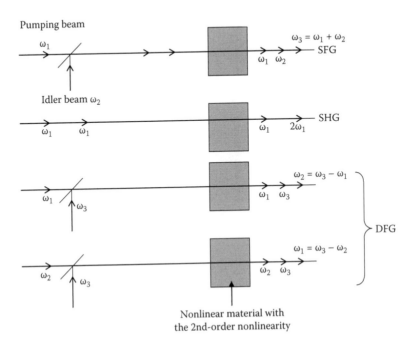

FIGURE 3.1 Arrangements of second harmonic generation (SHG), sum frequency generation (SFG), and difference frequency generation (DFG) systems.

required for SHG, SFG, or DFG. The second-order nonlinear phenomena do not appear in the crystals that have the inversion symmetry in the structure, because the even-order electric susceptibilities are zero when the material has the inversion symmetry. Details on this point will be described in the next section. Another condition to get SHG, SFG, or DFG is satisfaction of Equation (3.17). Now, omitting the derivation process, the calculated result of the conversion efficiency of SHG, φ_{SHG}, at the z point from the incident plane is shown as follows:

$$\varphi_{SHG} = \frac{I_2}{I_1} = 2 < \sinh^2(G \sin \varphi_i) >= G^2 + \frac{1}{4}G^4 + \frac{1}{36}G^5 + \cdots \tag{3.18}$$

where

$$G = \frac{\chi^{(2)}\omega_1 z}{2cn(\omega_1)} \sqrt{\frac{2\omega_1}{n(2\omega_1)}} A_0 \frac{\sin(z\Delta K / 2)}{z\Delta K / 2}, \tag{3.19}$$

$$\varphi_i = 2\varphi_1 - \varphi_3 - (z\Delta K / 2). \tag{3.20}$$

If $G < 1$, φ_{SHG} can be approximated with only the first term, G^2:

$$\varphi_{SHG} \cong \frac{(\chi^{(2)})^2 \omega_1^2 l^2}{c^3 \varepsilon_0 n^2(\omega_1) n(2\omega_1)} I_1 \left[\frac{\sin^2\left(\dfrac{z\Delta K}{2}\right)}{\left(\dfrac{z\Delta K}{2}\right)^2} \right] \tag{3.21}$$

where I_1 is the intensity of the pumping wave, l is the optical path length in the nonlinear material, and $\Delta K = K_3 - (2K_1)$. φ_{SHG} depends on the [ ] term in Equation (3.21), except for the case of $\Delta K=0$. When $\Delta K=0$, the following relation of Equation (3.22) holds, and Equation (3.21) is independent of z.

$$\left[\frac{\sin^2\left(\dfrac{z\Delta K}{2}\right)}{\left(\dfrac{z\Delta K}{2}\right)^2} \right] = \sin c^2\left(\frac{z\Delta K}{2}\right) = 1 \tag{3.22}$$

This condition is the momentum conservation denoted in Equation (3.17) and called the *phase-matching condition* between pumping and signal rays. The conversion efficiency φ_{SHG} depends also on gain constant $M=(\chi^{(2)})^2/n(\omega_1)^2 n(2\omega_1)$. In addition, φ_{SHG} is proportional to the square of the frequency of the pumping wave, ω_1^2, and the intensity of the pumping wave, $I(\omega_1)$. In general, the condition $\Delta K=0$ can be written including polarization directions of pumping, idler, and signal rays (α, β, and γ) as follows:

$$\Delta K = K_3^\gamma(2\omega_1) - (K_1^\alpha(\omega_1) + K_2^\beta(\omega_1)) = 0 \tag{3.23}$$

When the light beam with the wavelength λ_0 comes into a medium having the refractive index $n^\sigma(\omega)$ and propagates in the medium, the wavelength changes from λ_0 to $\lambda = \lambda_0 / n^\sigma(\omega)$, so that the wavevector K_0 in vacuum changes to K in the media.

$$K^\sigma = \left(\frac{2\pi}{\lambda}\right) K_0 = \left(\frac{\omega}{c}\right) n^\sigma(\omega) K_0 \tag{3.24}$$

Equation (3.23) can be rewritten as follows:

$$\Delta K = (1/c)\left\{\omega_3 n^{\gamma}(\omega_3)K_{03} - \left[\omega_2 n^{\beta}(\omega_1)K_{02} + \omega_1 n^{\alpha}(\omega_1)K_{01}\right]\right\} = 0 \qquad (3.25)$$

Equation (3.25) is the phase-matching condition. When $\Delta K=0$, the phase of the SHW (second harmonic wave) generated at $z = l$, and the phase of SHW generated at $z = l_2$ agree with each other. Therefore, the SHW can propagate with no attenuation. When $\Delta K \neq 0$, the intensity of the SHW damps with $\sin c^2(z\Delta K/2)$ as a function of the propagation length z. The z value that corresponds to the peak of $\sin c^2(z\Delta K/2)=1$ gives the peak position of the intensity of the SHW. The function $\sin c^2(z\Delta K/2)$ decreases rapidly after $z\Delta K/2=\pi/2$. The length from the incident surface to $z\Delta K/2=\pi/2$ is a critical length, and it is called the *coherent length*, l_c, in the nonlinear phenomenon. To get large efficiency in the SHG, the direction and polarizing plane of the incident beams should be selected to satisfy the phase- or velocity-matching condition using ordinary and extraordinary rays.

3.3 MATERIALS FOR NONLINEAR OPTICS

3.3.1 SINGLE CRYSTALS

As mentioned before, the second-order nonlinear phenomena do not appear in the crystals that have the inversion symmetry in the structures, because the even-order electric susceptibilities are zero. Therefore, the crystals that show the second-order nonlinear optical phenomena belong to the noncenter of the symmetry group. The crystal structures, properties of symmetry, and optical properties are listed in Table 3.1. There are 32 point groups, and 11 point groups in all of them have center of symmetry. As a result, materials belonging to 21 point groups can be used for the even-order nonlinear optics. However, all tensor elements in the second-order susceptibility $\chi_{ijk}^{(2)}$ are zero in the point group O(432). So, it cannot be used for second-order nonlinear optics. As will be shown in a later section, all of the crystals can be used for the odd-order nonlinear optics.

Now, we will consider the second-order electro-susceptibility $\chi_{ijk}^{(2)}(\omega_s, \omega_p, \omega_q)$, where ω_s, ω_p, and ω_q are the angular frequencies of the incident beams p and q, and the generated signal beam s, respectively.[3,4,8] The subscripts i, j, and k denote the polarization vectors of the light. $\chi_{ijk}^{(2)}$ has the following symmetries:

1. Characteristic exchange symmetry

$$\chi_{ijk}^{(2)}(\omega_s, \omega_p, \omega_q) = \chi_{ikj}^{(2)}(\omega_s, \omega_q, \omega_p), \qquad (\omega_s = \omega_p + \omega_q) \qquad (3.26)$$

2. Exchange symmetry in nonresonance
   When we can ignore damping,

$$\chi_{ijk}^{(2)}(\omega_s, \omega_p, \omega_q) = \chi_{jki}^{(2)}(\omega_p, -\omega_q, \omega_s) = \chi_{kil}^{(2)}(\omega_q, \omega_s, -\omega_p) \qquad (3.27)$$

In addition, by applying characteristic exchange symmetry,

TABLE 3.1

Crystal Structures and Optical Properties

Crystal Structure	Point Group	Center of Symmetry Yes○ No×	Number of Symmetry Axis 2(π)	3(3π/2)	4(π/2)	6(2π/3)	Number of Symmetry Face	Optical Property
Cubic	O_h	○	6	4	3		9	Isotropic
	O	×	6	4				
	T_d	×	3	4			6	
	T_h	○	3	4			3	
	T	×	3	4				
Tetragonal	D_{4h}	○	4		1		5	Anisotropic
	D_4	×	4		1			
	C_{4r}	×			1		4	
	C_{4h}	○			1		1	
	C_4	×			1			
	D_{2d}	×	3				2	
	S_4	×	1					
Hexagonal	D_{6h}	○	6			1	7	Optically mono-axial
	D_6	×	6			1		
	C_{6v}	×				1	6	
	C_{6h}	○				1	1	
	C_6	×				1		
	D_{3h}	×	3	1				
	C_{3h}	×		1			1	

(Continued)

TABLE 3.1 (Continued)
Crystal Structures and Optical Properties

Crystal Structure	Point Group	Center of Symmetry Yes○ No×	Number of Symmetry Axis				Number of Symmetry Face	Optical Property
			$2(\pi)$	$3(3\pi/2)$	$4(\pi/2)$	$6(2\pi/3)$		
Trigonal	D_3	×	3	1				
	D_{3d}	○	3	1			3	Optically mono-axial
	C_{3v}	×		1			3	
	C_{3i}	○		1				
	C_3	×		1				
Orthorhombic	D_{2h}	○	3				3	
	D_2	×	3					Anisotropic
	C_{2v}	×	1				2	
Monoclinic	C_{2h}	○	1				1	Optically biaxial
	C_2	×	1					
	Cs	×					1	
Triclinic	C_i	○						
	C_1	×						

$$\chi_{ijk}^{(2)}(\omega_s,\omega_p,\omega_q) = \chi_{ikj}^{(2)}(\omega_s,\omega_q,\omega_p) = \chi_{jik}^{(2)}(\omega_p,\omega_s,-\omega_q) = \chi_{kji}^{(2)}(\omega_q,-\omega_p,\omega_s). \quad (3.28)$$

Here, $\omega_s = \omega_p + \omega_q$ or $\omega_p + \omega_q - \omega_s = 0$.

3. Symmetry of Kleinman[9]

If the material has no dispersion for the incident lights, $\chi_{ijk}^{(2)}$ does not depend on the order of i, j, and k. That is,

$$\chi_{ijk}^{(2)}(\omega_s,\omega_p,\omega_q) = \chi_{jki}^{(2)}(\omega_s,\omega_p,\omega_q) = \chi_{kij}^{(2)}(\omega_s,\omega_p,\omega_q) = \chi_{kji}^{(2)}(\omega_s,\omega_p,\omega_q)$$
$$= \chi_{ikj}^{(2)}(\omega_s,\omega_p,\omega_q) = \chi_{jik}^{(2)}(\omega_s,\omega_p,\omega_q) = \chi_{kji}^{(2)}(\omega_s,\omega_p,\omega_q). \quad (3.29)$$

Tensors of $\chi_{ijk}^{(2)}(\omega_s,\omega_p,\omega_q)$ are classified into six groups by the exchange of ω_s, ω_p, and ω_q, and each group is also classified into six groups by the exchange of i, j, and k. Furthermore, i, j, and k can take any of x, y, and z, independently, which gives nine cases. As a result, total numbers of $\chi_{ijk}^{(2)}(\omega_s,\omega_p,\omega_q)$ are $6 \times 6 \times 9 = 324$.

However, these numbers decrease for the following reasons:

1. Polarizabilities must be real numbers.

$$P_i^{(2)}(-(\omega_p+\omega_q),-(k_p+k_q)) = P_i^{(2)}((\omega_p+\omega_q),(k_p+k_q))^*. \quad (3.30)$$

If there is no absorption in the medium, electric fields have the following relation:

$$E_j(-\omega_p,-k_p) = E_j(\omega_p,k_p)^* \quad (3.31)$$

and

$$E_k(-\omega_q,-k_p) = E_k(\omega_q,k_q)^* \quad (3.32)$$

As a result, the next relation holds:

$$\chi_{ijk}^{(2)}(-\omega_s,-\omega_p,-\omega_q) = \chi_{ijk}^{(2)}(\omega_s,\omega_p,\omega_q)^* \quad (3.33)$$

So that the numbers of independent $\chi_{i,j,k}^{(2)}$ reduce to $(324 \div 2) = 162$.

2. From the characteristic exchange symmetry, the numbers of independent $\chi_{i,j,k}^{(2)}$ reduce to $162 \div 2 = 81$.

3. From the exchange symmetry in nonresonance, the numbers of independent $\chi_{i,j,k}^{(2)}$ further reduce to $81 \div 3 = 27$.

4. Furthermore, when the symmetry of the crystal structure is taken into consideration, the numbers of independence $\chi_{ijk}^{(2)}$ decrease more. Before describing the relation between the structure symmetry and $\chi_{ijk}^{(2)}$, the convenient method to express $\chi_{ijk}^{(2)}$ will be described. The subscript $i = (x, y, z)$ is first selected and the remaining subscripts (j, k) are expressed by one character m. So the tensor $\chi_{ijk}^{(2)}$ is decribed as $\chi_{im}^{(2)}$. Now, m is expressed by the combination of $x(=1)$, $y(=2)$, and $z(=3)$. The relation between m and 1, 2, 3 is listed as follows[7,9]:

jk→	xx	yy	zz	yz (zy)	xz (zx)	xy (yx)
↓	11	22	33	23 (32)	13 (31)	12 (21)
m→	1	2	3	4 ($\bar{4}$)	5 ($\bar{5}$)	6 ($\bar{6}$)

The polarizations and the susceptibilities induced by the incident light are expressed as in Equations (3.34), (3.35), and (3.36) using the symbols presented above:

$$
\begin{pmatrix} P_x = (\omega_p + \omega_q) \\ P_y = (\omega_p + \omega_q) \\ P_z = (\omega_p + \omega_q) \end{pmatrix} = \varepsilon_0 [\chi] \begin{bmatrix} E_x(\omega_p)E_x(\omega_q) \\ E_y(\omega_p)E_y(\omega_q) \\ E_z(\omega_p)E_z(\omega_q) \\ E_y(\omega_p)E_z(\omega_q) + E_z(\omega_p)E_y(\omega_q) \\ E_z(\omega_p)E_x(\omega_q) + E_x(\omega_p)E_z(\omega_q) \\ E_x(\omega_p)E_y(\omega_q) + E_y(\omega_p)E_x(\omega_q) \end{bmatrix} \tag{3.34}
$$

$$
[\chi] = \begin{bmatrix} \chi_{xxx} & \chi_{xyy} & \chi_{xzz}\ \chi_{xyz} & (\chi_{xzy}) & \chi_{xxz}\ (\chi_{xzx}) & \chi_{xxy} & (\chi_{xyx}) \\ \chi_{yxx} & \chi_{yyy} & \chi_{yxz}\ \chi_{yyz} & (\chi_{yzy}) & \chi_{yxz}\ (\chi_{yzx}) & \chi_{yxy} & (\chi_{yyx}) \\ \chi_{zxx} & \chi_{zyy} & \chi_{zzz}\ \chi_{zyz} & (\chi_{zzy}) & \chi_{zxz}\ (\chi_{zzx}) & \chi_{zxy} & (\chi_{zyx}) \end{bmatrix} \tag{3.35}
$$

$$
= \begin{bmatrix} \chi_{11}\chi_{12}\chi_{13}\chi_{14}(\overline{\chi_{41}})\chi_{15}(\overline{\chi_{51}})\chi_{16}(\overline{\chi_{61}}) \\ \chi_{21}\chi_{22}\chi_{23}\chi_{24}(\overline{\chi_{42}})\chi_{25}(\overline{\chi_{52}})\chi_{26}(\overline{\chi_{62}}) \\ \chi_{31}\chi_{32}\chi_{33}\chi_{34}(\overline{\chi_{43}})\chi_{35}(\overline{\chi_{53}})\chi_{36}(\overline{\chi_{63}}) \end{bmatrix} \tag{3.36}
$$

If the exchange symmetry in nonresonance is satisfied, $\chi_{\alpha\beta}$ equals $\overline{\chi}_{\alpha\beta}$ in Equation (3.36), and the numbers of elements of $\chi_{ijk}^{(2)}$ reduce to 18. Furthermore, if Kleinman's symmetry exists, the numbers of elements decreases to 10 as shown below, where only the elements without parentheses are independent.

$$
\chi_{ijk}^{(2)} = \begin{bmatrix} \chi_{11} & \chi_{12} & \chi_{13} & \chi_{14} & \chi_{15} & \chi_{16} \\ (\chi_{16}) & \chi_{22} & \chi_{23} & \chi_{24} & (\chi_{14}) & (\chi_{12}) \\ (\chi_{15}) & (\chi_{24}) & \chi_{33} & (\chi_{23}) & (\chi_{13}) & (\chi_{14}) \end{bmatrix} \tag{3.37}
$$

The independent elements of each crystal can be decided by operating the elements of structure symmetries like S_4^2, C_2 and σ_d to the polarizability and comparing the obtained polarizability with the former one. From this operation, we know the elements of zero. Some examples of χ_{im} are shown in Table 3.2,[2] where the nonlinear optical coefficient tensor d_{im} is half of χ_{im} —that is, $d_{im} = \dfrac{1}{2}\chi_{im}$.

TABLE 3.2

Nonlinear Coefficient for the Second Harmonic Generation (SHG)

[Hexagonal]

$$C_{3h}(\bar{6}), [d_{ij}] = \begin{bmatrix} d_{11} & -d_{11} & 0 & 0 & 0 & -d_{22} \\ -d_{22} & d_{22} & 0 & 0 & 0 & -d_{11} \\ 0 & 0 & 0 & 0 & 0 & 0 \end{bmatrix}$$

$$C_{6}(6), [d_{ij}] = \begin{bmatrix} 0 & 0 & 0 & d_{14} & d_{15} & 0 \\ 0 & 0 & 0 & d_{15} & -d_{14} & 0 \\ d_{31} & d_{31} & d_{33} & 0 & 0 & 0 \end{bmatrix}$$

$$D_{6}(6222), [d_{ij}] = \begin{bmatrix} 0 & 0 & 0 & -d_{14} & 0 & 0 \\ 0 & 0 & 0 & 0 & -d_{14} & 0 \\ 0 & 0 & 0 & 0 & 0 & 0 \end{bmatrix}$$

$$D_{6}(6222), [d_{ij}] = \begin{bmatrix} 0 & 0 & 0 & -d_{14} & 0 & 0 \\ 0 & 0 & 0 & 0 & -d_{14} & 0 \\ 0 & 0 & 0 & 0 & 0 & 0 \end{bmatrix}$$

$$C_{6v}(6mm), [d_{ij}] = \begin{bmatrix} 0 & 0 & 0 & 0 & d_{15} & 0 \\ 0 & 0 & 0 & d_{15} & 0 & 0 \\ d_{31} & d_{31} & d_{33} & 0 & 0 & 0 \end{bmatrix}$$

$$D_{3h}(\bar{6}m2), [d_{ij}] = \begin{bmatrix} 0 & 0 & 0 & 0 & 0 & -d_{22} \\ -d_{22} & d_{22} & 0 & 0 & 0 & 0 \\ 0 & 0 & 0 & 0 & 0 & 0 \end{bmatrix}$$

$m \perp X$

$$D_{3h}(6m2), [d_{ij}] = \begin{bmatrix} d_{11} & -d_{11} & 0 & 0 & 0 & 0 \\ 0 & 0 & 0 & 0 & 0 & -d_{11} \\ 0 & 0 & 0 & 0 & 0 & 0 \end{bmatrix}$$

$m \perp Y$

[Cubic]

$$T(23), [d_{ij}] = \begin{bmatrix} 0 & 0 & 0 & d_{14} & 0 & 0 \\ 0 & 0 & 0 & 0 & d_{14} & 0 \\ 0 & 0 & 0 & 0 & 0 & d_{14} \end{bmatrix}$$

$$T_{d}(\bar{4}3m), [d_{ij}] = \begin{bmatrix} 0 & 0 & 0 & d_{14} & 0 & 0 \\ 0 & 0 & 0 & 0 & d_{14} & 0 \\ 0 & 0 & 0 & 0 & 0 & d_{14} \end{bmatrix}$$

Source: From R. L. Byer: *Nonlinear Optics* (eds. P. G. Harper and B. S. Wherrett), pp. 47–160, Academic Press, New York (1977). With permission.

The materials used for nonlinear optics should have the large second-order susceptibility. They must also be transparent at the frequencies used such as ω_s, ω_p, and ω_q. In addition, they must be birefringence crystals to satisfy the phase matching. The optical properties of some materials able to be used in SHG, SFG, and DFG are listed in Table 3.3.[3,4,7,10]

There are some other second-order nonlinear optic phenomena. When the energy difference E_d between two energy levels equals the energy of two times the incident photon energy, $2E_{pn}$, two-photon absorption or emission occurs. It is similar to the Raman scattering process. In Raman scattering, one photon is scattered by the fluctuation of elementary excitations existing in the material. This type of Raman scattering is called *spontaneous Raman scattering*. On the other hand, the scattering is called stimulated Raman scattering if the scattering occurs by the fluctuation created due to the incident photon. Super-radiation is also one of the second-order nonlinear phenomena. Usually, optical absorption and emission occur in individual atoms on molecules. So, the intensities of these processes in N-atom or N-molecule systems are N times the intensities in the individual atom or molecule. In super-radiation, the radiation field and N-atom system interact with each other. So, the intensity of emitted light is proportional to N^2. This phenomenon is known as Dich's super-radiation. Other super-radiation occurs in the organics with linear chain structures as will be described in detail.[11] In those materials, the bound state of several Frenkel excitons is stable, when the exciton is accompanied by a static dipole moment larger than the transition dipole moment along the chain direction.

3.3.2 OTHER MATERIALS FOR THE SECOND-ORDER NONLINEAR OPTICS

Two-dimensional organics, like porphyrins and phthalocyanines, and the structural flexible polymers are the useful second-order nonlinear optical materials. They have the advantages of high performance and processing flexibility. For example, porphyrins form different types of crystalline phases by the selected preparation condition. Some crystalline phases exhibit the red-shifted Q-band absorption and the others exhibit the blue-shifted Q-band absorption. Former excitons are called *J-aggregates* and *H-aggregates*.

Organic materials with appropriate donors and acceptors in their π-conjugated system exhibit the large second-order optical nonlinearity, and π-electron systems responsive to electric fields of light offer low dielectric constants. These features can be utilized for highly efficient frequency conversions. Electric-field poled polymers have been extensively developed, but organic crystals do not attract so much attention. Only a few organics are known as second-order nonlinear optical materials. Organic 2-methyl-4-nitroaniline (MNA) thin-film crystal is one of the materials for second-order nonlinear optics. The CT axis of molecules of MNA arranges parallel to the electro-principal axis. So, when the light propagates along the direction perpendicular to the electro-principal axis, the maximum second-order nonlinear susceptibility coefficient χ_{11} is obtained for the 1.06 μm laser beam.

The fiber from 3,5-dimethyl-1-4 nitrophenyl pyrazole also can be used for SHG in the Cherenkov radiation arrangement.[12] However, accomplishing phase matching in the Cherenkov arrangement is very difficult. In this case, quasi-phase

TABLE 3.3

Second-Order Nonlinear Coefficients $(d_{ij} = \chi_{ij}^{(2)}/2)^7$

Materials	Crystal Symmetry	Wemple[10] d_{ij}(10⁻⁹esu) (Fundamental Wavelength (μm))	Boyd[4] d_{ij}(10⁻⁹esu) (Fundamental Wavelength (μm))	Shen[3] d_{ij}[(3/4π) × 10⁻⁸esu] (Fundamental Wavelength (μm))
GaAs	Td(43 m)	d_{14} 334	—	d_{14} 337 (10.6)
GaP	"	d_{14} 200	—	d_{14} 70 (3.39)
InSb	"	d_{14} 1265	—	—
CdSe	C_{6v}(6 mm)	d_{33} 69 (1.83)	d_{15} 74, d_{31} 68	d_{33} 109; d_{31} 57, d_{15} 62 (10.6)
KDP (KH$_2$PO$_4$)	D_{2d}(42 m)	$d_{14}(=d_{36})$ 1.14 (0.266)	$d_{14}(=d_{36})$ 1.2	d_{14} 0.98; d_{36} 0.94 (1.0582)
ADP (NH$_4$H$_2$PO$_4$)	D_{2d}(42 m)	$d_{14}(=d_{36})$ 1.38 (0.266)	—	d_{14} 0.96; d_{36} 0.97 (0.6943)
ZnGeP$_2$	"	$d_{14}(=d_{36})$ 286 (1.83)	—	—
CdGeAs$_2$	"	d_{36} 563 (5.83)	d_{36} 1090	—
AgGaSe$_2$	"	d_{36} 78.8 (1.83)	d_{36} 81	—
AgGaS$_2$	"	d_{36} 28.6 (0.946)	—	—
LiIO$_3$	C_6(6)	d_{33}-14.3, d_{14}-13.6 (0.694)	d_{35}-13, d_{36}-10	d_{33}-8.4; d_{35}-11.2 (1.0642)
BNN (Ba$_2$NaNb$_5$O$_{15}$)	C_{2v}(2 mm)	d_{33}-47.7	d_{31}-35, d_{32}-35, d_{33}-48	d_{31}-29.1; d_{32}-29.1, d_{33}-40 (1.0642)
Tl$_3$AsS$_3$	C_{3v}(3 m)	(−) 95.5 (1.83)	—	d_{22} 6.14; d_{31}-11.6, d_{33} 81.4 (1.0582)
LiNbO$_3$	"	d_{33} -85.9 (0.532)	d_{22} 7.4, d_{31} 14, d_{33} 98	d_{14} 0.018 (1.0582)
LiTaO$_5$	"	d_{33} -47.7	—	—
α-SiO$_2$	D_3(32)	d_{11} 0.88	d_{11} 0.96	d_{11} 0.8
Te	"	d_{11} 2196	—	d_{11} 10⁴ (10.6)
BaTiO$_3$	C_{4v}(4 mm)	$d_{13}(=d_{15})$ -43, d_{33}-16 (5.38)	d_{15}-4.1, d_{31}-43, d_{33}-16	d_{15}-34.4; d_{31}-36, d_{33}-13.2 (1.0582)
PbTiO$_3$	"	d_{33} 20.3	—	—
β-BaB$_2$O$_4$	—	—	d_{11} 46	—

Note: $\chi^{(2)}(\text{esu}) = (3/4\pi) \times 10^4 \chi^{(2)}$ (m/V).

matching is used—that is, a period of polarization inversion is matched to the coherent length.

3.4 THIRD-ORDER NONLINEAR OPTICS

3.4.1 BASIC MATTERS

The third-order nonlinear effects are observed in all materials including glasses and liquids. The same frequency generation by the degenerate for-wave mixing (DFWM) method is one of most useful technique in optics, because the generated wave is the conjugate wave of the pumping wave. The coherent anti-Stokes Raman scattering (CARS) is a good method with which to research a material. The self-focusing on dispersion (optical Kerr) effect is important in the optical transmission using a fiber.

In the third-order process, the energy conservation law is written as

$$\omega_s = \omega_p \pm \omega_q \pm \omega_r \tag{3.38}$$

The momentum conservation is also written as

$$\boldsymbol{K}_s = \pm \boldsymbol{K}_p \pm \boldsymbol{K}_q \pm \boldsymbol{K}_r \tag{3.39}$$

The third-order nonlinear electronic polarizability $P_i^{(3)}(\omega_s)$ is expressed as follows:

$$P_i^{(3)}(\omega_s) = \varepsilon_0 \sum_{jkl} \chi_{ijkl}^{(3)}(\omega_s, \omega_p, \omega_q, \omega_r) E_{jkl}(\omega, \boldsymbol{K}) \tag{3.40}$$

where

$$E_{jkl}(\omega, \boldsymbol{K}) = E_j(\omega_p, \boldsymbol{K}_p) E_k(\omega_q, \boldsymbol{K}_q) E_l(\omega_r, \boldsymbol{K}_r), \tag{3.41}$$

here $(i,j,k,l) \equiv x,y,z$.

$\chi^{(3)}$ is calculated quantum mechanically for the electron transition in the system. As a result, the third-order electronic susceptibility can be written as follows:

$$\varepsilon_0 \chi_{ijkl}^{(3)}(\omega_s, \omega_p, \omega_q, \omega_r) = \frac{N}{\hbar^3} \sum_{abmn} (M_1 + M_2 + \cdots + M_8). \tag{3.42}$$

Each M can be written for each transition process by the following forms:

$$M_1 \equiv \rho_{0aa} \frac{\mu_{mn}^i \mu_{an}^j \mu_{ba}^k \mu_{mb}^l}{W_{mn} V_{bn} U_{an}},$$

$$M_2 \equiv \rho_{0aa} \frac{\mu_{an}^l \mu_{ab}^j \mu_{bm}^k \mu_{mn}^i}{W_{na} V_{ma} U_{ba}}, \tag{3.43}$$

and so on,

$$M_8 \equiv \cdots.$$

Each M term has six cases by exchange among ω_p, ω_q and ω_r, and i, j, k, and l can take directions x, y, and z, independently. So, the numbers of elements $\chi_{i,j,k,l}^{(3)}$ become $8 \times 6 \times 3^4 = 3{,}888$. However, the numbers of elements decrease most by considering the symmetry of crystal structures as the same as the second-order nonlinear case. In addition, when the special term in M-terms is large compared with other M-terms, for example in the case of resonance between two levels, other terms can be treated as a constant, approximately. Therefore, $\chi_{i,j,k,l}^{(3)}$ can be divided into the resonance term and the nonresonance term.

$$\chi_{i,j,k,l}^{(3)} = \chi_R^{(3)} + \chi_{NR}^{(3)} = \chi_R^{(3)} + const. \tag{3.44}$$

The $\chi^{(3)}$ elements allowed in 32 crystal symmetry groups have been given by Butcher et al.,[13,14] and the measured values of $\chi^{(3)}$ on some materials have been shown in Table 3.4.[4,15–17]

The third-order electronic susceptibility is given by measuring the absorption band as follows[18]:

$$|\chi^{(3)}| = const \times M\left(\frac{\alpha_p\Gamma^2}{E_p}\right) / [(E_p - h\omega)^2 + \Gamma^2]^{1.5} \tag{3.45}$$

where E_p, α_p, Γ, and M are the lowest excitation energy, the peak value of the absorption coefficient, the half-width of the absorption band, and the molecule density in numbers, respectively. The physical origins of $\chi^{(3)}$ are given in Table 3.5. The magnitudes and the response times are also shown in Table 3.5.

TABLE 3.4
Values of $\chi^{(3)}$ of Some Typical Materials

Material	$\chi^{(3)}$ (esu) and Response Time	Literature (Reference Number)
Air (20°C)	$\chi_{1111}^{(3)} = 1.2 \times 10^{-17}$	4
CSZ	$\chi_{1111}^{(3)} = 1.9 \times 10^{-12}$, ~2ps	4
Glass	$\chi_{1111}^{(3)} = (1 \sim 100) \times 10^{-8}$	4
CdSe doped glass	$\chi_{1111}^{(3)} = 1.4 \times 10^{-14}$ at 1.06 μm, ~30ps	
LiF	$\chi_{1111}^{(3)} = 0.6 \times 10^{-11}$ at 1.06 μm	4
NaCl	$\chi_{1111}^{(3)} = 0.6 \times 10^{-11}$	15, 16
Ge (300°C)	$\chi_{1111}^{(3)} = 2.5 \times 10^{-11}$ at 1.06 μm	15, 16
GaAs (300°C)	$\chi_{1111}^{(3)} = 1.2 \times 10^{-10}$ at 1.06 μm	15, 16
n-GaAs×	$\chi_{1111}^{(3)} = 0.9 \times 10^{-10}$ at 1.06 μm	15, 16
GaP (300K)	$\chi_{1122}^{(3)} = (2.3 \pm 0.3) \times 10^{-16}$ at 1.06 μm, ~8ps	16, 17
ZnS (300K)	$\chi_{1111}^{(3)} = (9.2 \pm 1.9) \times 10^{-10}$ at 1.06 μm, ~1.5ps	16, 17

TABLE 3.5

Physical Mechanisms of Nonlinear Refractive Index Change and the Characteristic Values

Physical Mechanism	Nonlinear Refractive Index $\chi^{(3)}$ (esu)	Response Time (sec)
Electronic hyperpolarizability	$10^{-13} \sim 10^{-14}$	10^{-15}
Molecular libration	$10^{-12} \sim 10^{-13}$	$\sim 2 \times 10^{-13}$
Molecular reorientation	$10^{-11} \sim 10^{-12}$	$10^{-11} \sim 10^{-12}$
Electrostriction	$10^{-11} \sim 10^{-12}$	$10^{-8} \sim 10^{-9}$
Thermal change	$10^{-5} \sim 10^{-4}$	$10^{-1} \sim 1$

3.4.2   SOME OF THE THIRD-ORDER NONLINEAR PHENOMENA

3.4.2.1   Four-Wave Mixing (FWM)

When three waves come into a material at the same time, new waves are generated by mixing the polarizabilities of three waves. When the same frequencies are used for the three incident waves, the wave of the same frequency is generated. This phenomenon is called *degenerated four-wave mixing*. In this case, two pumping waves come into the material from opposite sides, K_1, $-K_1$, and the third pumping wave K_3 comes into the material with suitable angle to the two wave directions. The momentum conservation satisfies in $K_4 = -K_3$, $K_4' = K_3-2K_1$ and $K_4'' = K_3 + 2K_1$. The generated three waves propagate in different directions. If the intensity of the third pumping wave is not so strong, the generated wave with K_4 is the phase-conjugate wave of the third pumping wave. The conjugate wave has the following properties. When the original wave is written as $E_0(r,t) = e_0 A_0 e^{i(K_0 \cdot r - \omega t)}$, the conjugate wave can be written as $E_0(r,t) = e_0^* A_0^* e^{i(K_0 \cdot r - \omega t)}$.

1. The unit vector of the original wave, e_0, changes to e_0^*. When the original wave is a left-handed circularly polarized wave, the reflected wave is also a left-handed circularly polarized wave.
2. $A_0(x, y, z)$ changes to $A_0^*(x, y, z)$. This means that the wave fronts are inverse to each other.
3. $K_c = -K_0$ means that the propagation directions of both waves are just opposite.

In the DFWM (degenerate four-wave mixing), with the intensity of the generated wave with wave vector $-K_3$, it is possible to get a larger intensity than that of the original wave.

3.4.2.2   Coherent Anti-Stokes Raman Scattering (CARS)

CARS is one of the four-wave mixing effects. The two pumping waves with frequencies ω_1 and ω_2 come into the medium, respectively, and the wave with the differential frequency $(\omega_1 - \omega_2)$ is generated for $\omega_1 > \omega_2$. Further, if the ω_1 wave interacts with the polarization of the differential wave, the third polarization with the frequency $\omega_4 = 2\omega_1 - \omega_2$ is generated, and the coherent wave with frequency ω_4 radiates. The wave with ω_2 is usually used for the excitation of the material.

3.4.2.3 Self-Focusing and Self-Divergence

The refractive index under intensive light irradiation is written as follows:

$$n = n^{linear} + n^{non-linear}$$

$$= n_0 + \Delta n. \tag{3.46}$$

Here,

$$n_0^2 = 1 + \chi^{(1)}, \tag{3.47}$$

$$2n_0 \Delta n' = \chi^{(2)} <E>_t + \chi^{(3)} <E^2>_t + \cdots \tag{3.48}$$

$<E>_t$ means the time average. The Δn is the time average of $\Delta n'$. Therefore, the odd power terms of E are zero in the isotropic materials and the crystals with a central symmetry. So, Δn is given by Equation (3.49):

$$\Delta n = (\frac{1}{2}n_0)\{\chi^{(3)} <E>_t^2 + \chi^{(5)} <E>_t^4 + \cdots\}. \tag{3.49}$$

If we can ignore higher terms than $<E>_t^2$ and assume no absorption, the refractive index n is the parabolic function of the electric field E. This is a similar phenomenon to the Kerr effect. So, this effect is called the optical Kerr effect. The radial distribution of laser intensity decreases usually from center to outside. The refractive index changes as a parabolic function of a radius if the intensity of the beam changes linearly toward the radial direction. As a result, the self-focusing or the self-divergence of the beam occurs. This is very convenient for optical communication.

3.4.3 MATERIALS FOR THE THIRD NONLINEAR OPTICS

Inorganic materials are useful for the third nonlinear optics as described in Table 3.4. In addition, organics, polymers, and low-dimensional materials are very useful for the third nonlinear optics.

3.4.3.1 Organics

Organic polymers have the flexibility in the structure and the possibility to fabricate at large scale. From these points, the linear chain-type organic polymers are particularly promising materials for the third nonlinear optics. The characteristic property of the organic materials used for the third nonlinear optics is to have π-electrons bound loosely to molecules. The transition momentum of the molecule in these materials concentrates upon the lowest excited electron state. These cause the appearance of large nonlinear coefficients in these materials. The nonlinear optical response increases by use of the collective excitation like the super-radiation as described before. The collective excitation can be induced by Frenkel excitons. It cannot occur through Winner excitons except for exciton molecules. In any way, the excited energy in organics with π-electrons can migrate resonantly among molecules. If the excitonic state created by optical absorption at the molecule migrates in the chain on the aggregate, many molecules respond only to one photon absorption.

Therefore, in the linear chain of a π-electron conjugate system, the coherent length is N in molecular units. If the dipole moment of the molecule is μ_m, the dipole moment μ_s of the system becomes $N^{1/2}$ times μ_m. The spectral width, W_s, of the system also becomes $1/N^{1/2}$ times that of molecule, W_m, and the intensity of radiation I_s from the system becomes N times that of molecule, I_m. As a result, the $\chi_s^{(3)}$ of the system N_s becomes $N_m^x (x \geq 2)$ times that of molecule, χ_m. These relations are summarized in the following:

$$\mu_s = \mu_m \times N^{1/2} \tag{3.50}$$

$$W_s = W_m \times 1/N^{1/2} \tag{3.51}$$

$$I_s = I_m \times N, \tag{3.52}$$

$$\chi_s^{(3)} = \chi_m^{(3)x}, \quad x \geq 2 \tag{3.53}$$

Many trials to make large conjugate systems have been done. For our purpose, the main bones, chain, were connected by the bridges of substituted side group such as basics, atoms, or molecules. If π-conjugation between the main chain and the side group is made, the $\chi^{(3)}$ should become larger than the $\chi^{(3)}$ of side group materials such as dyes, halogen slots, sulfur, and transition metals.

There is one question that a collective excitation of one-dimensional charge transfer cannot exist in two-dimensional organics, like porphyrins and phthalocyanines. Those organic systems are made by the organic molecular beam epitaxy. They exhibit red- and blue-shifted Q-band absorption peaks, as compared with the monomers in solution. These shifts depend on the preparation condition. The polycrystalline phase that appears as the red-shifted excitons is called J-aggregates, and the polycrystalline phase that appears as the blue-shifted excitons is called H-aggregates. The photo-induced absorption on the blue side of the exciton absorption peak and bleaching of the exciton absorption peak are seen at 760 nm for ClIn Phythalo-cyanines. These phenomena can be interpreted to mean that the photo-induced absorption peak is the transition to the bi-excitons in this two-dimensional system. This indicates the collective excitation of one-dimensional charge transfer can be excited in two-dimensional polymers.

Polydiacetylene (Poly-DCHP), polythiophene, and polyacetylene are used frequently as main chain bone materials. The monomer and the polymer structures of polydiacetylene are shown in Figure 3.2. The molecules of these materials have large ionicity by introducing side bands easily. For example, in acetylene, poly [(0-trimethylulyl) phenyl]] acetylene, poly [4-tert-butyl-2, 6-dimethylphenyl] acetylene and poly [0-isopropyl pheny] acetylene were synthesized and tested for the third-order nonlinear optics.

The pseudoisocynanin (PIC) was discovered for the first time by Jelly[19] and Scheibe.[20] After that time, many aggregates of this polymer have been tested to find large $\chi^{(3)}$ materials. The dynamical processes of the excited state in the J-aggregates pseudoisocynanin-halogen base (PIC:Br) were studied by a pump-probe method. The free induction decay time of these aggregates was 290 fsec. Such a long free

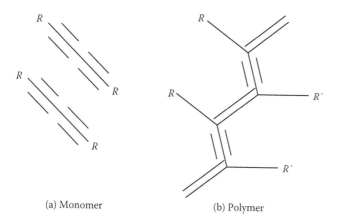

(a) Monomer (b) Polymer

FIGURE 3.2 Monomer and polymer structures of polydiocetylene.

induction time is caused by the aggregation. The magnitude of $\chi^{(3)}$ of the PIC (dye)-J-aggregates is approximately 0.6×10^{-7} esu.

The chain polymer [polydiacetylene (poly-DCHD)] was also studied extensively. The $\chi^{(3)}$ along the main chain is ~10^{-6} esu, and the figure of merit, $\chi^{(3)}$/abs. coeff., is 10^{-18} esu cm at the wavelength 650 nm. The $\chi^{(3)}$ on the perpendicular direction to the main chain of this polymer is 10^{-7} esu and the figure of merit is 10^{-11} esu cm at the wavelength 720 nm. Several types of polydiacetylene aggregates with many kinds of side bonds, like hetero atoms, were researched. For example, the $\chi^{(3)}$ of some diphenyl-substituted polydiacetylenes (poly-MADA) was one order smaller than that of poly-DCHD, but the absorption coefficient is one order smaller than that of poly-DCHD at near the THG (third harmonic generation) wavelength. Therefore, the figure of merit was about the same in both materials. In the poly-MADA, the refractive index as a function of light intensity I can be written as $n = n_0 + \gamma I$. The proportional constant, γ, is -4×10^{-3} cm^2/GW. This is 30 times as great as that of the standard material CS_2.

As other examples, one-dimensional metal compounds have been studied. Dionedioximes d^8 metal has a planar structure. The existence of hydrogen bonding between the ligands offers electrical neutrality in the molecules. The films made by the vacuum evaporation of the molecules have linear metal chains. Many kinds of metals such as Cu, Ni, Pd, and Pt were tested to check the influence of the number of electrons in metals.

The $\chi^{(3)}$ values obtained by substituting the metal increase proportionally to the number of electrons in the metal. However, the distance between metals increases proportionally to the electron numbers in metals. Therefore, the figure of merit does not depend on the kind of doping metal. The magnitude of the $\chi^{(3)}/\alpha$ was 10^{-8} esu cm at 1064 nm in Pt (dmg)$_2$, where α is the absorption coefficient.

One of the other nonlinear optical effects is the induced circular dichroism in chiral. The chiral has anisotropy to left- and right-handed circular polarized lights and shows the optical rotation of the linear polarized light. The optical nonlinearity in chiral was larger than that of helicenes by a factor of about 100 at the resonance wavelength region. Therefore, chiral should be used as an optical shutter.

3.4.3.2 Complex Materials Dispersed with Metal-, Semiconductor-, and Organic Microcrystals

The magnitude of the third-order nonlinear optical constant in complex materials dispersed with microclusters shown in Figure 3.3 depends on the cluster size and the kind of matrix. The color of the stained glass dispersed with metal clusters is created by the resonance between the surface plasmon of the metal and the light. When the complex dielectric constants of metal and matrix are $\varepsilon_m(\omega)$ and $\varepsilon_d(\omega)$, respectively, the complex dielectric constant of the medium $\varepsilon(\omega)$ is given by

$$\varepsilon(\omega) = \varepsilon_d + 3P\varepsilon_d(\omega)\frac{\varepsilon_m(\omega) - \varepsilon_d(\omega)}{\varepsilon_m(\omega) + 2\varepsilon_d(\omega)}, \tag{3.54}$$

where P is a volume fraction of nano-crystals. The metal clusters feel the local electric field E_L. E_L is given by Equation (3.56):

$$E_L = \frac{3\varepsilon_d(\omega)}{\varepsilon_m(\omega) + 2\varepsilon_d(\omega)}E_0 = f_i(\omega)E_0, \tag{3.55}$$

where E_0 is the electric field of incident light, and f_i is a factor called the local electric field factor. The local electric field increases at the resonance frequency of the surface plasmon. This condition is expressed as $\varepsilon_m^{'}(\omega) + 2\varepsilon_d(\omega) = 0$, where $\varepsilon_m^{'}(\omega)$ is the real part of metal clusters. The third-order electric susceptibility $\chi^{(3)}$ under the resonance condition is given as follows:

$$\chi^{(3)} = pf_i^4 \times \chi_m^{(3)} \tag{3.56}$$

The $\chi^{(3)}$ of the medium dispersed with metal clusters is proportional to both the fourth power of the local field factor and the occupation rate. The increasing ratios of f_i^4 in the oxide glass dispersed with Cu, Au, and Ag microclusters are around 5, 5, and 4×10^4 times, respectively. The figure of merit is given by $(nc/\omega_s)(|f_1|^2/|\varepsilon_m^{''}|) \times |\chi_m^{(3)}|$, where ω_s is the frequency of surface plasmon. c and $\varepsilon_m^{''}$ are the optical velocity in vacuum and the imaginary part of the dielectric constant of metals, respectively. The $\chi^{(3)}$s of the media doped with Cu, Au, and Ag are approximately 2×10^{-6} esu,

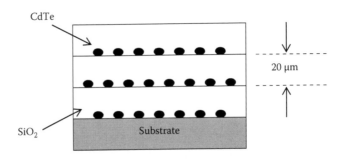

FIGURE 3.3 Film dispersed with microcluster fabricated using the alternate deposition technique.

4×10^{-8} esu, and 3×10^{-9} esu, respectively. These values depend on the occupation rate and slightly on the matrix glass as is seen from Equations (3.54) and (3.56). The response time measured by the relaxation of nonlinear absorption is composed of two components. The response time of the fast component is 3ps and that of the slow component is 200ps at the pumping intensity 4 WJ/cm². These values decrease with decreasing pumping intensity. The relaxation time is determined by the strength of electron-phonon coupling.

The excitons in a nano-size cluster are confined quantum mechanically. When the size of a cluster is less than approximately 2.5 times the exciton size, the internal motion of the exciton is confined. When the size of a cluster is larger than approximately 2.5 times of the exciton size, the translational motion of the exciton is confined. In the latter case, the oscillator strength depends on the crystallite volume because the giant oscillator strength effect occurs the same as in the impurity-trapped exciton case. The coherent length of the weak confined exciton covers the whole region of the crystal. Therefore, the oscillator strength f_x for the weak confined exciton is given by the following equation:

$$f_x = \frac{8}{\pi} \times \frac{R^3}{V} f_0, \tag{3.57}$$

where f_0, R, and V are the oscillator strength of bulk crystal, the radius of the microcrystal, and the volume of the unit cell, respectively. f_0 is given by

$$f_0 = \frac{2m_0\varpi}{\hbar} |\mu_{eg}|^2 \frac{1}{\pi a_b^3} \tag{3.58}$$

where $|\mu_{eg}|$ is the dipole moment of the transition between e and g states. If one assumes the spherical-shaped microcrystal surrounded by an infinite potential, the oscillator strength f_x is proportional to the volume of the crystal. If only two levels exist in the crystal, the imaginary part of the third nonlinear susceptibility in a nearly resonant energy region is given by the following equation:

$$Im\chi^{(3)} \propto f_x^2 NT_1 / \Gamma_h^2, \tag{3.59}$$

where, N, T_1, and Γ_h are the number's density of crystals, the lifetime of excitons in a nonradiative process, and the homogeneous line width, respectively. $|\chi^{(3)}|$ can be approximated by $Im\chi^{(3)}$. Therefore, $\chi^{(3)}$ is proportional to R^6. However, the lifetime of exciton radiation depends on the crystal size according to the following equation:

$$\tau_\gamma = \frac{3m_\beta c^3}{16e^2 n\varpi^2 f_0} \left(\frac{V}{R^3}\right) \propto \frac{1}{R^3}. \tag{3.60}$$

When the lifetime is determined by τ_γ, the T_1 is nearly equal to τ_γ. So, $\chi^{(3)}$ is proportional to R^3. When $\chi^{(3)}$ is measured by DFWM, the pumping light is absorbed by only certain sized crystals. Therefore, $\alpha\Gamma_h$ is proportional to $f_x N$, where α is an absorption coefficient. As a result, $|\chi^{(3)}|$ is given by the following equation:

$$|\chi^{(3)}| \propto f_x \alpha T_1 / \Gamma_h \tag{3.61}$$

In the above treatment, the local electric field effect on the dielectric constant was neglected, because it does not bring so large an effect in semiconductors due to the magnitudes in combinations of dielectric constants of semiconductor and matrix.

The size dependences of oxide glasses doped with CuCl and CuBr have been studied by Nakamura et al.[21] and others.[22–24] The oscillator strength increases with $R^{2.2}$ in CuCl and with $R^{2.0}$ in CuBr.[21] The saturation in the increment has been observed at approximately 4 nm in CuCl. The f_x decreases in the region larger than 4 nm in size. No saturation effect has been observed in CuBr. The difference may be caused by the different structure of the valence band. The exciton radii of CuBr and CuCl are 1.25 nm and 0.68 nm, respectively.

The stacking layers structure constituted by CdTe particles and oxide glasses as shown in Figure 3.3 has been tested for the third nonlinear optical element. $\chi^{(3)}$ was not so large, but the relaxation time was 1.1 psec.

Multiple quantum wells also have been researched for third nonlinear elements. For example, ZnCdSe/ZnSSe multiple quantum wells exhibited sharp excitonic absorption, and the $\chi^{(3)}$ of it was approximately 2×10^{-4} esu at the near absorption region. The longitudinal relaxation time, T_1, showed biexponential behavior. The fast decay time was 2ps and the slow one was 40ps. The transverse relaxation time, T_2, was 1.7ps.

Organic microcrystals also have been tested. The shift of the exciton absorption due to the size occurs at the region of one order larger than that of metal and semiconductor. $\chi^{(3)}$ values of some organic microcrystals are shown in Table 3.6.[25] The polymers dispersed with semiconductor particles are also materials for the third nonlinear optics. For example, CdS/polymer showed $\chi^{(3)}$ larger than 10^{-7} esu, the absorption less than 10^2 cm^{-1}, and τ less than 10 psec. The radius of exciton in CdS is large, and it

TABLE 3.6

$\chi^{(3)}$ Values of Various Organic Microcrystals Dispersed in Water[25]

Sample Material	Sample Size (mm)	$\chi^{(3)}$ (esu)	$\chi^{(3)}/\alpha$ (esu · cm)
a) J-aggregated MCSe-C18	2 ~ 20	2×10^{-13}	1×10^{-9}
b) J-aggregated MCSe-C18	6.2 ~ 20	1.1×10^{-12}	1.8×10^{-9}
c) Poly(14-8ADA)	10 ~ 300	8.6×10^{-13}	8.6×10^{-10}
d) Polydiacetylene [poly(DCHD)]	23 ~ 70	1.4×10^{-12}	6.1×10^{-10}
e) Poly(DCHD)	23 ~ 100	2.2×10^{-12}	9.3×10^{-10}
f) Poly(DCHD)	16 ~ 150	1.6×10^{-12}	1×10^{-9}

	Sample a)	Sample b)	Sample c)	Sample d)	Sample e)	Sample f)
Dispersion concentration($\times 10^{-4}$M)	2	6.2	10	23	23	16
Crystal size (nm)	20	20	300	70	100	150

Source: H. Kassi et al.: *Jpn. J. Appl. Phys.*, 31, L1132 (1992). With permission.

TABLE 3.7

$\chi^{(3)}$ Values of Some Standard Materials Measured by THG, Degenerate Four-Wave Mixing (DFWM), z-Scan, and Kerr-Shutter Methods

Method	THG	DFWM	Z-scan	Z-scan	Kerr
λ(μm)	2	0. 79	0. 79	0. 79	0. 79
Repetition (Hz)	10	10	1 K	100M	100M
Pulse width	5 ns	150 fs	150 fs	100 fs	100 fs
Fused silica	1	1.2	1.2		
TaFD43 glass[a]	15	38	38	38	38
AOT-5 glass[a]	27	78		63	68
CSz	5	38	29	−18	114
BBO crystal	170	88		98	

[a] Made by Hoya Crystal Co.

TABLE 3.8

Summary of the Size-Dependent Excitonic Absorption of Various Materials

	Compound	Size (nm)	λmax of Exciton Peak (nm)	Energy Shift ΔE(cm^{-1})	Measured Temperature	Literature (Reference Number)
Inorganics	CuCl	Bulk	386		77K	21
		3.3	384	100		
		2.1	382	300		
	CdSe	Bulk	674		2K	
		12	664	300		
		1.8	507	4900		
Organics	NTCDA/	Bulk	560		R. T.	22
	PTCDA	4	558	40		
	(superlattice)	1	555	160		
	PPY/	Bulk			R. T.	23
	PBT(Polymer	9		80		
	superlattice)	3		260		
	Perylene	Bulk	480		R. T.	24
		200	470	500		
		50	450	1400		
	PDA(DCHD)	Bulk	665		R. T.	
		150	652	300		
		50	635	700		

is a Winner-type exciton. So, the origin of large $\chi^{(3)}$ may not be the giant oscillator effect. The magnitude of $\chi^{(3)}$ and the response time of a tin dichlorophthalocyanine film formed by vapor deposition were 4×10^{-8} esu and less than 1ps, respectively.

Measurements of the absolute value of $\chi^{(3)}$ are not so easy. Therefore, the relative measurements are usually used. Some standard materials and $\chi^{(3)}$ values have been listed in Table 3.7. The size-dependent excitonic absorptions are summarized in Table 3.8.

The characteristic values of nonlinear optical phenomena of some materials are listed in Tables 3.4 through 3.8.

The data used in this section are mainly picked up from *Mesoscopic Materials and Clusters*, edited by Arai et al. (Kodansha; Springer, 1999), the *Pre-proceedings of the Symposium on Non-linear Photonic Materials,* Tokyo, 1996 (in Japanese), and *Fundamental Quantum Optics*, by Kudo and Wakaki (Tokai University Press, 2009) (in Japanese).

REFERENCES

Further details of the basic theory can be found in the literature. Refer to References 1 through 5 concerning degenerate four-wave mixing

1. A. Yariv: *Introduction to Optical Electronics*, Holt, Rinehart and Winston (1971).
2. R. L. Byer: *Nonlinear Optics* (eds. P. G. Harper and B. S. Wherrett), pp. 47–160, Academic Press, New York (1977).
3. Y. R. Shen: *The Principles of Nonlinear Optics*, Wiley (1984).
4. R. W. Boyd: *Nonlinear Optics*, Academic Press (1992).
5. N. Bloembergen: *Nonlinear Optics*, W. A. Benjamin, New York (1965).
6. M. Fox: *Optical Properties of Solids*, Oxford University Press (2001).
7. K. Kudo and M. Wakaki: *Fundamentals of Quantum Optics*, pp. 258–333, Tokai University Press (2009) (in Japanese).
8. M. Weissbluth: *Photon-Atom Interactions*, pp. 360–397, Academic Press (1989).
9. D. A. Kleinman: Nonlinear Dielectric Polarization in Optical Media, *Phys. Rev.*, 126, 1977 (1962).
10. S. W. Wemple and M. DiDomenico, Jr.: *Applied Solid State Science* (ed. R. Wolfe), Vol. 3, Academic Press, New York (1972).
11. M. K. Gonokami, N. Peyghambarion, K. Meissner, B. Flugel, Y. Sato, K. Ema, R. Shimano, S. Hazumdar, F. Guo, T. Takihiro, H. Ezaki, and E. Hanamura: Exciton strings in an organic charge-transfer crystal, *Nature,* 367, 47 (1994).
12. A. Harada, Y. Okazaki, K. Kamiyama, and S. Umegaki: Generation of blue coherent light from a continuous-wave semiconductor laser using an organic crystal-cored fiber, *Appl. Phys. Lett.*, 59, 1535 (1991).
13. P. N. Butcher: *Nonlinear: Optical Phenomena*, Ohio State University Press, Columbus (1965).
14. P. N. Butcher and D. Cotter: *The Elements of Nonlinear Optics*, Cambridge University Press (1990).
15. R. K. Jain and M. B. Klein: *Optical Phase Conjugation* (ed. R. A. Fischer), Academic Press, New York, p. 307 (1983).
16. N. Ockman, W. Wang, and R. R. Alfano: Application Of Ultrafast Laser Spectroscopy to The Study of Semiconductor Physics, *Int'l. J. Modern Phys.*, B5, 20, 3165 (1991).
17. W. E. Bron, J. Kuhl, and B. K. Rhee: Picosecond-laser-induced transient dynamics of phonon in GaP and ZnSe, *Phys. Rev.* B34, 6961 (1986).
18. T. Yoshimura: Estimation of enhancement in third-order nonlinear susceptibility induced by a sharpening absorption band, *Optics Commun.*, 70, 535 (1989).
19. E. E. Jelly: Spectral Absorption and Fluorescence of Dye in the Molecular State, *Nature,* 138, 1009 (1936).

20. G. Scheibe: Richstreffen Der Deutchen Chemiker In Verbindung Mit Der 49. Hauptversammlung Des Vereins Deutscher Chemiker Vom 7.-11. Juli 1936 in Munchen, *Angew. Chem.*, 49, 563 (1936).
21. A. Nakamura: *Hyomen* (*Surface*, in Japanese), 30, 330 (1992).
22. F. F. So, S. R. Forrest, Y. Q. Shi and W. H. Staser: Quasi-epitaxial growth of organicultiple quantum well structures by organic molecular beam deposition, *Appl. Phys. Lett.*, 56, 674 (1990).
23. M. Fujitsuka , R. Nakahara, T. Iyoda, T. Shimizu and H. Tsuchiya: Optical properties of conjugated polymer superlattices prepared by potential-programmed electropolymerization, *J. Appl. Phys.*, 74, 1283 (1993).
24. H. Nakanishi: *Proceedings of the fifth symposium on Nonlinear Photonic Materials* (1996) at Yamaha Hall, Tokyo, Japan, p. 87.
25. H. Kassi, H. S. Nalwa, H. Oikawa, S. Okada, H. Matsuda, N. Minami, A. Katsuta, K. Ono, A. Mukoh and H. Nakanishi: A Novel Preparation Method of Organic Microcrystals, *Jpn. J. Appl. Phys.*, 31, L1132 (1992).

4 Materials for Solid-State Lasers

Taro Itatani

CONTENTS

4.1   DEVELOPMENT OF LASERS AND LASER MATERIALS

Laser action was first described [1] by Schawlow and Townes in 1958. The first demonstration of laser oscillation was achieved [2] for a ruby laser ($Cr^{3+}:Al_2O_3$) by Maiman in 1960. The first CW (continuous wave) operation was reported [3] for a crystal of $Nd^{3+}:CaWO_4$ in 1961. Since then, various kinds of laser oscillation have been reported for divalent ions (e.g., Sm^{2+}, Dy^{2+}, Tm^{2+}, Ni^{2+}, Co^{2+}, and V^{2+}) and trivalent ions (Nd^{3+}, Er^{3+}, Ho^{3+}, Ce^{3+}, Tm^{3+}, Pr^{3+}, Gd^{3+}, Eu^{3+}, Yb^{3+}, Cr^{3+}, and Ti^{3+}), which are found in rare earth and transition metals and summarized [4] in Table 4.1.

Host materials for lasers are classified into three groups: crystals, glasses, and ceramics. Host materials should be uniform and large enough for laser action volume, and have excellent optical mechanical and thermal properties. Uniform optical properties (e.g., gain and refractive index) are necessary to prevent inhomogeneous propagation of optical beams. Birefringence should also be suppressed to maintain polarization for laser oscillation. Mechanical and thermal stabilities are important for stable laser oscillation under strong pumping and thermal stress. Thermal conductivity is also important for dissipating heat generated by pumping, and the thermal expansion coefficient is also important for keeping good beam properties. The robustness is required for thermal shock by optical pumping, and high melting temperature is also required for reliability for laser action of host materials.

TABLE 4.1
Typical Laser-Active Ions, Host Materials, and Emission Wavelengths

Laser-Active Ion	Host Materials	Lasing Wavelength
Titanium (Ti^{3+})	Al_2O_3 (sapphire)	700 nm to 1050 nm
Chromium (Cr^{2+})	ZnS_xSe_{1-x}	1900 nm to 3400 nm
Chromium (Cr^{3+})	Al_2O_3 (ruby), LiSaF, LiCaF, BeAl$_2$O$_4$ (Alexandrite)	700 nm to 900 nm
Chromium (Cr^{4+})	YAG, $MgSiO_4$ (forsterite)	1100 nm to 1600 nm
Neodymium (Nd^{3+})	YAG, YVO_4, YLF, glass	1000 nm to 1300 nm
Ytterbium (Yb^{3+})	YAG, glass	1000 nm to 1100 nm
Erbium (Er^{3+})	YAG, glass	1500 nm to 1600 nm
Thulium (Tm^{3+})	YAG, glass	2000 nm to 2010 nm
Holmium (Ho^{3+})	YAG, YLF, glass	2050 nm to 2100 nm
Praseodymium (Pr^{3+})	Silica-fiber	1300 nm
Cerium (Ce^{3+})	YLF, LiCAF, LiSAF	280 nm to 330 nm

4.2 POPULATION INVERSION AND ENERGY LEVELS

Laser oscillation is analyzed on a model using the concepts of population inversion and energy levels in lasing materials. Population inversion and energy levels are explained using a simple model of a two-level system (Figure 4.1). According to the Boltzmann distribution, the atomic density (N_2) in the upper energy level E_2 is less than the density (N_1) in the ground-state energy E_1 at thermal equilibrium.

$$N_2/N_1 = \exp(-(E_2 - E_1)/kT) \tag{4.1}$$

As a result, population difference $N_1 - N_2$ is always positive. Atoms at lower energy levels are pumped to higher energy levels as the laser materials are excited. Under strong pumping, the population difference becomes negative, and population inversion is obtained (Figure 4.1b). Laser oscillation is achieved under this condition.

4.2.1 THREE-LEVEL SYSTEM MODEL

The three-level system model (Figure 4.2) is a simple but realistic model for various kinds of lasers. There are three energy levels. E_1 is the ground state energy level, E_2 is

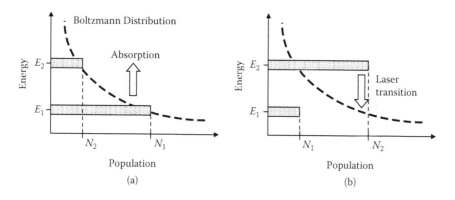

FIGURE 4.1 (a) Thermal equivalent condition. (b) Population inversion.

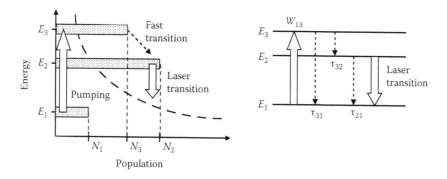

FIGURE 4.2 Energy diagram for three-level system.

the intermediate energy level for laser transition, and E_3 is the upper state energy level for pumping from the ground state E_1. Initially, most atoms are at E_1 as determined by the Boltzmann distribution. Following pumping at the rate of W_{13} from E_1 to E_3, the distribution of each level is determined by τ_{31} (lifetime of E_3 to E_1), τ_{32} (lifetime of E_3 to E_2), and τ_{21} (lifetime of E_2 to E_1). Excited atoms at E_3 transfer to the intermediate energy level E_2. In general, this process is nonradiative and much faster than τ_{21}, which is determined by the fluorescence rate. When the pumping rate is below the laser threshold, the transition from E_2 to E_1 is determined by the spontaneous emission and some drain by fluorescence. When the pumping exceeds the laser threshold, the laser emission is dominant from E_2 to E_1. To obtain efficient laser oscillation, τ_{32} should be fast.

$$\tau_{32} \ll \tau_{21}, \tau_{31} \tag{4.2}$$

The population N_3 is negligible compared to N_1 and N_2. Therefore,

$$N_1 + N_2 \doteqdot N_{tot}. \tag{4.3}$$

The critical condition for population inversion is defined as

$$N_2 = N_1 = N_{tot} / 2 \tag{4.4}$$

The population of the intermediate level, N_2, should be larger than that of the ground level, N_1, to maintain amplification. The pumping power for population inversion (i.e., $N_2 > N_1$) is necessary for more than half of the ground state density (N_{tot}). This is a disadvantage for a three-level system, which needs stronger pumping for laser oscillation compared to a four-level system.

4.2.2 FOUR-LEVEL SYSTEM MODEL

A four-level system model (Figure 4.3) is applied for many practical lasers. Laser transition occurs between E_2 and E_1. In contrast with the three-level model, the ground level is not the lowest energy level. Population inversion can be easily obtained under

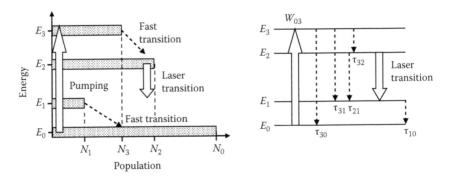

FIGURE 4.3 Energy diagram for four-level system.

the conditions given in the following relations, and more efficient laser oscillation can be achieved compared with the three-level system.

$$\tau_{32}, \tau_{10} \ll \tau_{21}, \tau_{31} \qquad (4.5)$$

$$N_1/N_0 = \exp(-(E_1 - E_0)/kT) \ll 1 \qquad (4.6)$$

τ_{21} is the fluorescent lifetime. If energy difference between E_0 and E_1 exceeds kT, then $N_1/N_0 \ll 1$ and the ground state is almost empty. Some of this laser material, such as Yb:YAG, is explained as a four-level system by cooling the laser crystal to reach the condition of $N_1/N_0 \ll 1$.

4.3 LASER CRYSTALS

Laser science and laser technology have been developed with much effort to synthesize new laser crystals composed of laser-active ions and host materials. Various kinds of laser crystals have been demonstrated [5,6]. Solid-state lasers have become popular because a wide range of coherent light is obtained by utilizing nonlinear crystals for wavelength conversion and semiconductor lasers for pumping. Diode-pumped solid-state lasers (DPSSLs) are widely used in semiconductor, display, and solar panel manufacturing industries.

4.3.1 ND:YAG LASER

Nd:YAG lasers are important because of their high gain and good thermal and mechanical properties of the host crystal.

4.3.1.1 Material Properties

Yttrium aluminum garnet (YAG), discovered in 1969, is one of the best host materials and possesses good optical, thermal, and mechanical properties [5,6]. The chemical formula for YAG is $Y_3Al_5O_{12}$. YAG has a cubic crystal structure of garnet. Its material properties are summarized [7–9] in Table 4.2. The radius of the Nd ion is different by 3% from that of the Y ion. Therefore, the Nd^{3+} ion gets stress from the

---

TABLE 4.2

Typical Laser-Active Ions, Host Materials, and Emission Wavelengths

Chemical formula	$Y_3Al_5O_{12}$
Melting point [°C]	1970
Mohs hardness	8.5
Density [g/cm³]	4.56
Heat capacity [J/gK]	0.625
Thermal conductivity [W/cmK]	0.14
Thermal expansion coefficient	7.5×10^{-6}
$\delta n/\delta T [K^{-1}]$	7.3×10^{-6}

lattice, and this stress influences the energy levels of Nd^{3+}, resulting in additional absorption and birefringence. The maximum doping of Nd is empirically limited to 1.5%. Higher doping shortens the fluorescence lifetime and broadens the absorption spectrum, resulting in higher pumping intensity.

4.3.1.2 Laser Performances

This laser operates as a four-level system (Figure 4.4), and its laser oscillation was demonstrated [10] by Geusic et al. in 1964. Its laser properties are summarized in Table 4.3. The upper level $^4F_{3/2}$ has a long fluorescence lifetime of 230 μs, with pumping efficiency exceeding 95% [11]. The lower level for laser transition is $^4I_{11/2}$, with an efficiency of 60% [12]. The absorption spectrum for Nd:YAG is presented [13] in Figure 4.5. A Nd:YAG crystal is grown by replacing Y^{3+} with Nd^{3+} in a YAG crystal. The output power reached 124 W for continuous wave (CW) operation, and the output energy of 11.2 mJ was obtained with the peak power of 93 kW in Q-switched operation at 10 Hz by side-pumping to the laser crystal and using a V-shaped double resonator (Figure 4.6) [14].

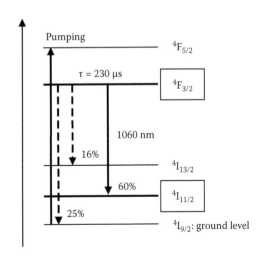

FIGURE 4.4 Energy level of Nd:YAG laser.

TABLE 4.3

Laser Properties of Nd:YAG

Absorbing wavelength (spectrum width) [nm]	~800 (<10)
Lasing wavelength [nm]	1060
Ratio of (absorbing/lasing) wavelength	0.76
Fluorescence lifetime [μs]	230
Stimulated emission cross section [cm^{-2}]	2.8×10^{-19}
Refractive index	1.82
Scatter loss [cm^{-1}]	0.002

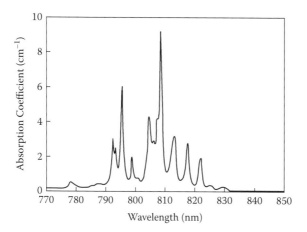

FIGURE 4.5   Absorption spectrum for Nd:YAG.

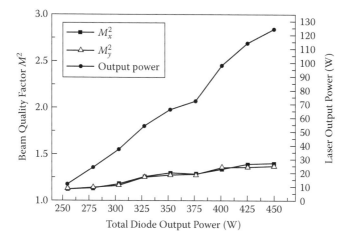

FIGURE 4.6 M^2 and laser output power for Nd:YAG laser.

4.3.2   Nd:YLF Laser

4.3.2.1   Material Properties

Yttrium lithium fluoride (YLF) with Nd^{3+} ion offers a laser transition of 1053 nm, which matches the pumping of Nd-doped phosphate and fluorophosphate glasses. The chemical formula for YLF is $LiYF_4$. YLF has a tetragonal crystal structure. Its material properties are summarized [5,6] in Table 4.4. This crystal has a natural birefringence that can eliminate its depolarization thermally. The maximum doping of Nd is empirically limited to 3 wt%.

TABLE 4.4

Material Properties of Yttrium Lithium Fluoride (YLF) at Room Temperature

Chemical formula	$LiYF_4$
Melting point [°C]	825
Mohs hardness	4 ~ 5
Density [g/cm³]	3.99
Heat capacity [J/gK]	0.79
Thermal conductivity [W/cmK]	0.06
Thermal expansion coefficient	a-axis: 13×10^{-6}, c-axis: 8×10^{-6}
$\delta n/\delta T[K^{-1}]$	$n_0 = 4.3 \times 10^{-6}$, $n_e = 2 \times 10^{-6}$

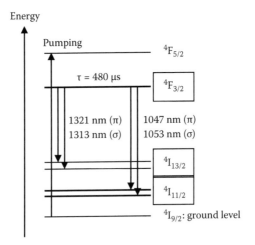

FIGURE 4.7   Energy level of Nd:YLF laser.

4.3.2.2 Laser Performances

This laser operates as a four-level system (Figure 4.7). Its laser properties are summarized in Table 4.5. Its energy diagram is basically the same as that of Nd:YAG, although the laser emissions take a slightly different wavelength for σ and π polarizations. The lasing polarization is selected by a polarization control element (e.g., an electro-optic modulator or a polarizing-beam splitter), and the wavelength is determined by cavity mirrors with dielectric coating. The upper level $^4F_{3/2}$ has a long fluorescence lifetime of 480 µs, which is double that of Nd:YAG. The absorption spectrum [15] for Nd:YLF is shown in Figure 4.8. Nd:YLF is generally used with high-power Q-switching because of its long upper lifetime and weak thermal lens effect. The output energy reaches 379 mJ with a pulse duration of 8.7 ns and repetition of 10 Hz using an electro-optic modulator of $La_3Ga_5SiO_{14}$ for Q-switching. The schematic diagram of the resonator is presented in Figure 4.9. The output energy over 400 mJ has demonstrated [16] at the modulation voltage of 1000 V, which is shown in Figure 4.10.

TABLE 4.5

Laser Properties of Nd:YLF

Absorbing wavelength (spectrum width) [nm]	792 (<5), 797 (<5)
Lasing wavelength [nm]	1047(σ), 1053(π)
Ratio of (absorbing/lasing) wavelength	0.76
Fluorescence lifetime [μs]	480
Stimulated emission cross section [cm^{-2}]	1.2×10^{19}
Refractive index	$n_0 = 1.4481$, $n_e = 1.4704$
Scatter loss [cm^{-1}]	0.002

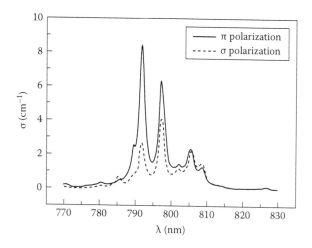

FIGURE 4.8 Yttrium lithium fluoride (YLF) absorption spectrum for a Nd_3^+ concentration of 0.8 mol%.

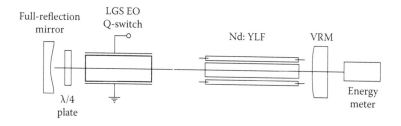

FIGURE 4.9 Nf:YLF laser cavity with a Q-switch of LGS crystal.

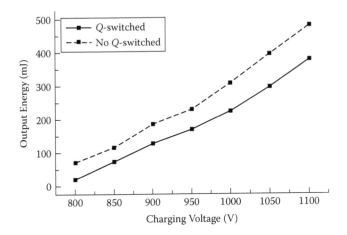

FIGURE 4.10 Output energy versus charging voltage of the capacitor.

4.3.3 Nd:YVO$_4$ Laser

4.3.3.1 Material Properties

YVO$_4$ has a Zircon tetragonal crystal structure. Its material properties are summarized [5,6] in Table 4.6. Yttrium vanadate (YVO$_4$) has several drawbacks. It has a shorter lifetime than Nd:YAG, and its thermal conductivity is less than half that of Nd:YAG and Nd:YLF. Typical doping of Nd is 1 atom%.

4.3.3.2 Laser Performances

This laser operates as a four-level system (Figure 4.11). Even though the potential for an efficient laser was recognized [17] in 1966, this crystal was not industrialized until the 1990s, after development of growth technology that eliminated light scattering and color centers. The upper level $^4F_{3/2}$ has a rather short lifetime of 100 μs, though the stimulated cross section is four times larger than that of Nd:YAG in the π-direction. The absorption spectrum is broader and smoother than the spectrum for Nd:YAG crystal as shown in Figure 4.12. Its laser properties are summarized in Table 4.7. These properties are suitable for multimode diode laser pumping when the emission wavelength is controlled by the temperature of the diodes. Nd:YVO$_4$ is often pumped by fiber-coupled diode lasers in the end-pumping scheme, which is shown in Figure 4.13. Resonant pumping at 888 nm has been introduced [18] for high efficiency and good beam quality. This wavelength provides low isotropic absorption, with equal absorption coefficients for c-polarized light, which leads to smoother uniform absorption for a long crystal with a reduced front-face heat load and temperature, and polarization-insensitive absorption suitable for fiber pumping. A long (30 mm) YVO$_4$ crystal with Nd doping of 0.4% is used for the cavity configuration with optical-fiber-guided pumping that produces a long and smooth pumping volume in the laser active medium. An output power of 60 W was

TABLE 4.6
Material Properties for YVO4 at Room Temperature.

Chemical formula	YVO_4
Melting point [°C]	1810
Mohs hardness	5
Density [g/cm³]	4.22
Heat capacity [J/gK]	0.565
Thermal conductivity [W/cmK]	0.05
Thermal expansion coefficient	a-axis: 4.4×10^{-6}, c-axis: 11×10^{-6}
$\delta n / \delta T [K^{-1}]$	a-axis: 8.5×10^{-6}, c-axis: 3×10^{-6}

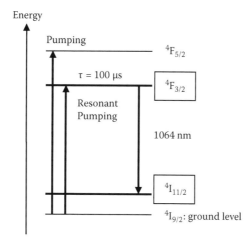

FIGURE 4.11 Energy level of Nd:YVO$_4$ laser.

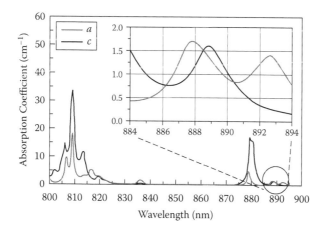

FIGURE 4.12 Nd:YVO$_4$ absorption spectrum for a-axis and c-axis polarization.

TABLE 4.7

Laser Properties of Nd:YVO₄

Absorbing wavelength (spectrum width) [nm]	809 (<1)
Lasing wavelength [nm]	1064
Ratio of (absorbing/lasing) wavelength	0.76
Fluorescence lifetime [μs]	100
Stimulated emission cross section [cm⁻²]	11.4×10^{-19}
Refractive index	$n_0 = 1.96$, $n_e = 2.17$ (c direction)
Scatter loss [cm⁻¹]	0.002

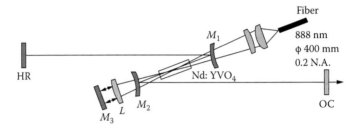

FIGURE 4.13   Typical cavity configuration for Nd:YVO₄ laser.

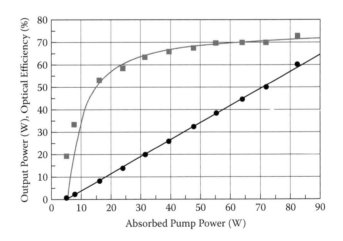

FIGURE 4.14   Output properties of the Q-switched Nd:YVO₄ laser.

demonstrated [19] with pump power of 108 W corresponding to 55% efficiency, which is shown in Figure 4.14. The shape of the output beam is round and diffraction limited with $M^2 = 1.05$. Many applications by using, for example, a cavity-damped, Q-switched oscillator up to 60 W with an optical efficiency of 73%, were realized with this system.

4.3.4   Yb:YAG Laser

4.3.4.1   Material Properties

See Section 4.3.1.

4.3.4.2   Laser Performances

Ytterbium³⁺ (Yb³⁺) is used as a laser-active ion, different from Nd³⁺. This ion has only one excited state ($^2F_{2/5}$) and a ground state ($^2F_{2/7}$). These states are spectrally broadened by a phonon-induced transition within the Stark level manifolds. The energy diagram for this laser is depicted in Figure 4.15. This laser operates as a quasi-three-level system, so higher pumping is necessary compared with the four-level system. The laser operates as a nearly four-level system beyond 1080 nm.

Yb³⁺ has a longer lifetime of 1 ms, and its absorption bandwidth of 4 nm is much wider than that of Nd³⁺. Thus, laser action with higher output is expected. Its absorption spectrum is broad and smooth at room temperature as shown in Figure 4.16 [20]. The spectrum at 78 K becomes sharp and is suitable for diode pumping. Its laser properties are summarized in Table 4.8. The configuration of a cryogenically cooled Yb:YAG laser [21] is presented in Figure 4.17, and CW output power up to 550 W has been achieved as shown in Figure 4.18. The laser crystal is cooled with liquid nitrogen for improved efficiency, and the laser medium with multiple thin discs and bonded to transparent thermally conductive sapphire plates enables efficient cooling to obtain good beam quality at high power operation. Bonding technology will be explained in Section 4.7.

4.3.5   Ti:Sapphire Laser

Since the laser action of Ti:sapphire was first reported [22] in 1982, this laser has been extensively investigated and widely used for tunable lasers and ultrafast lasers.

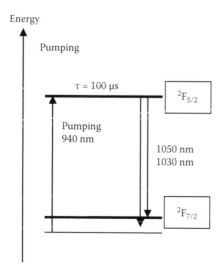

FIGURE 4.15   Energy level of Yb:YAG laser.

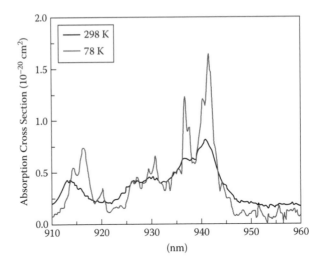

FIGURE 4.16 Yb:YAG (crystal) absorption spectrum for two temperatures.

TABLE 4.8
Laser Properties of Yb:YAG

Absorbing wavelength (spectrum width) [nm]	940, 970
Lasing wavelength [nm]	1030
Ratio of (absorbing/lasing) wavelength	0.91 for 940 nm, 0.94 for 970 nm
Fluorescence lifetime [μs]	1200
Stimulated emission cross section [cm^{-2}]	2×10^{-20}
Refractive index	1.82
Scatter loss [cm^{-1}]	0.003

4.3.5.1 Material Properties

Monocrystalline Al_2O_3 (sapphire) is one of the best host materials. It possesses good optical, thermal, and mechanical properties and was used for pioneering demonstrations [2] of a ruby laser.

The chemical formula for sapphire is Al_2O_3. Sapphire has a hexagonal crystal structure. Its material properties are summarized [6] in Table 4.9. A Ti^{3+} ion is substituted for an Al^{3+} ion. The radius of the Ti^{3+} (0.76Å) ion is much different from that of the Al^{3+} (0.51Å) ion. This laser crystal is made by the Czochralski method with the typical Ti concentration of 0.1wt%.

4.3.5.2 Laser Performances

Ti:sapphire, also known as a vibronic laser, has a strong interaction of electronic states with lattice vibrations (i.e., phonons). This vibrational-electronic (vibronic) interaction leads to strong homogeneous broadening, resulting in a high gain bandwidth. The upper level (^2E) has a long fluorescence lifetime of 3.2 μs. The absorption

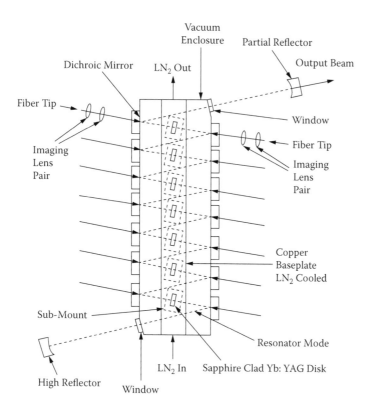

FIGURE 4.17 Continuous wave diode pumped Yb:YAG cryogenic laser.

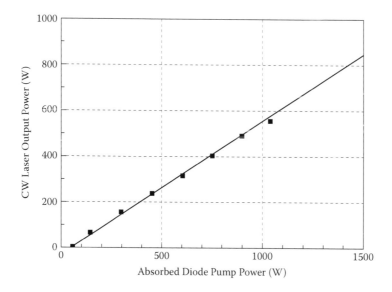

FIGURE 4.18 Output power of the Yb:YAG cryogenic laser.

TABLE 4.9

Material Properties for Sapphire at Room Temperature

Chemical formula	Al_2O_3
Melting point [°C]	2040
Mohs hardness	9
Density [g/cm³]	3.98
Heat capacity [J/gK]	0.75
Thermal conductivity [W/cmK]	0.33
Thermal expansion coefficient	5×10^{-6}
δn/δT [K⁻¹]	13×10^{-6}

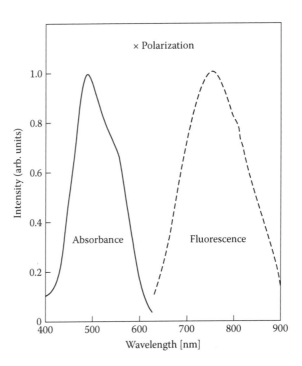

FIGURE 4.19   Absorption and fluorescence spectra for Ti:Al₂O₃ crystal.

and fluorescence spectra are shown [23] in Figure 4.19. This laser operates as a four-level system with the upper level ^2E and the lower level $^2T_2$. Its laser properties are summarized in Table 4.10. This system was introduced [22] by Moulton in 1986. A tuning range from 700 nm to 1050 nm has been achieved as depicted in Figure 4.20. This wide bandwidth is also useful for generating short pulses. Many laser systems based on Ti:sapphire lasers (e.g., generation by passive mode-locking and amplification by chirped pulse) have been commercialized and widely used.

TABLE 4.10

Laser Properties of Ti:Sapphire

Absorbing wavelength (spectrum width) [nm]	488 (<200)
Lasing wavelength [nm]	800
Ratio of (absorbing/lasing) wavelength	0.76
Fluorescence lifetime [μs]	3.2 μs
Stimulated emission cross section [cm⁻²]	$4.1 \times 10^{-19}/cm^2$ CASIX
Refractive index	1.76

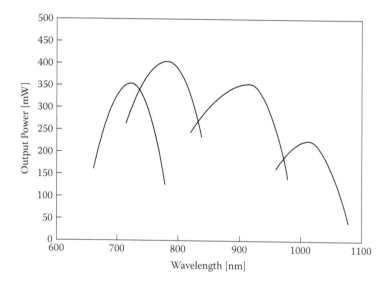

FIGURE 4.20 Tuning range for Ti:Al₂O₃ crystal.

A noncollinear optical parametric amplifier for chirped optical pulses based on the ultrafast Ti:Al₂O₃ laser system has demonstrated [24] 7.6 fs pulse generation as shown in Figure 4.21. This system produces the peak power of 2 TW at the 30 Hz repetition rate. The output spectrum and reconstructed pulse width and phase are shown in Figure 4.22. Average output power up to 220 W was achieved [25] by improving the pumping configuration.

4.3.6 ALEXANDRITE LASER

4.3.6.1 Material Properties

Alexandrite ($Cr^{3+}:BeAl_2O_4$) is a common name for chromium-doped Chrysoberyl with an orthorhombic structure. This crystal is grown by the Czochralski method, like a ruby. Chromium is doped to replace Al^{3+} ions in the crystal with Cr^{3+} ions. The material properties of Alexandrite are presented [26] in Table 4.11. The optical

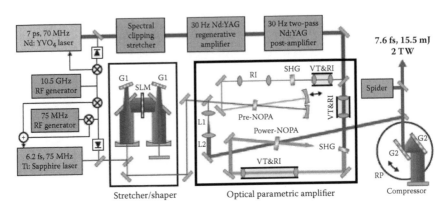

FIGURE 4.21 The 2 TW 7.6 fs laser system. G1 and G2 are grating units; L1 and L2 are lenses.

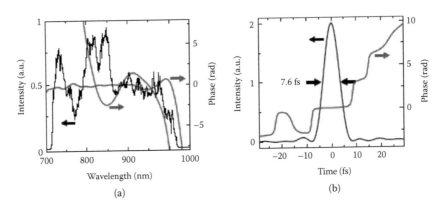

FIGURE 4.22 (a) Spectrum intensity and spectral phase. (b) Reconstructed pulse intensity and phase in the time domain.

TABLE 4.11
Material Properties for Alexandrite at Room Temperature

Chemical formula	$BeAl_2O_4$
Melting point [°C]	1870
Mohs hardness	8.5
Density [g/cm³]	3.7
Heat capacity [J/gK]	0.83
Thermal conductivity [W/cmK]	0.23
Thermal expansion coefficient	a: 5.9, b: 6.1, c: 6.7 × 10⁻⁶
δn/δT[K⁻¹]	8 × 10⁻⁶

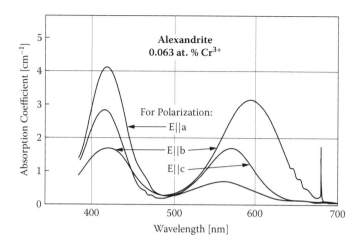

FIGURE 4.23   Absorption spectrum for Alexandrite.

and mechanical properties of Alexandrite are similar to those of the ruby. Thus, this material is a good host laser crystal with high thermal conductivity, hardness, and chemical stability. The Alexandrite crystal exhibits birefringence due to its ortho-rhombic structure. Laser emission polarized to the *b*-axis is 10 times stronger than that under the alternate polarization condition.

4.3.6.2   Laser Performances

This laser operates as either a three-level system or a four-level system with the upper level ^2E and the lower level $^4A_2$. The absorption spectrum of the crystal doped at 0.063 atom% Cr^{3+} is presented in Figure 4.23. This laser is also a vibronic laser like Ti:sapphire lasers and is used as a wavelength tunable laser in the visible region. The upper level ^2E has a long fluorescence lifetime of 260 μs. This long lifetime is advantageous for cheap flash lamp pumping and Q-switched operation. Alexandrite lasers have been widely used for medical and cosmetic applications (e.g., skin reju-venation and hair removal). Although less absorption by melanin for ruby laser light, the wavelength of 755 nm from an Alexandrite laser offers the best balance between efficiency for all hair types and epidermal safety for darker-skinned patients.

4.4   GLASS MATERIALS FOR LASER

Various kinds of glasses have been investigated for laser applications [4]. Higher and more uniform doping is possible with glasses than with laser crystals, and glasses have optically isotropic properties. The main advantages of the glass medium are size capability and shape flexibility. The size of the glass exceeds 1 m for high-power applications, and single-mode fibers with micrometer scale have been achieved. Nd-doped glass lasers are used as driver lasers for laser fusion at the National Ignition Facility (NIF), and Er-doped glass fiber amplifiers are widely used in opti-cal communication networks.

4.4.1 MATERIAL PROPERTIES

Silicate and phosphate glasses are commonly used because they have sufficient quality and are widely available. Phosphate glasses are used mainly for high-energy applications, and silicate glasses are used mainly for optical communications. The material properties for silicate glasses (fused silica) and phosphate glasses are summarized [27,28] in Table 4.12.

4.4.1.1 Laser Performances

An Nd:glass laser operates as a four-level system as shown in Figure 4.24. The upper level is the lower-lying $^4F_{3/2}$ multiplet, and the lower terminal level is the lower-lying $^4I_{11/2}$ multiplet for laser transition. The upper level $^4F_{3/2}$ has a long fluorescence lifetime exceeding 300 μs. The terminal level $^4I_{11/2}$ is emptied to the ground state $^4I_{9/2}$ by

TABLE 4.12

Material Properties of Silicate and Phosphate Glasses at Room Temperature

	Fused Silica (SiO$_2$)	Phosphate Glass Q-246 (Kigre, Inc.)	Phosphate Glass Q-88 (Kigre, Inc.)
Transition temperature [°C]	616	470	366
Mohs hardness	~7	~6	~5
Density [g/cm³]	2.2	2.55	2.71
Specific heat [J/gK]	0.75	0.93	0.81
Thermal conductivity [W/cmK]	0.014	0.013	0.0084
Thermal expansion coefficient [/K]	0.51×10^{-6}	9×10^{-6}	10×10^{-6}
δn/δT [/K]	2×10^{-8}	2.9×10^{-6}	-0.5×10^{-6}

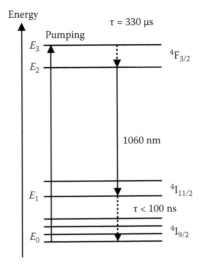

FIGURE 4.24 Energy level of Nd: Glass laser.

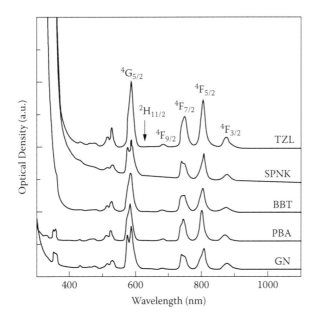

FIGURE 4.25 Absorption spectra of Nd: Glass (GN for Germante, PBA for phosphate, BBT for borate, SPNK for lead silicate, and TBL for tellurate).

the fast nonradiation transition of less than 100 ns. Because the energy separation between $^4I_{11/2}$ and $^4I_{9/2}$ is much greater than kT (27 meV), the population density of terminal level $^4I_{11/2}$ is almost zero for the heated condition with laser pumping.

The absorption spectra for various kinds of oxide glasses are presented [29] in Figure 4.25. The stimulated emission cross section of a Nd:phosphate glass is larger than that of a Nd:silicate glass. A Nd:phosphate glass has been adopted [30] in the high-power laser system at the NIF (Figure 4.26). The lasing properties are governed by the absorption, storage, and release of the pumping energy. Energy storage capability is not influenced much by the host glass materials. However, release capability is limited by the host glass properties. Typical properties of a Nd:phosphate glass and a Nd:silicate glass are indicated in Table 4.13. The fluorescence spectrum is not sensitive to temperature. Therefore, the operation range is wide (–100°C to 100°C in general).

In high-power laser systems, the main limiting factors are nonlinear effects and shock resistance. The nonlinear properties of the refractive index and the two-photon absorption should be suppressed. A high shock parameter of 12 W/cm, which exceeds 30 times that of the Nd:phosphate glass LHG8, has been reported [31] for the Nd:silicate glass.

There are important differences between glass lasers and crystal lasers with regard to the thermal conductivity and the emission linewidth. Specifically, glass lasers have poor thermal conductivities and broader emission linewidth. The low thermal conductivity limits the high-power operation, and the broader linewidth results in the amplification for a shorter pulse width and higher storage energy capability. Doping of active ions more than 3 wt% is easily obtained [29].

FIGURE 4.26   Nd:phosphate glass for the National Ignition Facility (NIF).

TABLE 4.13
Laser Properties of Nd:Phosphate Glass and Silicate Glass

	Phosphate Glass Q-246 (Kigre)	Silicate Glass Q-88 (Kigre)
Absorbing wavelength (spectrum width) [nm]	810 (10)	810 (10)
Lasing wavelength [nm]	1054	1062
Ratio of (absorbing/lasing) wavelength	0.77	0.76
Fluorescence lifetime [μs]	330	330
Stimulated emission cross section [cm^{-2}]	4×10^{-20}	2.9×10^{-20}
Refractive index	1.538	1.581
Scatter loss [%cm^{-1}] at lasing wavelength	0.0008	0.0008

4.5   CERAMIC LASERS

Laser action for transparent oxide ceramics doped with Nd was achieved [32] with an optical loss of 5 to 7%/cm. Vacuum sintering of Y_2O_3 and Al_2O_3 was performed to achieve [33] the CW operation of Nd:ceramic YAG. The advantages of ceramics over glasses are higher thermal conductivity and thermal shock resistance. On the other hand, the advantages of ceramics over laser crystals are the lower cost and ease in obtaining a larger gain medium.

4.5.1   MATERIAL PROPERTIES

The chemical formula for ceramic YAG is $Y_3Al_5O_{12}$, the same as that for the YAG single crystal. The advantages of ceramic YAG over laser crystals are

FIGURE 4.27 Polycrystalline Nd: YAG ceramic laser rods.

(a) SEM image (b) TEM image

FIGURE 4.28 Yttrium aluminum garnet (YAG) ceramic consists of small YAG crystal with grain boundaries less than 1 nm.

the lower cost and ease in obtaining a larger volume of gain medium. The main disadvantage of ceramic materials is their optical loss due to the interface scattering and nonuniformity of the material. An improved method of casting wet-synthesized powder was developed [34] to demonstrate a laser power exceeding 1 kW. Neodymium-doped YAG nano-crystalline ceramics [35] have the very low pore volume concentration of 1 ppm. The pictures of ceramic laser rods are shown in Figure 4.27. Scanning electron microscopy (SEM) and transmission electron microscopy (TEM) images (Figure 4.28) clearly indicate that the boundary areas are less than 1 nm. The ceramic YAG has good thermal conductivity (12 W/mK), comparable to that of the YAG single crystal. Its material properties are summarized [36] in Table 4.14. The maximum doping concentration of Nd ions exceeds 6 wt%.

TABLE 4.14

Material Properties for Ceramic Yttrium Aluminum Garnet (YAG) at Room Temperature

Chemical formula	$Y_3Al_2O_{12}$
Melting point [°C]	1970
Hardness	15 GPa
Density [g/cm³]	4.55
Heat capacity [J/gK]	0.602
Thermal conductivity [W/cmK]	10.7 W/mK
Thermal expansion coefficient	8×10^{-6}
δn/δT[K⁻¹]	9.2×10^{-6}

4.5.2 LASER PERFORMANCES

This laser operates as a four-level system, the same as Nd:YAG (single crystal). The absorption and fluorescence spectra [37] for Nd:ceramic YAG is shown in Figure 4.29.

The high-power outputs from a ceramic YAG and a crystal YAG are shown in Figure 4.30. The slope efficiency is 41.2% for the ceramic YAG laser, which is slightly higher than that for the crystal YAG laser (38.4%). The output powers reached 110 W for the ceramic YAG laser and 103 W for the crystal YAG laser. Ceramic Nd:YAG lasers with a slab-and-disk configuration have demonstrated [38] the average power of 67 kW with a 20% duty cycle.

4.6 SEMICONDUCTOR LASERS

Semiconductor lasers have been extensively investigated since 1961 and widely distributed in worldwide markets for optical communications and data storages. Light-emitting diode (LED)–based illumination and displays are also prevailing rapidly. Semiconductor lasers differ from other lasers especially in their small size (typically less than 1 mm) and the direct pumping by electrical current.

4.6.1 MATERIAL PROPERTIES

Semiconductor laser material has started from GaAs. Subsequently, a variety of semiconductor materials have joined the line-up such as InP for longer wavelength emissions and GaP for shorter wavelength emissions. A simplified model of energy states and density of states is illustrated in Figure 4.31. E_c and E_v are the energy levels of the conduction and valence bands, respectively, and E_g is the band gap of the semiconductor. Without current pumping, the valence band is filled with electrons, and the conduction band is empty (Figure 4.31a). With current pumping, electrons and holes are injected into the conduction and valence bands, respectively (Figure 4.31b). The laser transition occurs with this population inversion. Operations of the semiconductor lasers (e.g., calculation of density of states, energy levels, and gain of

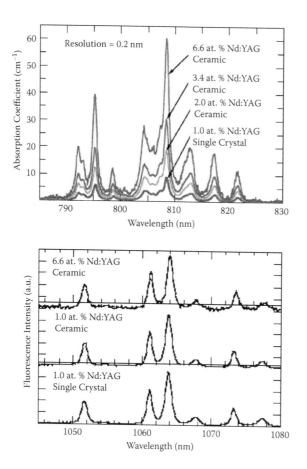

FIGURE 4.29 Absorption and fluorescence spectra for ceramic Nd:YAG.

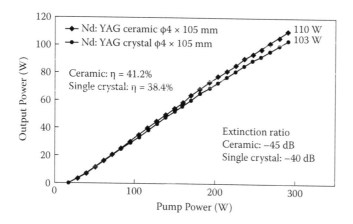

FIGURE 4.30 High-power outputs from a ceramic yttrium aluminum garnet (YAG) laser and a crystal YAG laser.

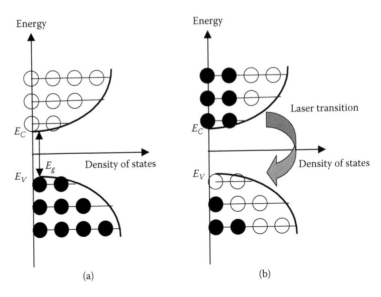

FIGURE 4.31 Energy and density of states for semiconductor lasers. (a) Density of states without current pumping, (b) density of states with current pumping.

TABLE 4.15
Material Properties for Semiconductors

	Si	GaN	GaP	GaAs	InP
Band structure	Indirect	Direct	Direct	Direct	Direct
Band gap	1.12 eV	3.39	1.42 eV	1.34 eV	2.27 eV
	(1107 nm)	(365 nm)	(873 nm)	(925 nm)	(546 nm)
Melting point (decomposition)[°C]	1410	2500	1467	1240	1062
Mohs hardness	6.5 to 7	9	4 to 5	3.5 to 4.5	3
Density [g/cm³]	2.33	6.1	4.13	5.32	4.79
Specific heat [J/gK]	0.7	0.42	0.44	0.35	0.33
Thermal conductivity [W/cmK]	1.31	1.5	1.1	0.46	0.67
Thermal expansion coefficient [/K]	2.6×10^{-6}	a: 3.2×10^{-6} b: 5.6×10^{-6}	5.3×10^{-6}	5×10^{-6}	4.5×10^{-6}
Refractive index	3.42	$n_o = 2$ $n_e = 2.2$	3.66	3.45	3.45
$\delta n/\delta T[/K]$	1.66×10^{-4}	0.61×10^{-4}	2×10^{-4}	0.61×10^{-4}	0.83×10^{-4}

the medium) are described in detail in the references [39]. The physical properties for typical semiconductor materials are summarized [40,41] in Table 4.15. Silicon, which is the typical semiconductor material for electronics applications, has an indirect band gap structure and is not a lasing material. Typical lasing materials (e.g., GaN, GaP, GaAs, and InP) which have a direct band gap structure are classified as

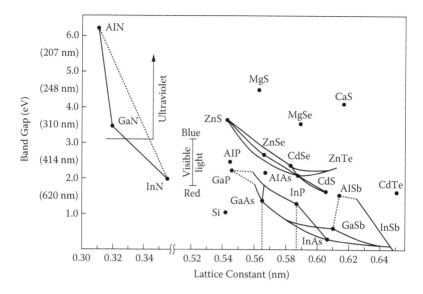

FIGURE 4.32   Relation between band gap and lattice constant.

III-V compound semiconductors. Their thermal conductivities exceed that of YAG crystal. The relationship between band gap energies and lattice constants for III-V compound semiconductors is shown [41] in Figure 4.32. By changing the composition, we can adjust the lasing wavelength. The most important III-V compound alloy system is $In_xGa_{1-x}As_yP_{1-y}$ for optical communication. In particular, these compounds are used as high-power laser diodes for optical pumping of solid-state lasers. Long device lifetime more than 10,000 hours has been achieved by using an aluminum-free structure to prevent oxidation at the emitting surface.

4.6.2   LASER PERFORMANCES

Typical band diagram of a semiconductor laser is depicted in Figure 4.33. The laser has a p-i-n junction structure and an optical waveguide to confine the light. Electrons and holes are injected into the i-GaAs layer and recombined to emit the laser radiation. It is important that the semiconductor gain layer has a confinement structure in order to achieve efficient laser oscillation. The AlGaAs compound forms a clad layer, which has a lower refractive index than that of the GaAs core layer. Typical commercial semiconductor lasers are summarized in Table 4.16. Semiconductor lasers have realized lasing wavelength ranging from ultraviolet to infrared.

4.7   FUNCTIONAL MATERIALS

4.7.1   BONDED MATERIALS

Bonding technology for laser crystals is important for improving the damage threshold and reducing the thermal effect in the gain medium. Laser-induced damage

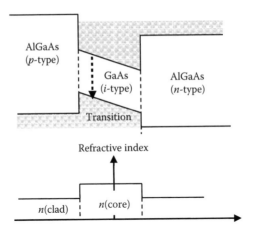

FIGURE 4.33   Typical semiconductor laser structure.

TABLE 4.16
Semiconductor Lasers with Different Wavelengths and Applications

Wavelength (nm)	Lasing Material	Substrate	Applications
405	InN	Sapphire	Blue-ray disc, violet imaging
445	InGaN	Sapphire	Projector
650	AlGaInP	GaP	DVD
670	AlGaInP	GaP	Pointers (red)
785	AlGaAs	GaAs	CD
980	InGaAs	GaAs	Pumping for YAG lasers
1310	InGaAsP	InP	Optical fiber communication with minimum dispersion
1550	InGaAsP	InP	Optical fiber communication with minimum loss

occurs on the surface of the gain medium without bonding. The damage threshold on the surface is very low compared to the damage level for materials in general. Bonded composite crystals are important for improving laser performance. Thermal load on the gain surface can be reduced, and the damage threshold can also be reduced by covering the surface of the gain medium with a nondoped medium.

The typical configuration of a Nd:YAG laser [42] with a bonded crystal is shown in Figure 4.34. This Nd:YAG composite rod consists of a 4-mm-long and 1.1 wt% Nd-doped rod in the middle and two 4-mm-long nondoped end caps with a diameter of 3 mm. The total length is 12 mm with two 4-mm-long nondoped YAG caps and one 4-mm-long Nd:YAG gain rod. The output power and the beam quality are indicated in Figure 4.35. The maximum output power of 15.2 W was achieved with a slope efficiency of 45% (Figure 4.35). The maximum output power without a bonding structure is limited to 8.3 W.

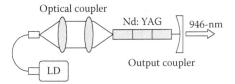

FIGURE 4.34　The high-power diode-end-pumped 946 nm Nd:YAG laser.

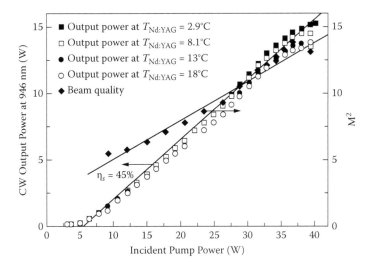

FIGURE 4.35　Laser output and beam quality for different pumping powers.

4.7.2　Semiconductor Saturable Absorber

The semiconductor saturable absorber (SESAM) is a key component for ultrafast lasers with passive mode locking, in which an ultrashort pulse train can be obtained by repeated gain saturation and absorption saturation, but this system needs additional self-starting and self-sustaining conditions. These conditions are satisfied by introducing a SESAM and are demonstrated [43] for various lasers of Nd:YAG, Nd:YLF, Nd:YVO$_4$, Yb:YAG, Ti:Al$_2$O$_3$, Nd:glass, and so on.

A schematic image of a SESAM is presented in Figure 4.36. It is made of a single quantum well absorber and a Bragg reflector of GaAs and AlAs on a GaAs substrate. As the incident optical pulses reach into the SESAM, the front edges of the pulses are absorbed. Then absorption becomes saturated, and the pulses are reflected back from SESAM. An advanced passive mode-locked laser has been demonstrated to generate an average power of 60 W in a 810 fs pulse train for a thin-disk Yb:YAG laser (Figure 4.37) [44]. A broadband SESAM has been demonstrated [45] for a Forsterite laser, and 19.4 fs pulses were obtained in 1300 nm region. The structure for the broadband SESAM is illustrated in Figure 4.38a, and the fringe-resolved autocorrelation results at 19.4 fs are presented in Figure 4.38b.

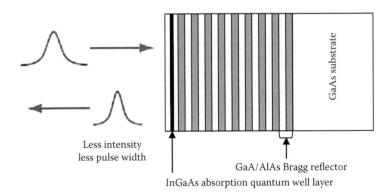

FIGURE 4.36   Structure of semiconductor saturable absorber (SESAM).

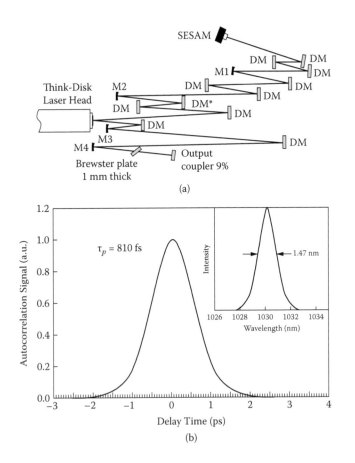

FIGURE 4.37   (a) Configuration of thin-disk Yb:YAG laser with a semiconductor saturable absorber (SESAM). DM: dispersive mirror, M: mirror. (b) Autocorrelation trace and spectrum of the 810-fs pulses.

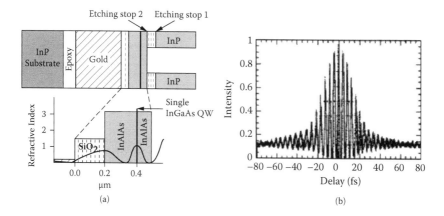

FIGURE 4.38 (a) Structure of the broadband semiconductor saturable absorber (SESAM) using a metal reflector. QW: quantum well. (b) Fringe-resolved autocorrelation trace with pulse width of 19.4 fs.

REFERENCES

1. A. L. Schawlow and C. H. Townes: "Infrared and Optical Masers", *Phys. Rev.*, 112, 1940 (1958).
2. T. H. Maiman: *Nature,* 187, 493 (1960).
3. L. F. Johnson and L. Nassau: *Proc. IRE,* 49, 1704 (1961).
4. Marvin J. Weber: *Handbook of Laser Wavelength*, CRC Press (1998).
5. M. J. Weber: *Handbook of Laser Science and Technology Supplement 2: Optical Materials*, CRC Press (1995).
6. M. Bass: *Handbook of Optics*, Vol. II, section 33.53 properties of crystals and glasses (2011).
7. O. C. Cronemyer: "Optical Absorption Characteristics of Pink Ruby", *J. Opt. Soc. Am.*, 56, 1703 (1966).
8. D. M. Dodd, D. J. L. Wood, and R. L. Barns: "Spectrophotometric Determination of Chromium Concentration in Ruby", *J. Appl. Phys.*, 35, 1183–1186 (1964).
9. T. H. Maiman, R. H. Hoskins, I. J. D'Haenens, C. K. Asawa, and V. Evtuhov: "Stimulated Optical Emission in Fluorescent Solids. II. Spectroscopy and Stimulated Emission in Ruby", *Phys. Rev.*, 123, 1151 (1961).
10. J. E. Geusic, H. M. Marcos, and L. G. Van Uitert: "Laser Oscillations In Nd doped Yttrium Aluminum, Yttrium Gallium and Gadolinium Garnets", *Appl. Phys. Lett.*, 4, 182–184 (1964).
11. T. Kushida, and J. E. Geusic: "Optical Refrigeration in Nd-Doped Yttrium Aluminum Garnet", *Phys. Rev. Lett.*, 21, 1172 (1968).
12. T. Kushida, H. M. Marcos, and J. E. Geusic: "Laser Transition Cross Section and Fluorescence Branching Ratio for Nd^{3+} in Yttrium Aluminum Garnet", *Phys. Rev.*, 167, 289 (1968).
13. J. Lu, M. Prabhu, J. Song, C. Li, J. Xu, K. Ueda, A. A. Kaminskii, H. Yagi, and T. Yanagitani: "Optical properties and highly efficient laser oscillation of Nd:YAG ceramics", *Appl. Phys.*, B71, 469–473 (2000).
14. J. Ma, G. Zhong, J.Wu, Z. Qiao, G. Wang, and Q. Duanmu: "124 W diode-side-pumped Nd:YAG laser in dynamic fundamental mode", *Opt. Laser Technol.*, 42, 552–555 (2010).

15. N. U. Wetter, E. C. Sousa, F. A. Camargo, I. M. Ranieri, and S. L. Baldochi: "Efficient and compact diode-sidepumped Nd:YLF laser operating at 1053 nm with high beam quality", *J. Opt. A: Pure Appl. Opt. Col.*, 10, 104013 (2008).

16. S. Zhang, Q. Wang, Z. Tian, X. Yin, H. Zhang, Y. Li, and S. Li: "Performance of Nd:YLF laser by using La$_3$Ga$_5$SiO$_{14}$ crystal electrooptic Q-switch", *Opt. Laser Technol.*, 37, 608–611 (2005).

17. J. R. O'Connor: "Unusual crystal - Field energy levels and efficient laser properties of YVO$_4$:Nd", *Appl. Phys. Lett.*, 9, 407–409 (1966).

18. SPIE. Newsroom 10.1117/2.1200707.0708 (2007).

19. L. McDonagh, R. Wallenstein, R. Knappe, and A. Nebel: "High-efficiency 60 W TEMoo Nd:YVO$_4$ oscillator pumped at 888 nm", *Opt. Lett.*, 31, 3297–3299 (2006).

20. D. J. Ripin, J. R. Ochoa, R. L. Aggarwal, and T. Y. Fan: "165-W cryogenically cooled Yb:YAG laser", *Opt. Lett*, 29, 2154–2156 (2004).

21. D. C. Brown, J. M. Singley, E. Yager, K. Kowalewski, L. Guelzow, and J. K. Kuper: "Kilowatt class high-power CW Yb:YAG cryogenic laser", *Proc. SPIE*, 6952, 69520K, Laser source Technology for Defence and Security IV (2008).

22. P. F. Moulton: "Ti-Doped Sapphire: Tunable Solid-state Laser", *Opt. News*, 8, 9 (1982).

23. P. F. Moulton: "Spectroscopic and laser characteristics of Ti:Al$_2$O$_3$", *J. Opt. Soc. Am. B*, 3(1), 125–133 (1986).

24. R. Rao, G. Vaillancourt, H. S. Kwok, and C. P. Khattak: *OSA Proc. on Solid State Lasers*, 5, 39 (1989).

25. S. Witte, R. Th. Zinkstok, A. L. Wolf, W. Hogcrvorst, W. Ubachs, and K. S. E. Eikema: "A source of 2 teravvatt, 2.7 cycle laser pulses based on noncollinear optical parametric chirped pulse amplification". *Opt. Express*, 14, 8168–8177 (2006).

26. H. Samelson, J. C. Walling, and D. F. Heller: *Proc. SPIE*, 335, 85 (1982).

27. D. B. Leviton and B. J. Frey: "Temperature-dependent absolute refractive index measurements of synthetic fused silica", *Proc. SPIE*, 6273, 62732K (2006).

28. Kigre Inc., Home page, www.kigre.com.

29. Z. Jiang, J. Yang, and S. Dai: "Optical spectroscopy and gain properties of Nd^{3+}-doped oxide glasses", *J. Opt. Soc. Am. B*, 21(4), 739–743 (2004).

30. Lawrence Livermore National Laboratory, National Ignition Facility. lasers.llnl.gov.

31. Y. Fujimoto, H. Yoshida, M. Nkatsuka, T. Ueda, and A. Fujinoki: "Development of Nd-doped Optical Gain Material Based on Silica Glass with High Thermal Shock Parameter for High-Average-Power Laser", *Jap. J. Appl. Phys.*, 44(4 A), 1764–1770 (2005).

32. C. Greskovitch and J. P. Chernoch: "Polycrystalline ceramic lasers", *J. Appl. Phys.*, 44, 4599–4606 (1973).

33. A. Ikesue, T. Kinoshita, K. Kamata, and K. Yoshida: *J. Am. Ceram. Soc*, 78, 1033 (1995).

34. J. Lu, J. Song, M. Prabhu, J. Xu, K. Ueda, H. Yagi, T. Yanagitani, and A. A. Kaminskii: *Jpn. J. Appl. Phys.*, 39, LI048 (2000).

35. J. Lu, K. Ueda, H. Yagi, T. Yanagitani, Y. Akiyama, A. A. Kaminskii: "Neodymium doped yttrium aluminium garnet (Y$_3$Al$_5$O$_{12}$) nanocrystalline ceramics – a new generation of solid state laser and optical materials", *J. Alloys and Compounds*, 341, 220–225 (2002).

36. Transparent Polycrystalline YAG Ceramics, http://www.konoshima.co.jp/en/ceramics/PDF/YAG_HP.pdf.

37. T. Taira, A. Ikesue, et al.: *OSA TOPS on Advanced Solid-State Lasers*, 19(3), 430–432 (1998).

38. R. M. Yamamoto, B. S. Bhachu, K. P. Cutter, S. N. Fochs, S. A. Lets, C. W. Parks, M. D. Rotter, and T. F. Soules: In *Conference on Advanced Solid-State Photonics* (Optical Society of America, 2008), paper WC5(2008).

39. A. Yariv and P. Yeh: *Photonics Optical Electronics in Modern Communications*, Oxford University Press (2006).

40. S. M. Sze: *Technology of Semiconductor Devices*, Wiley (2002).

41. H. K. Toenshoff and M. Hartmann: *Forschung im Ingenieurwesen*, 63, 1434 (1997).

42. R. Zhou, E. Li, H. Li, P. Wang, and J. Yao: "Continuous-wave, 15.2 W diode-end-pumped Nd:YAG laser operating at 946 nm". *Opt, Lett*, 31, 1869–1871 (2006).

43. U. Keller, K. J. Weingarten, F. X. Kartner, D. Kopf, B. Braun, S. D. Jung, R. Fluck, C. Hoenninger, N. Matuschek, and J. Aus der Au: *IEEE, J. Selected Topics in QE*, 2, 435 (1996).

44. E. Innerhofer, T. Sudmeyer, F. Brunner, R. Haring, A. Aschqwanden, R. Paschotta, C. Honninger, M. Kumkar, and U. Keller: "60-W average power in 810-fs pulses from a thin-disk Yb:YAG laser", *Opt. Lett.*, 28, 367–369 (2003).

45. Z. Zhang, K. Torizuka, T. Itatani, K. Kobayashi, T. Sugaya, T. Nakagawa, and H. Takahashi: "Broadband semiconductor saturable-absorber mirror for a self-starting mode-locked Cn:forsterite laser", *Opt. Lett.*, 23, 1465–1467 (1998).

5 Materials for Optical Waveguides

Bishnu P. Pal

CONTENTS

5.1 INTRODUCTION

Optical waveguides form the building blocks of integrated optics as well as fiber optics—the two fields that could together be broadly classified as guided wave optics. Optical fibers essentially function as optical transmission media for optical signals [1,2]; they are of dielectric materials and hence are functionally passive, as they cannot respond to external voltages or electric currents. In view of its passive characteristic, optical fibers are unable to find applications as functional devices for several other functions required in an optical communication network. Integrated optical waveguides on the other hand could be built around passive as well as active materials like semiconductors, electro-optic materials, and so forth [3–5]. In this chapter we focus on optical waveguides of integrated optical variety only. During the 1980s though a lot of effort was spent on developing guided wave components based on semiconductor materials, technologies based on $LiNbO_3$, glass, polymers, and silica developed at a much faster pace. Several high-performance discrete functional devices—passive as well as active—are now available based on these materials. In Section 5.2 we discuss the basic physics underlying wave guidance in a slab/planar waveguide and the power carried by a mode in Section 5.3. This is followed by a discussion in Section 5.4 on modeling wave guidance in 3D-waveguide geometries, in which light is confined in two dimensions. Section 5.5 is devoted to a matrix technique to model wave guidance in multilayer waveguides (e.g., anti-resonant reflecting optical waveguide [ARROW]) based on silica on silicon material systems. In Sections 5.6, 5.7, and 5.8 we discuss physics of electro-optic, semiconductor, and polymer waveguides. Section 5.9 describes various technologies for fabricating integrated optical waveguides followed by a conclusion.

5.2 SLAB/PLANAR WAVEGUIDES: PHYSICS AND ANALYSIS

Fundamental to understanding wave guidance in optical waveguides is the concept of modes supported by the waveguide. The physics of an optical waveguide in terms of its mode(s) is best illustrated and understood through a slab/planar geometry characterized by three layers (shown in Figure. 5.1a). In its simplest geometry, it is an optical structure in which the refractive index is solely a function of one variable (e.g., x). The central core region of the uniform refractive index n_1 is sandwiched between two cladding layers, each of uniform refractive index n_c and n_s, and each of which is less than n_1. If $n_c = n_s = n_2$, such a structure is known as a *symmetric* waveguide. We may assume the refractive index to vary along x as follows:

$$n^2(x) = n_c^2 \quad\quad x \geq d \quad\quad \text{(cover)}$$

$$= n_1^2 \quad\quad 0 \prec x \prec d \quad \text{(core)} \quad\quad\quad (5.1)$$

$$= n_s^2 \quad\quad x \leq d \quad\quad \text{(substrate)}$$

Such a refractive index profile (RIP) is shown in Figure 5.1b. The wave equation, which governs propagation in such a composite dielectric structure, is given by

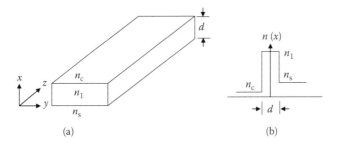

FIGURE 5.1 (a) The planar/slab waveguide geometry; (b) refractive index profile of the waveguide. For a symmetric waveguide $n_c = n_s = n_2$.

$$\nabla^2 \Psi = \varepsilon_0 \mu_0 n^2(r) \frac{\partial^2 \Psi}{\partial t^2} \tag{5.2}$$

where ε_0 and μ_0 are dielectric permittivity and magnetic permeability of free space, respectively, and Ψ represents either the electric field E (x, y, z, t) or the magnetic field H (x, y, z, t) associated with the propagating electromagnetic wave. For the above-mentioned RIP and choosing z as the direction of propagation, without any loss of generality, spatial variation of the fields would be confined to the xz-plane. If we choose $e^{i\omega t}$ as the temporal dependence of Ψ, the solution to Equation (5.2) for the electric field could be written as

$$E(x, y, z, t) = E_j(x) e^{i(\omega t - \beta z)} \qquad j = x, y, z \tag{5.3}$$

where ω represents frequency of the propagating wave and β the corresponding propagation constant, whose value is dictated by the parameters of the waveguide. For example, for a multimode waveguide, there would be several possible β, while for a single-mode waveguide there would be only one β allowed. The parameters that characterize a given waveguide are refractive index contrast between the core and its surrounding medium, core dimensions (either full width or half width), and the wavelength of the propagating light. These parameters could be clubbed within a unique parameter known as the V-parameter/normalized frequency of a waveguide, which is defined for a symmetrical waveguide as

$$V = \frac{2\pi}{\lambda_0} \frac{d}{2} \left(n_1^2 - n_2^2 \right)^{\frac{1}{2}} \tag{5.4}$$

where d is the width of the core; for an asymmetric waveguide in which $n_c \neq n_s$, n_2 is replaced by n_s, and λ_0 represents wavelength at which the waveguide is operated. At a given V number, an optical waveguide could admit only one or more than one value of β; the precise number is dictated by the electromagnetic boundary conditions. It can be shown that the boundary conditions allow only a discrete set of βs. The transverse fields associated with each of these βs constitutes a guided mode of the waveguide. A mode is thus formally defined as a transverse electric field distribution, which propagates through the waveguide without any change in its distribution except for its accumulated phase. It can be shown (see later in the chapter) that at a given V-number, a waveguide could support either one or more than one mode; in the former case the waveguide is classified as a single-mode waveguide, while it would be called a multimoded waveguide in the latter case. An arbitrary optical field incident on a waveguide could be expanded into a sum over the allowed guided modes and a continuum of unguided radiation modes as [6]

$$E(x,y) = \sum_m a_m E_m(x) e^{i(\omega t - \beta z)} + \int a(\beta) E(\beta) e^{i(\omega t - \beta z)} d\beta \qquad (5.5)$$

In Equation (5.5), the coefficient a_m is such that power in the mth mode is proportional to $|a_m|^2$. Further, these guided modes are mutually orthogonal and are normalized such that they satisfy the orthonormality condition

$$\int_{-\infty}^{\infty} E_m^*(x) E_n(x) dx = \partial_{mn} \qquad (5.6)$$

where $\partial_{mn} = 0$ for $m \neq n$ and equals 1 for $m = n$; it represents the Kronecker delta function. The orthonormality condition normalizes the power carried by each mode to unity. The radiation modes also form an orthogonal set, although the orthonormality condition is required to be defined appropriately in terms of Dirac delta function [6]. We would ignore radiation modes in our discussions in this chapter.

For the RIP given by Equation (5.1), the waveguide could normally support both transverse electric (TE) and transverse magnetic (TM) modes. For the TE modes, the only nonzero field components are E_y, H_x, and H_z, while for the TM modes, the corresponding components are H_y, E_x, and E_z. By substituting Equation (5.2) for E (and also the corresponding equation for H) in the two curl equations of Maxwell, one can obtain the following wave equation satisfied by the TE mode [4,6,7]:

$$\frac{d^2 E_y(x)}{dx^2} + \kappa_m^2 E_y(x) = 0 \qquad (5.7)$$

where

$$\kappa_m = \sqrt{k_0^2 n^2(x) - \beta_m^2} \qquad (5.8)$$

which represents transverse components of the plane wave vector $k_0 n(x)$. The solutions of the wave equation (Equation 5.7) for TE modes would be given in the three distinct regions of an asymmetric waveguide as

$$\begin{aligned}
E_y(x) &= A e^{-\gamma_c x} & x \geq d & \quad \text{cover} \\
&= B e^{-i\kappa_m x} + C e^{+i\kappa_m x} & 0 \leq x \leq d & \quad \text{core} \\
&= D e^{+\gamma_s x} & x \leq 0 & \quad \text{substrate}
\end{aligned} \qquad (5.9)$$

where

$$\gamma_{c,s} = \sqrt{\beta_m^2 - k_0^2 n_{c,s}^2} \quad \text{and} \quad \kappa_m = \sqrt{k_0^2 n_1^2 - \beta^2} \tag{5.10}$$

Because for a guided mode its field must necessarily be oscillatory inside the core and *evanescent* (i.e., decaying exponentially) in the cover and the substrate, $\gamma_{c,s}$ and κ_m are necessarily all real and positive quantities. Thus for a guided mode, its propagation constant must satisfy the following inequality:

$$k_0 n_1 > \beta_m > k_0 n_s > k_0 n_c \tag{5.11}$$

where it is assumed that $n_s > n_c$. It is apparent from Equation (5.9) that the guided mode power decays to $1/e$ of its value at $\gamma_{c,s}^{-1}$ distances, respectively, from the interfaces of core-cover and core-substrate. The closer the value of β to $k_0 n_{c,s}$, the greater would be the penetration of the evanescent tail into the cladding, and it would be more confined when β is closer to $k_0 n_1$. The electromagnetic boundary conditions require that the tangential components E_y and $H_z \left(\propto \dfrac{dE_y}{dx} \right)$ across the core-substrate and the core-cover interfaces be continuous. By making use of these conditions, one could get two expressions for the ratio B/C, which through algebraic manipulation would finally yield the following eigenvalue equation:

$$\tan(\kappa_m d) = \frac{\gamma_c/\kappa_m + \gamma_s/\kappa_m}{1 - \dfrac{\gamma_c \gamma_s}{\kappa_m^2}} \tag{5.12}$$

The same equation could be rewritten as

$$\kappa_m = (\varphi_s + \varphi_c + m\pi)/d; \quad m = 0,1,2,3, ... \tag{5.13}$$

where

$$\tan \varphi_{s,c} = \frac{\gamma_{s,c}}{\kappa_m} \tag{5.14}$$

The parameter β appears on both sides of the equal sign in Equations (5.12) and (5.13). These are transcendental equations, either of which could be solved graphically or numerically. Further, because m can take only discrete values, for a given waveguide (i.e., at a given V-number), only *discrete* values of β would be yielded by solution of either Equation (5.12) or Equation (5.13). Values of β for $m = 0, 1, 2, ...$ correspond to propagation constants of TE_0, TE_1, TE_2, ... modes. For symmetric waveguides, because $n_c = n_s = n_2$, Equation (5.12) can be recast as either

$$\tan\left(\frac{\kappa_m d}{2}\right) = \frac{\gamma_2}{\kappa_m} \tag{5.15a}$$

or

$$\cot\left(\frac{\kappa_m d}{2}\right) = -\frac{\gamma_2}{\kappa_m} \qquad (5.15b)$$

where $\gamma_2 = \sqrt{\beta^2 - k_0^2 n_2^2}$. If we introduce a normalized parameter b defined through

$$b = \frac{\beta^2 - k_0^2 n_2^2}{n_1^2 - n_2^2} \qquad (5.16)$$

which is known as the normalized propagation constant, the allowed guided mode regime (Equation 5.11) in terms of the parameter b would satisfy the following inequality:

$$0 \le b \le 1 \qquad (5.17)$$

Equations (5.15a) and (5.15b) can thus be rewritten in terms of the dimensionless parameters V and b as

$$\tan\left(V\sqrt{1-b}\right) = \sqrt{\frac{b}{1-b}} \qquad (5.18a)$$

and

$$\tan\left(V\sqrt{1-b}\right) = -\sqrt{\frac{1-b}{b}} \qquad (5.18b)$$

These two equations can be conveniently solved graphically or numerically. For determining the modal field associated with a particular mode having a particular β as yielded by the solution of Equations (5.18a) and (5.18b), the modal fields for a symmetric waveguide are obtained from Equation (5.9) by eliminating three out of the four unknown constants A, B, C, and D, which can be determined through the electromagnetic boundary conditions (mentioned earlier). This procedure yields the fields as [7]

$$E_y(x) = \frac{D}{\cos\left(\dfrac{\kappa_m d}{2}\right)} \cos\left[\kappa_m\left(x - \frac{d}{2}\right)\right] \qquad (5.19)$$

in which β_m contained in the expression for κ_m is given by the solution of Equation (5.18a). Likewise, β_m obtained through solution of Equation (5.18b) yields the field as [7]

$$E_y(x) = \frac{D}{\sin\left(\dfrac{\kappa_m d}{2}\right)} \sin\left[\kappa_m\left(\frac{d}{2} - x\right)\right] \qquad (5.20)$$

It can be seen from Equation (5.19) that E_y is symmetric in x about the mid-plane of the core, while in Equation (5.20) the field E_y is antisymmetric in x with respect to the mid-plane of the waveguide. Thus, in a symmetric waveguide, the guided TE modes will, in general, consist of *symmetric* (Equation 5.19) and *antisymmetric* (Equation 5.20) modes; these are also called *even* and *odd* modes, respectively. For a given waveguide because V would be known at a given operating wavelength, a plot of the left- and right-hand sides of Equations (5.18a) and (5.18b) as a function of b within the range defined by Equation (5.17) on the same figure will yield b (hence β) through their intersections. The number of intersections also determines the number of modes. As an example, consider a symmetric planar waveguide formed from silicon oxynitride (SiON) having refractive index (r.i.) n_1 as 2.01 as the core layer, which is surrounded by silica (SiO$_2$ of r.i. $n_2 = 1.458$) as the cladding at the operating wavelength of He-Ne laser (0.6328 μm). Figure 5.2 shows universal curves (full curves) for b as a function of V for the TE modes in a symmetric planar waveguide; values of b at different values of V are obtained by solving Equations (5.18a) and (5.18b). For $V = 1$, all TE modes except the TE$_0$ modes are cut off in a symmetric planar waveguide. By definition the mth mode is *cut off* in a waveguide when its propagation constant β_m equals $k_0 n_2$, implying $b = 0$. Thus at a mode cutoff, Equations (5.18a) and (5.18b) become

$$\tan\left(V_c^{(TE_m)}\right) = 0 \Rightarrow V_c^{(TE_m)} = m\pi; \quad m = 0,1,2,\ldots\ldots \tag{5.21}$$

where $V_c^{(TE_m)}$ stands for normalized cutoff frequency (i.e., the V-number at which mth TE-mode is cut off in a particular waveguide). Even values of m including $m = 0$ correspond to symmetric TE modes, and odd values to antisymmetric modes. Because λ_0, n_1, and n_2 are fixed, different values of V would correspond to different widths (d) of the waveguide. These universal curves could be readily used as follows:

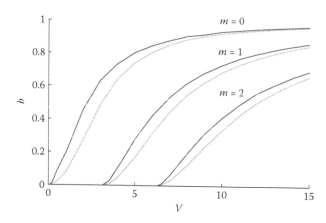

FIGURE 5.2 Universal plot showing normalized propagation constant b versus normalized frequency V for various TE$_m$ (full curves) in a symmetrical planar waveguide. The dashed curves correspond to TM$_m$ modes in a planar optical waveguide; for the TM modes the ratio between the core and cladding refractive indices was taken as 1.38 assuming the waveguide to be a symmetric guide of SiON/SiO$_2$.

for example, for a $SiO_2/SiON/SiO_2$ planar waveguide, from Figure 5.2 it can be read that for $V = 1$, which corresponds to a $d \sim 0.14$ µm, silicon oxynitride waveguide will support only one TE mode at the 0.6328 µm wavelength. If d is increased by about fivefold to ~ 0.73 µm, V increases to ≈ 5, at which the silicon oxynitride waveguide would support two symmetric and one antisymmetric TE modes.

It is evident from Equation (5.21) that the TE_0 mode is never cut off in a symmetric planar waveguide; it would be supported for arbitrarily small core width or refractive index contrast. For design purposes the cutoff condition, Equation (5.21), can be rewritten in a more useful form [4]

$$\frac{d}{\lambda} = \frac{m}{2\sqrt{n_1^2 - n_2^2}} \tag{5.22}$$

This condition implies that the smallest ratio (d/λ) for the mth mode to be supported in a waveguide is given by Equation (5.22). For example, for any d/λ less than 0.35, a silicon oxynitride planar waveguide will function as a single-mode (i.e., support only TE_0 mode) waveguide.

By repeating the above procedures for the TM modes of an asymmetric planar waveguide, the eigenvalue equation equivalent to Equation (5.12) for determining βs of the possible TM modes in a given waveguide can be shown to be [6]

$$\tan(\kappa_m d) = \frac{\left(\dfrac{n_1^2}{n_s^2}\right)\dfrac{\gamma_s}{\kappa_m} + \left(\dfrac{n_1^2}{n_c^2}\right)\dfrac{\gamma_c}{\kappa_m}}{1 - \left(\dfrac{n_1^4}{n_c^2 n_s^2}\right)\dfrac{\gamma_c \gamma_s}{\kappa_m^2}} \tag{5.23}$$

For a symmetric planar waveguide the eigenvalue equations equivalent to Equation (5.18) for the TM modes in terms of V and b are [4]

$$\tan\left(V\sqrt{1-b}\right) = \left(\frac{n_1^2}{n_2^2}\right)\sqrt{\frac{b}{1-b}} \qquad \text{(symmetric modes)} \tag{5.24a}$$

$$\tan\left(V\sqrt{1-b}\right) = -\left(\frac{n_1^2}{n_2^2}\right)\sqrt{\frac{1-b}{b}} \qquad \text{(antisymmetric modes)} \tag{5.24b}$$

which show that in the TM case the ratio n_1^2/n_2^2 as a multiplier (on the right-hand side), essentially differentiates the eigenvalue equations for the TM modes from those for the TE modes. The b-V dispersion curves for the TM modes of a silicon oxynitride planar waveguide are also shown in Figure 5.2 as dashed curves; ratio of the core-to-cladding refractive index at 0.6328 µm wavelength is ~ 1.38 in such a waveguide. For an asymmetric waveguide, $V_c^{(TM_m)} > V_c^{(TE_m)}$. For the lowest-order mode (i.e., for $m = 0$),

$$V_c^{TE_0} = \tan^{-1}\left(\sqrt{a}\right) \quad \text{and} \quad V_c^{TM_0} = \tan^{-1}\left(\frac{n_1^2}{n_s^2}\sqrt{a}\right) \tag{5.25}$$

where a represents the asymmetry parameter, and it is defined through

$$a = \frac{\left(n_s^2 - n_c^2\right)}{\left(n_1^2 - n_s^2\right)} \tag{5.26}$$

Thus in an asymmetric waveguide both TE_0 and TM_0 modes have finite but different cutoff frequencies, which implies that for

$$\tan^{-1}\left(\sqrt{a}\right) < V < \tan^{-1}\left(\frac{n_1^2}{n_s^2}\sqrt{a}\right) \tag{5.27}$$

only the TE_0 mode is supported in an asymmetric planar waveguide. Such a wave-guide, in which all other modes except the TE_0 mode are cut off, is called a *single-polarization* single-mode waveguide [8]. In the case of weakly guiding waveguides, for which $n_1 \approx n_s$, TE and TM modes are nearly degenerate (i.e., the n_{eff} of TE_m and TM_m modes are almost the same).

5.3 POWER CARRIED BY A GUIDED MODE IN A PLANAR WAVEGUIDE

The time average of the Poynting vector is given by

$$\langle S \rangle = \langle E \times H \rangle = \frac{1}{2}\mathrm{Re}\left(E \times H\right) \tag{5.28}$$

where <....> implies time average. It can be shown that the net power carried by a symmetric TE mode along z-direction per unit length along y is given by [7]

$$P_z = \frac{1}{2}\int_{-\infty}^{\infty}\mathrm{Re}\left(E \times H\right)\cdot\hat{z}\,dx$$

$$= \frac{\beta_m D^2}{4\omega\mu_0\cos^2\xi}\left[d + \frac{2}{\gamma_2}\right] \tag{5.29}$$

where $\xi = \frac{1}{2}\kappa_m d$ and $\hat{z}$ is the unit vector along z. It is evident from Equation (5.29) that the guided-mode power is confined to an *effective guide half-width* of $[d/2 + 1/\gamma_2]$; the quantity $1/\gamma_2$ represents the characteristic decay length of the evanescent portion of the modal field. Thus one can ascribe a confinement factor Γ to each mode through the following [7]:

$$\Gamma = \frac{\text{Modal power carried inside the core}}{\text{Total power carried by the mode}}$$

$$= \frac{\xi + \sin\xi\cos\xi}{\xi + \kappa_m / \gamma_2} \tag{5.30}$$

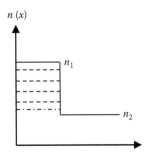

FIGURE 5.3   Dotted lines correspond to n_{eff} of different guided modes; the one shown as dashed-dotted at the bottom is close to its cutoff.

At the mode cutoff, $\beta_m = k_0 n_2$ (i.e., when the mth mode reaches its cutoff, e.g., through variation of the operating wavelength) in a given waveguide, $\gamma_2 = 0$, and hence $\Gamma = 0$; the mode, thus, ceases to be guided inside the core. This can be understood by referring to Figure 5.3, where mode effective indices $n_{\text{eff}} (= \beta/k_0)$ (in the form of dashed lines) of guided modes are shown relative to the refractive index profile (full line) of the waveguide, which is assumed here to be a step-index waveguide. With increase in magnitude of the operating wavelength, it can be seen from Equation (5.4) that the operating V-number decreases and hence n_{eff} of a guided mode would become closer to the cladding index as the mode would see the signature of the cladding more strongly. As the value of n_{eff} drops below the cladding refractive index at a still longer wavelength, the modal field transforms to an oscillatory field everywhere (see Equations 5.9 and 5.10; the parameter $\gamma_{c,s}$ becomes an imaginary number) in the structure, and hence it ceases to be a guided mode.

5.4   THREE-DIMENSIONAL OPTICAL WAVEGUIDE STRUCTURES

In Section 5.2, we discussed analysis of waveguiding in a planar structure. However, the density of guided-wave components on a substrate can be significantly increased by confining the guided optical energy in both the x and y directions [9]. In contrast to the planar geometry, three-dimensional waveguides (several examples are shown in Figures 5.4a through 5.4d) consist of rectangular or near-rectangular cores, which are difficult to analyze analytically. Studies of propagation effects in them generally require extensive numerical analyses [10,11]. Hocker and Burns [12] have, however, proposed a relatively simple and approximate approach, which is called the effective-index method. This method can be illustrated through the example of an embedded strip waveguide (cf. Figure 5.4a), in which the core's dimension is $d_a \times d_b$ and its refractive index is n_1, which is surrounded on its three sides by a medium of refractive index n_3. It is assumed that the core is covered with a medium of refractive index n_2.

The method starts with the assumption that the waveguide extends infinitely along y. This makes it an asymmetric planar waveguide with refractive index varying only along the x direction as shown in Figure 5.5b. For such an asymmetric planar guide having a core of width d_a and refractive index n_1, which is sandwiched between two media of refractive indices n_2 and n_3, its modes are determined through the procedures

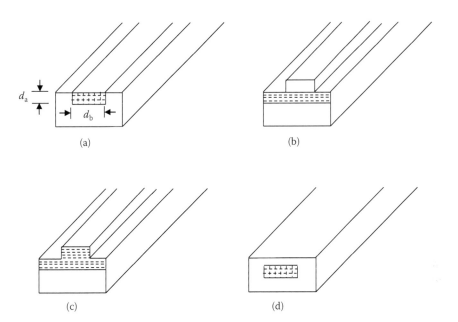

FIGURE 5.4   (a) Embedded strip guide; (b) strip-loaded guide; (c) rib guide; (d) immersed guide.

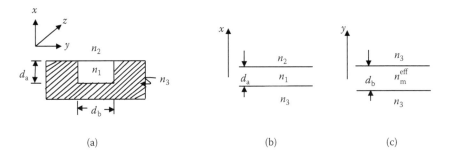

FIGURE 5.5   (a) An embedded strip waveguide of dimension $d_a \times d_b$ having a core of r.i. n_1 surrounded by a material of r.i. n_3 on its three sides and has a cover of r.i. n_2; (b) planar guide along x; (c) planar guide along y with core of r.i. n_{eff}.

outlined in Section 5.2. Depending on polarization of the input beam, either TE (electric field along y) or TM (magnetic field along y) modes will be excited. An effective index n_m^{eff} can be associated with the mth mode of this waveguide. At the next step the method assumes that a core material of refractive index n_m^{eff} can replace the entire asymmetric waveguide along the x direction. In the yz plane, we thus obtain a symmetric planar pseudo-waveguide of core refractive index n_m^{eff} of width d_b surrounded by a medium of index n_3 (see Figure 5.5c). Thus, we can study propagation of such a symmetric pseudo-waveguide by finding the propagation constants β_m of different possible guided modes from the universal mode dispersion curve (shown in Figure 5.2).

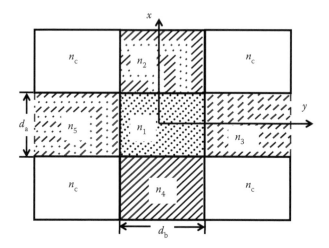

FIGURE 5.6 Most general rectangular waveguide in which the core of r.i. n_1 is surrounded by different r.i. materials on its four sides except for the four corners where the r.i. is n_c.

For each value of m, there will be n solutions for the effective waveguide structure in the yz plane. Thus, in the effective-index model, each mode is designated with a pair of subscripts: m and n. According to [12], there is agreement for the low-order modes and for large aspect ratios in the values of β_{mn} found by this *effective-index* method and extensive numerical methods [10,11]. In particular, the agreement occurs far from cutoff. In view of this and its simplicity, the effective-index method is used extensively in the literature to model propagation in three-dimensional waveguides.

Figure 5.6 represents the most general form of a rectangular waveguide. In the mid-1980s, an alternate technique [13] to the effective index method based on perturbation theory akin to that used in solving quantum mechanical problems was proposed to obtain modal propagation constants with rectangular waveguide geometries. This technique was applied to a number of integrated and fiber optical waveguide geometries and devices [14–16]. It relies on choosing a fictitious rectangular optical waveguide, the index profile of which is separable in x and y coordinates, and which closely resembles the actual waveguide except at the corners. Because the index profile is separable in x and y, the modal solution to the fictitious waveguide becomes simple. Furthermore, because the real index profile differs little from the fictitious profile, simple perturbation theory is then applied to obtain the correct propagation characteristics, and hence the b-V dispersion curves of the real waveguide. Details of this method could be found in the literature [13,14].

5.5 MULTILAYER OPTICAL WAVEGUIDES

5.5.1 SILICA WAVEGUIDES

Optical waveguides are sometimes difficult to form on substrates like silicon (r.i. = 3.5), which is so widely studied from the point of view of integrated circuits in electronics, due to the lack of another suitable transparent medium of refractive index higher than that of silicon. In such cases, another material layer could be grown as a buffer layer on silicon before deposition/formation of the guiding layers. Thus, the typical refractive index (r.i.) profile of such a composite structure can be represented as shown in Figure 5.7. Because of the high-index silicon substrate, the waveguide behaves as a leaky structure unless the buffer layer is thick enough. To reduce mode leakage loss in such a waveguide, the silica buffer layer is grown thick enough to ensure that the evanescent tail of the guided field is negligible at the interface between the silica buffer layer and the silicon substrate. To achieve this, typically the silica layer thickness must be greater than 4 μm, requiring a long deposition time. The problem of long deposition time can be overcome in a novel waveguide configuration [17,18]. It involves a multilayer planar configuration known as anti-resonant reflecting optical waveguides (ARROW). The layered structure of the waveguide and the corresponding refractive index profile are shown in Figures 5.8a and 5.8b, respectively. The bottom silica layer (~2 μm) is called the *second cladding layer*, whereas the top silica layer (~4 μm) forms the *core* of the waveguide. The intermediate high index layer (typically examples of which are poly-Si, Titania) about 0.1 μm thick between these two regions is called the first *cladding*. Two independent physical phenomena are exploited in this waveguide geometry. The silica/doped silica-air interface at the top provides a total internal reflecting surface, whereas the high refractive index layer sandwiched between the core and the second cladding region serves as a highly reflecting (>99%) interface.

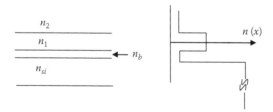

FIGURE 5.7 An optical waveguide along with its refractive index profile in which Si is used as a substrate to form the waveguide with a buffer layer of intermediate r.i. (n_b) and sufficient thickness to isolate the Si substrate from the core (of r.i. n_1).

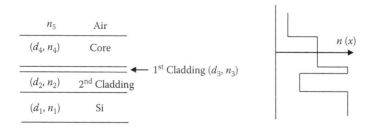

FIGURE 5.8 An anti-resonant reflecting optical waveguide (ARROW) wave along with its refractive index (r.i.) profile.

Thus an ARROW on silicon uses silica/doped silica as the core like that in an optical fiber. The thickness of the first cladding layer is chosen to be small to act as a Fabry-Perot resonator and closely matched the anti-resonant conditions of the resonator. Anti-resonances in a Fabry-Perot etalon are spectrally broad. From the Fabry-Perot analogy the waveguide will work over a wide spectral range. Thus the fabrication tolerance is comfortable. Under optimum conditions, reflectivity could be almost 99.96% from the set of two interfaces of poly-Si/TiO_2-SiO_2 and SiO_2-Si [17]. Approximate expressions for optimum thicknesses of the two reflecting layers are given by [17,18]

$$d_3^{opt} = \frac{\lambda}{4n_3}(2N+1)\left[1-\left(\frac{n_4}{n_3}\right)^2+\left(\frac{\lambda}{4n_3 d_{eff}}\right)^2\right]^{-1/2}, \quad N=0,1,2,.... \quad (5.31)$$

and

$$d_2^{opt} \cong \frac{1}{2}d_{eff}(2M+1), \quad M=0,1,2,..... \quad (5.32)$$

where

$$d_{eff} \cong d_4 \times \varsigma \frac{\lambda}{2\pi\sqrt{n_4^2-n_5^2}} \quad (5.33)$$

and

$$\varsigma = 1 \qquad \text{for TE modes}$$
$$= \left(\frac{n_5^2}{n_4^2}\right) \text{for TM modes} \quad (5.34)$$

while M, N stand for order of anti-resonances. Quantities n_i ($i = 3, 4, 5$) and d_i ($i = 2, 3, 4$) correspond to different layers as marked on Figure 5.8. The cover layer 5 is usually air. These results have been derived to yield minimum loss for the fundamental mode in an ARRO waveguide under anti-resonant conditions.

Propagation characteristics of ARROWs were also obtained through a matrix approach [19]. In Figure 5.8, the refractive indices of the silicon substrate, the second cladding, the first cladding, the core, and the cover (air) are represented by n_1, n_2, n_3, n_4, and n_5, respectively. The thicknesses of the corresponding regions are d_1, d_2, d_3, and d_4, except for the cover for which thickness is infinity. The structure has five homogeneous layers, four interfaces, and five different refractive indices. Its propagation characteristics can be calculated through a method known as the matrix method [20], whose tutorial analysis is presented in Chapter 24 of the work by Ghatak and Thyagarajan [6]. It is a general method that can be implemented for any leaky, absorbing, or guided structure, in which the core layer refractive index could be graded or uniform. In a guided structure shown in Figure 5.9, for a y-polarized

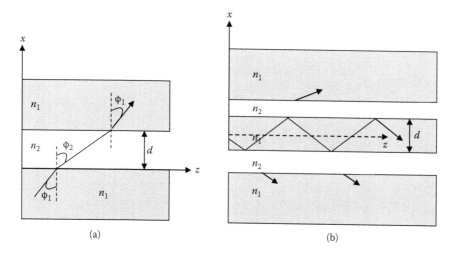

(a)   (b)

FIGURE 5.9 (a) A medium of r.i. n_2 in between two media each of same r.i. n_1 and (b) a leaky structure. (After [6] A. Ghatak and K. Thyagarajan, *Introduction to Fiber Optics*, Cambridge University Press, Cambridge (1998). ©1998 Cambridge University Press.)

propagating plane wave, the corresponding fields in the three individual refractive index regions could be written as [6]

$$E_1(x,z,t)=\left[e_1^+\exp^{i(\omega t-k_1\cos\varphi_1 x-\beta z)}+e_1^-\exp^{i(\omega t+k_1\cos\varphi_1 x-\beta z)}\right]$$

$$E_2(x,z,t)=\left[e_2^+\exp^{i(\omega t-k_2\cos\varphi_2 x-\beta z)}+e_2^-\exp^{i(\omega t+k_2\cos\varphi_2 x-\beta z)}\right] \qquad (5.35)$$

$$E_3(x,z,t)=\left[e_3^+\exp^{i(\omega t-k_3\cos\varphi_3 x-\beta z)}+e_3^-\exp^{i(\omega t+k_3\cos\varphi_3 x-\beta z)}\right]$$

where

$$k_i = k_0 n_i; \quad i=1,2,3$$

$$k_0 = \frac{\omega}{c} \qquad (5.36)$$

$$\frac{\beta}{k_0} = n_1\sin\varphi_1 = n_2\sin\varphi_2$$

Continuity of the fields and their derivatives at the interfaces $x = 0$ and $x = d$ would yield the following matrix equations relating the field in one layer to the one in the adjacent layer:

$$\begin{pmatrix} e_1^+ \\ e_1^- \end{pmatrix} = \frac{1}{2}\begin{pmatrix} \left(1+i\frac{\gamma}{\kappa}\right) & \left(1-i\frac{\gamma}{\kappa}\right) \\ \left(1-i\frac{\gamma}{\kappa}\right) & \left(1+i\frac{\gamma}{\kappa}\right) \end{pmatrix}\begin{pmatrix} e_2^+ \\ e_2^- \end{pmatrix} \qquad (5.37)$$

and

$$\begin{pmatrix} e_2^+ \\ e_2^- \end{pmatrix} = \frac{1}{2} \begin{pmatrix} \left(1-i\frac{\kappa}{\gamma}\right)e^{-\gamma d} & 0 \\ \left(1+i\frac{\kappa}{\gamma}\right)e^{+\gamma d} & 0 \end{pmatrix} \begin{pmatrix} e_3^+ \\ 0 \end{pmatrix} \tag{5.38}$$

where

$$\kappa = k_1 \cos\varphi_1 = \sqrt{k_0^2 n_1^2 - \beta^2} \text{ and}$$

$$i\gamma = k_2 \cos\varphi_2 = k_0 \sqrt{n_2^2 - n_2^2 \sin^2\varphi_2} \Rightarrow \gamma = \sqrt{\beta_2^2 - k_0^2 n_2^2} \tag{5.39}$$

By assuming $\exp(\gamma d) \gg 1$, it can be shown that

$$\left|\frac{e_3^+}{e_1^+}\right|^2 \approx 16\frac{\gamma^2 \kappa^2}{\delta^4} \exp(-2\gamma d); \quad \delta^2 = \kappa^2 + \gamma^2 = k_0^2\left(n_1^2 - n_2^2\right) \tag{5.40}$$

This ratio would represent the tunneling probability. For a guided mode, the tunneling probability $\Rightarrow$ leakage loss would be extremely small. This basic analysis could be extended to a leaky structure by surrounding the above-mentioned waveguide with a layer of n_1 (see Figure 5.9b) so that it becomes a leaky waveguide because of having a second cladding of refractive index larger than n_2. It can be shown that the power in such a leaky waveguide would decay as [6]

$$P = P_0 \exp(-2\Gamma z) \tag{5.41}$$

where

$$\Gamma = \frac{8b^{3/2}(1-b)^{3/2}V^2}{d^3\beta\gamma} \exp(-2\gamma d) \tag{5.42}$$

This analysis was originally developed to treat quasimodes of a leaky planar waveguide. For a multilayer structure the same procedure is continued. For example, for a five-layer planar structure, the mode excitation efficiency $\eta(\beta) = |E_5^+/E_1^+|^2$ is evaluated for the given φ_1 (i.e., for the given $\beta = k_0 n_1 \sin\varphi_1$). The process is repeated by scanning the β-space through a variation in φ_1. A plot of $\eta(\beta)$ reveals resonant peaks that closely resemble Lorentzian functions in shape. For a single-mode guide, there would be only one such peak. The value of β at which the peak appears corresponds to the real part of the propagation constant. The full-width-at-half-maximum (FWHM) (= 2Γ) of the Lorentzian represents the leakage power loss coefficient. With the so derived propagation constant, the fields throughout the system can be computed by evaluating appropriate matrices. The method can be applied to any multilayer structure in which one or more layers could have complex refractive indices [20,21]. In loss less waveguides the parameter $\Gamma \to 0$. This method was used to design and fabricate ARRO waveguides [4,19]. The method is amenable to both TE and TM polarizations. An important attribute of this matrix method is that, in addition to giving leakage loss and propagation constant, it allows computation of the corresponding modal field distributions. The method is equally applicable to ARRO-B waveguides [22], which use a layer of lower refractive index between the two silica layers, rather than a layer of higher refractive index layer. In contrast to the ARRO waveguides, an ARRO-B waveguide is polarization insensitive.

5.6   ELECTRO-OPTIC WAVEGUIDES

5.6.1   ELECTRO-OPTIC MATERIALS

Electro-optic materials like $LiNbO_3$ find extensive application as optical waveguide materials for configuring devices like high-speed modulators in optical communication networks. In fact, $LiNbO_3$-based electro-optic modulators and switches were perhaps the very first active optical waveguide components, which spurred the growth of the field of integrated optics. The phenomenon of electro-optic (e-o) effect means electric field–induced change in the refractive index of a material. Crystalline materials like $LiNbO_3$, KDP, $LiTaO_3$, and $BaTio_3$ could be cited as examples, which exhibit the e-o effect. All of these are anisotropic materials and hence crystal optics is required to model light propagation in these materials. If the change of refractive index is proportional to the magnitude of the applied electric field, it is known as linear e-o effect or Pockel's effect. On the other hand if the change in refractive index is proportional to the square of the applied electric field, the phenomenon is called *quadratic e-o effect* or *Kerr effect*. It can be shown that in an e-o material, applied electric field could induce different amounts of change in refractive index for the two orthogonally polarized propagating light waves through it. Accordingly, phase accumulated by the two differently polarized beams with propagation in an e-o material in the presence of an external electric field would be different. This effect of electric field–induced phase retardation between the two differently polarized beams with propagation in an e-o material can be exploited to achieve amplitude modulation of the propagating light. In fact, whenever the signal transmission rate is ≥ 10 Gbit/s, e-o modulators are almost invariably used in an optical communication network. At such high transmission rates, direct modulation of the drive current of the semiconductor laser often leads to light emission with a chirp, which is detrimental from the point of view of temporal dispersion in a fiber optic communication link. In the next section we discuss e-o effect in the most widely studied e-o material as a waveguide, namely, $LiNbO_3$.

5.6.2   ELECTRO-OPTIC EFFECTS IN LiNbO₃

Lithium niobate is optically anisotroptic and its displacement vector D and electric field vector E are related through the following tensor relation:

$$D = \varepsilon_0 \sum \varepsilon_{ij} E_j \tag{5.43}$$

However, $\varepsilon_{ij} = \varepsilon_{ji}$ from symmetry consideration in a lossless material that is not optically active. In a principal axis system, in which ε_{ij} is diagonal, ε_{ij} is a 3×3 diagonal matrix, such that

$$D_x = \varepsilon_{xx}E_x, \quad D_y = \varepsilon_{yy}E_y, \quad D_z = \varepsilon_{zz}E_z \tag{5.44}$$

Accordingly in a principal axes system, these components are often represented with only one subscript as follows:

$$D_x = \varepsilon_x E_x, \quad D_y = \varepsilon_y E_y, \quad D_z = \varepsilon_z E_z \tag{5.45}$$

Accordingly, principal refractive indices n_i ($i = x, y, z$) could be defined as

$$n_i = \left(\varepsilon_i / \varepsilon_0\right)^{1/2}; \quad i = x, y, z \tag{5.46}$$

where ε_0 is free space dielectric permittivity. In the principal axes system, the index ellipsoid of an anisotropic medium is expressed as

$$\frac{x^2 + y^2}{n_o^2} + \frac{z^2}{n_e^2} = 1 \tag{5.47}$$

LiNbO$_3$ is a uniaxial anisotropic medium, and its $n_x = n_y = n_o$, and $n_z = n_e$ where the subscripts o and e stand for ordinary and extraordinary refractive indices. In the presence of an externally applied electric field along an arbitrary direction, in general the index ellipsoid of an e-o material can be written as [9]

$$\frac{x^2}{n'^2_{xx}} + \frac{y^2}{n'^2_{yy}} + \frac{z^2}{n'^2_{zz}} + \frac{2yz}{n'^2_{yz}} + \frac{2xz}{n'^2_{xz}} + \frac{2xy}{n'^2_{xy}} = 1$$

$$\frac{1}{n'^2_{ij}} = \frac{1}{n^2_{ij}} + \Delta\left(\frac{1}{n^2}\right)_{ij}; \quad i, j = x, y, z$$

$$\tag{5.48}$$

$$\frac{1}{n^2_{ij}} = 0 \text{ for } i \neq j;$$

$$\Delta\left(\frac{1}{n^2}\right)_{ij} = \sum_{k=x,y,z} r_{ijk}E_k; \quad i, j = x, y, z$$

where $[r_{ijk}]$ represents e-o tensor, which is often written in a contracted notation with the first two subscripts as $[r_{ij}]$; $i = 1, 2, 3, \ldots, 6$. Thus for an electric field applied along z (i.e., along the optic axis of LiNbO$_3$) $\Rightarrow E_x = E_y = 0$, change in $\left(\frac{1}{n^2}\right)$ would be given by

$$\Delta\left(\frac{1}{n^2}\right) = \begin{bmatrix} r_{13}E_3 & 0 & 0 \\ 0 & r_{13}E_3 & 0 \\ 0 & 0 & r_{33}E_3 \end{bmatrix} \tag{5.49}$$

Accordingly in the presence of the electric field E_z (= E_3), its index ellipsoid could be written as

$$x^2\left(\frac{1}{n_o^2}+r_{13}E_3\right)+y^2\left(\frac{1}{n_o^2}+r_{13}E_3\right)+z^2\left(\frac{1}{n_e^2}+r_{33}E_3\right)=1 \tag{5.50}$$

It is evident from Equations (5.45) and (5.48) that principal indices of refraction have changed from ($n_{\{x\}}=n_{\{y\}}$) and $n_{\{e\}}$ ($=n_{\{z\}}$) to

$$n_x' = n_y' = n_o - \frac{n_o^3 r_{13}E_3}{2}; \quad n_z' = n_e - \frac{n_e^3 r_{33}E_3}{2} \tag{5.51}$$

Depending on a particular crystal's symmetry group [23], elements of the e.o. tensor [r] are decided. Because LiNbO$_3$ is a crystal with 3m symmetry, only eight elements of its [r] are nonzero, as could be seen from the following:

$$[r] = \begin{bmatrix} 0 & -r_{22} & r_{13} \\ 0 & r_{22} & r_{13} \\ 0 & 0 & r_{33} \\ 0 & r_{51} & 0 \\ r_{51} & 0 & 0 \\ -r_{22} & 0 & 0 \end{bmatrix} \tag{5.52}$$

where

$$r_{22}(\equiv r_{222})\cong 3.4\times 10^{-12}\,\text{m/volt}; \; r_{13}(\equiv r_{113})\cong 8.6\times 10^{-12}\,\text{m/volt};$$

$$r_{33}(\equiv r_{333})\cong 30.8\times 10^{-12}\,\text{m/volt}; \; r_{51}(\equiv r_{131})\cong 28\times 10^{-12}\,\text{m/volt}; \tag{5.53}$$

Thus, with r_{33} being the largest in magnitude, one often chooses orientation of the applied external electric field with respect to orientation of the LiNbO$_3$ crystal so as to exploit this particular e.o. coefficient. LiNbO$_3$ is also highly transparent at the optical communication wavelength window around 1550 nm. Its Curie temperature of 1100°C ~1180°C enables fabrication of metal like Ti-in-diffused optical wave-guides [24,25]. In a later section, we describe fabrication of optical waveguides by different methods.

5.7 SEMICONDUCTOR WAVEGUIDES

Semiconductor materials are attractive for formation of waveguides because one can obtain an electrically active waveguide through this route. III-V group (of the periodic table) compound semiconductors are used to form sources and detectors, whose emission/absorption wavelength coincides with the optical fiber communication wavelength window. Figure 5.10 shows lattice constant versus band gap energy/ band gap wavelength of several such compounds [26]. In a semiconductor waveguide two different materials having different band gap energies are used to realize two different refractive index semiconductors, whose lattice constants match better than 0.1%, in order to form a waveguide with no interface defects. Refractive index of a semiconductor is inversely proportional to the band gap energy. GaAs or InP could be used as a substrate on which one or more thin layers are deposited for forming the waveguide. For example, by replacing a fraction x of Ga atoms in GaAs with Al atoms, one can realize the ternary compound $Al_xGa_{1-x}As$, whose lattice constant is almost the same as that of GaAs but with an increased band gap energy and hence of a different refractive index. The band gap energy depends on the fraction x, which could be approximately described for $0 < x < 0.45$ through [26,27]

$$E_g(x) = 1.424 + 1.247x \tag{5.54}$$

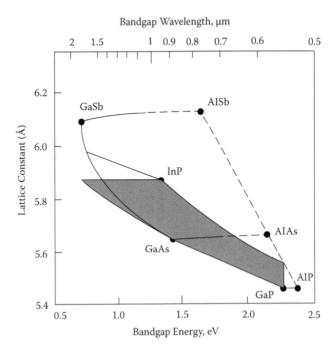

FIGURE 5.10 Lattice constant versus band gap energy/wavelength. (After bandgap [26]; ©2004 John Wiley & Sons, Inc.). Similar figures on bandgap engineering could be found on the Web, e.g. http://people.seas.harvard.edu/~jones.

Likewise band gap energy of InP could be significantly reduced by forming a quaternary compound $In_{1-x}Ga_xAs_yP_{1-y}$, whose lattice constant matches that of InP provided the ratio of x to y is restricted to 0.45. This material could be tailored to operate at any wavelength in the range 1 to 1.65 μm in which the optical telecommunication window 1.3 to 1.6 μm falls. Active devices like quantum well lasers and multiple quantum well devices require controlled, precise (thickness within 1 μm) deposition of different material layers to form the core and also buffer layers, when required in multiple quantum well cases. In some devices, known as strained quantum wells, slight material strain is introduced in a controlled manner in the form of thin layers intentionally within the core layer in order to improve device performance without creating defects [26]. These concepts are exploited in making waveguides through quantum dots and quantum wire materials [28,29]. In a multilayer semiconductor planar waveguide, due to the band gap difference between the layers and the refractive index difference, electrons and holes get confined in the central core layer, where they recombine to generate light in the form of guided wave and effectively function as a waveguide light source [26].

5.8   POLYMER WAVEGUIDES

The huge growth of optical communication industries in recent years has led to the development of newer technologies especially from the point of view of easy fabrication and economics with potentials for mass-scale production. Polymer integrated optics is one such platform, which has shown great promise in this direction [30–35]. The polymer waveguide technology platform is most suitable for high-component-count photonic circuits in planar waveguide geometry. Broadly the classes of polymers that were used have been acrylates, polyimides, and olefins. These materials, which could be thermoplastics, thermosets, or photopolymers, can readily yield planar single-mode, multimode, and micro-optical waveguide components of various dimensions. Typically, starting materials are either polymers or oligomers in solution or liquid monomers. Through halogenation, losses at the wavelengths of interest for optical communication in these materials could be drastically reduced. For example, in contrast to unfluorinated polymethyl methacrylate (PMMA), for which absorption loss is typically 0.5 dB/cm at 1550 nm wavelength, in 80% fluorinated PMMA it could be as low as 0.07 dB/cm [26]. A number of polymers have been shown to exhibit excellent environmental stability and have demonstrated capability in a variety of demanding applications [31]. Several passive and active components to realize a variety of functions like coupling, routing, filtering, and switching have been reported based on polymer-integrated optics. Research and development (R&D) on various adhesion schemes has enabled use of a variety of polymers on a variety of substrates. Through incorporation of chromophores [36] in polymers and subsequent poling for aligning their dipole moments, active devices like modulators have been realized.

5.9   OPTICAL WAVEGUIDE FABRICATION TECHNIQUES

The starting point in planar waveguide processing is a substrate, correctly oriented, polished flat and appropriately cleaned. Due to the requirement of very small dimensions, the possibility of contamination from dusts and so forth must be minimized, which may otherwise introduce scattering centers. Thus all processing is carried out in clean-room environments, in which constant temperature and humidity are maintained and filtered air is fed during the entire process of deposition/fabrication of the waveguide film in order to ensure process repeatability. Other services to the room like water, supply of process gases, and so forth, are also highly purified. The materials and substrates could be electro-optic organic polymers, semiconducting like GaAs or InP insulators like glass or silica. Modification of physical features like refractive index of the substrate near its surface can be achieved through ion exchange, ion implantation, or thermal diffusion. A variety of techniques are available for fabrication of optical waveguides—choice of the best technique for a particular purpose is dictated by a combination of the following considerations:

- Small attenuation in dB/cm
- Required core width d, core-cladding index difference Δn, and r.i. distribution $n(x, y)$
- In case of crystalline materials as substrates/hosts, setting of the optical axes
- Adhesiveness of the waveguiding material to the substrate
- Stability and reproducibility

The thin film deposition techniques are extensively exploited to realize optical waveguides. Examples for realizing amorphous material systems are spin or dip coating, vacuum evaporation, RF and DC sputtering, and chemical vapor deposition (CVD). On the other hand, for forming waveguides from crystalline materials on a crystalline substrate, either of the following strain-free epitaxial growth techniques could be followed: liquid-phase epitaxy (LPE) or vapor-phase epitaxy (VPE). Some of these techniques, mentioned above, are briefly described below.

5.9.1   SPIN-COATING TECHNIQUE

The waveguide materials are first dissolved in a solvent and a drop is placed on the planar face of the substrate, which is rapidly rotated (≥ 1000 rpm) in a spin-coating setup based on the centrifuge principle. Due to the centrifugal action, a thin film of this drop of the solution gets collected on the substrate to yield a step index waveguide. This is often used to create step-index polymer waveguides.

5.9.2   DIP-COATING TECHNIQUE

The substrate (typically a microscope glass slide) is dipped into the solution containing appropriate concentration of the solute and slowly lifted at a uniform rate to let a thin film of the solution adhere to it. The *rate* of lifting and *concentration* of the solution together decide waveguide parameters like thickness and refractive index. After deposition of the film, it is thermally cured or dried at an elevated temperature to increase film adhesiveness. Like spin coating, the dip-coating technique also yields step-index optical waveguides.

5.9.3 Vacuum Evaporation Technique

In this technique, both the substrate and the coating material, which would form the waveguide, are kept in a vacuum chamber at some distance apart. The substrates are typically preheated to a temperature of about $\sim200°C$ to enhance film adhesiveness to the substrate. The coating material is heated and the evaporated guiding film gets collected/deposited by vacuum evaporation at a low pressure of $<10^{-5}$ Torr on the substrate to yield a step-index optical waveguide.

5.9.4 Sputtering Technique

This technique is applicable to materials that cannot be evaporated by thermal heating due to their having relatively higher melting temperatures. Gases like Ar, Ne, or Kr (at a pressure of $\sim10^{-3}$ to 10^{-2} Torr) are introduced into a chamber containing an anode (substrate) and a cathode (target) to generate a high a.c. electric field of rf ~13.56 MHz in order to realize gaseous plasma in the space between the two electrodes. The Ar^{+} ions from the Argon plasma after collision with the cathode eject atoms randomly from the target. By the process known as *sputtering*, in which the target is bombarded with ions, having kinetic energies in the range ~10 eV $-$ 2 keV, several ejected atoms get collected on the nearby substrate placed on the anode, thereby depositing a thin film layer, which is slowly built up with time. The process usually produces a pure, uniform, durable, and optically low-loss amorphous or polycrystalline film.

5.9.5 Chemical Vapor Deposition (CVD) or Epitaxy Technique

This technique is somewhat similar to the telecommunication-grade optical fiber fabrication process and is widely employed in the semiconductor industry. A mixture of reactants is made to react in the gas phase at the hot substrate (placed in a static furnace), and gradually a film of suitable composition (to achieve the desired r.i.) is deposited on the substrate surface. Unreacted gases exit through the downstream. A number of variants of the standard CVD process also exist—for example, low-pressure CVD (LPCVD), plasma-enhanced CVD (PECVD), and metal organic CVD (MOCVD), in which metal alkalis form the mixing compounds for realizing a desired refractive index. Liquid-phase epitaxy (LPE), vapor-phase epitaxy (VPE), and the molecular beam epitaxy (MBE) techniques are also conceptually similar to CVD in which layer-by-layer deposition in a controlled manner is used to grow a specific waveguiding structure. The MBE and the MOCVD techniques, in particular, can yield extremely thin $\sim$ nm or sub-nm layers. In such small dimensions, the electrons and holes behave as if being confined within a quantum well. Multiple core layers separated in between by transparent barrier layers, all ~10 nm thicknesses, are called multiple quantum well devices [26,29]. Another variety is known as strained quantum well, in which an intentional strain is introduced within the core layers through introduction of an active layer with a slight mismatch in lattice constants without creating any defect center [26].

These precision techniques are applied when the waveguide film, whose crystalline structure (i.e., lattice constant) is similar in magnitude to that of the substrate crystal, is to be formed. The starting waveguide material could be either in a liquid form known as LPE, in a gaseous form known as VPE or CVD, or in molecular beam form called MBE. In order to form an active device like a laser diode, one forms one *p*- and one *n*-type cladding layer. When forward biased, due to reduction in the in-built junction electric field, the electrons and holes diffuse across the core-cladding interfaces to accumulate within the core region, where they recombine and generate light. Because the generated light is formed within a waveguide, it is guided as TE or TM mode within the device before exiting through the end-face [26].

5.10   POLYMER WAVEGUIDES

Polymer solutions like a mixture of PMMA and styrene acrylonitrile (SAN) copolymer or liquids like photoresist, polyurethane can be used and deposited on a planar substrate like a microscope glass slide by dip coating. Subsequently the deposited films are normally baked by heating at $60 \sim 100°C$ for times varying from 5 minutes to a few hours depending on the type of film. In order to form a 2D light confining waveguide [26,30], the process starts with the deposition of an ultraviolet (UV)- curable resin layer through spin coating on a copper-sputtered silicon substrate. After its cure through exposure to UV light, the cladding layer of deuterated and fluorinated PMMA is deposited followed by deposition of deuterated PMMA to form the core, which is finally coated with a photoresist. This photoresist is used to transfer the 2D waveguide pattern onto the core by reactive ion etching through UV light exposure of a suitable mask placed on the top, which contains the pattern of the waveguide, and fabricated separately through an electron beam writing machine. This layer is then overclad through spin coating, and a UV-cured photoresist is placed at the top as a protective layer. The sequence of these steps is shown in Figure 5.11.

In the plasma polymerization technique, an electric discharge is first created in *monomer* vapors, which are low-weight organic molecules. The electric discharge induces fragmentation of the monomers and subsequently forms *polymers* through cross-linking, which get deposited on the substrate. Due to the inherent nature of having many chemical bonds, mild acids, bases, or organic solvents do not affect these polymers, which exhibit good temperature stability. In case of certain materials, polymerization could also be induced through exposure to UV radiation or heat.

5.10.1   Ion-Exchange Technique for Graded Index Waveguides

In this technique, the substrate is placed in a suitable melt at an elevated temperature, which leads to an exchange of certain ions between the melt and the substrate leading to a high refractive index layer buried under the surface. For example, exchange of Na^+ from a glass substrate, placed in melts containing $Ag^+/K^+/Ti^+$ or Li^+ from $LiNbO_3$ and H^+ from benzoic acid yields a graded index waveguide. The r.i. profile of such a waveguide could be expressed through the equation

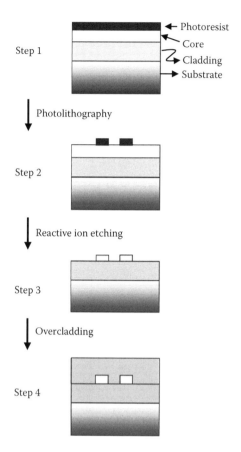

FIGURE 5.11   Fabrication steps for polymer waveguides. (After [35] R. Yoshimura, M. Hikita, S. Tomaru, and S. Imamura, *IEEE J. Lightwave Technol.* Vol. 16 (1998), 1030. ©1998 IEEE.)

$$n(x) = n_s + \Delta n \exp\left(-\frac{x^2}{d^2}\right)$$

with   (5.55)

$$d = 2\sqrt{Dt}$$

where D is the diffusion coefficient and t is time of diffusion.

In the proton exchange technique [37], Li^+ ions are exchanged with H^+ from an acid. This leads to a nearly step-index waveguide in the surface region of $LiNbO_3$ and $LiTaO_3$ substrates. Either of these substrates is placed in a solution of benzoic acid, which is heated to ~200°C. In equilibrium, the following reaction takes place:

$$C_6H_5COOH + Heat \rightarrow C_6H_5COO^- + H^+$$ (5.56)

The efficiency of this reaction depends on temperature. Protons (H^+) from the acid exchange with Li^+ ions under the following equilibrium reactions:

$$LiNbO_3 + xH^+ \rightarrow Li_{1-x}H_xNbO_3 + xLi^+ \tag{5.57}$$

and

$$LiTaO_3 + xH^+ \rightarrow Li_{1-x}H_xTaO_3 + xLi^+ \tag{5.58}$$

Due to the crystalline nature of these substrates, the optical properties are significantly influenced through the r.i. seen by waves having polarization components along the crystallographic Z-axis, for which Δn increase is ~0.12 at the He-Ne laser wavelength. A single-mode waveguide is formed within 5 minutes at a temperature $T \sim 249°C$. Loss (α) in these waveguides is ~0.5 dB/cm. The process may involve replacements of almost 70% Li^+ ions in the crystal surface; such high H^+ concentration may lead to damage to the crystal surface. Adding lithium salts like $LiNO_3$, Li_2O_3, can reduce this. The r.i. profile changes from a step type to a graded profile if the waveguide is annealed at ~400°C for 2 hours, which tends to a Gaussian profile after annealing for a few hours. The waveguides yielded by the proton exchange technique are inherently TM guiding. It may be worth mentioning that though it is a relatively easy fabrication technique, the e.o. and acousto-optic (a.o.) coefficients of $LiNbO_3$ get drastically reduced due to proton exchange in contrast to the other popular technique for $LiNbO_3$ waveguide fabrication by thermal in-diffusion technique. Therefore for modulators, metal-diffused $LiNbO_3$ waveguides are preferred for which e.o. and a.o. properties remain unaffected.

5.10.2 THERMAL DIFFUSION TECHNIQUE

This is often used for fabricating $LiNbO_3$ waveguides. By heating $LiNbO_3$ in a vacuum at ~1000°C, Li_2O is out-diffused from the crystal surface, providing a higher index layer near the surface. However, this process only changes the n_e index by an amount $\Delta n \sim 10^{-3}$.

On the other hand, in case of "in-diffusion" a metal film is deposited on $LiNbO_3$ followed by heating of the crystal in a flowing inert gas, such as Ar and N_2 or O_2 at ~1000°C. This leads to a metal (e.g., Ti) in-diffused layer with a higher r.i. near the surface. Both n_0 and n_e can be increased by almost the same amount—the actual index change can be controlled by metal-film thickness.

5.10.3 ION IMPLANTATION TECHNIQUE

In this technique, ions subjected to high ~20 to 300 kV potentials are accelerated before being directed to impinge on a substrate to induce lattice disorder leading to a high refractive index layer (e.g., implantation of B^+ into fused silica substrate results in the formation of an optical waveguide).

5.11   SILICON WAVEGUIDE TECHNOLOGY

5.11.1   SILICA ON SILICON WAVEGUIDES

In view of the high maturity of silicon-based electronic integrated circuit technology, silica on silicon offers an attractive technology for mass production of passive waveguide components [4 and references therein]. Though various process technologies have been proposed for this platform, one based on flame hydrolysis followed by reactive ion etching (RIE) is a popular process [33]. The flame hydrolysis process is also used in the fabrication of low-loss optical fibers by the outside vapor deposition technique. In its implementation for fabrication of silica on silicon waveguides, two porous layers of specific thickness of silica (SiO_2) and germania (GeO_2) doped silica are derived and deposited in sequence on a polished substrate of silicon wafer. The first layer effectively forms the lower cladding while germania-doped layers being of a higher refractive index form the core. These layers are porous in structure, which are converted to transparent continuous homogenous layers through a consolidation step, which involves the silicon wafer with the deposited layers having been subjected to heating on a stationary resistance heated furnace. Thereafter, a second coat of silica layers is again deposited through flame hydrolysis and consolidation as before to form the overcladding in order to realize a planar waveguide. In case of the ridge-type waveguides, the corresponding structural geometry is first formed through photolithography and RIE of the deposited core layer before depositing the overclad layers of silica. Due to the availability of good-quality silicon wafers of large size, one could make several components on one wafer and integrate these through waveguides as intermediate guided path joints between them so as to form integrated optical planar lightwave circuits, which is sometimes also referred to as silicon optical bench (SiOB). Through control of germania content in the core having silica as its claddings, one can achieve either low or high relative core cladding index difference $\Delta = \dfrac{n_1^2 - n_2^2}{n_1^2}$ in these waveguides. Low losses ~0.017 dB/cm have been demonstrated in waveguides having Δ of 0.45%, while in waveguides having higher Δ of 2%, the loss achieved was ~0.1 dB/cm.

5.11.2   SILICON OXYNITRIDE (SiON) WAVEGUIDES

SiON is a silicon alloy formed due to combination of silica and silicon nitride (Si_3N_4), the latter having a refractive index of 2.01. The advantage is that various core refractive indices between values greater than 1.45 and 2 are achievable in such waveguides without incurring additional loss, unlike the case of a germania-doped silica core [26]. Layer of SiON in the form of thin films can be deposited by PECVD to realize a refractive index up to 1.7, while for realizing a refractive index larger than 1.7, LPCVD is deployed. Starting materials in the two processes are different—in the case of PECVD, these are SiH_4, N_2O, and NH_3, while for LPCVD, SiH_2Cl_2, O_2, and NH_3 are required. Losses are typically less than 0.2 dB/cm, though it is higher in the 1.5 μm low-loss wavelength window of interest to the optical communication community, which however could be reduced through high-temperature annealing at about 1150°C postfabrication.

5.11.3 Silicon-on-Insulator (SOI) Waveguides

In this technology [38–40] a layer of insulating silica is buried inside a silicon wafer either through implantation of oxygen followed by high-temperature annealing or by oxidizing the top layer of the silicon wafer, which is bonded to a second silicon wafer. In both cases, the central portion of the upper silicon is thicker in order to form a rib waveguide [26]. The buried insulating layer could be formed by a two-step process in which oxygen is first implanted into the substrate, and then at the next step it is annealed at high temperature. The fabricated waveguide is typically multimoded; however, due to high differential loss of the higher-order modes relative to the fundamental mode, the waveguide essentially functions as a single-mode waveguide. In another technique [41], the waveguide is formed by in-diffusion of Ge (r.i. $\cong$ 4.3) into an epitaxial layer of doped silicon (lightly) on a heavily doped silicon substrate to form a Ge_xSi_{1-x} alloy. Alternatively, thin alternating layers of Ge and Si could be deposited. These approaches typically yield low losses (~4 dB/cm) in silicon waveguides.

5.11.4 Laser Written Waveguides

Fabrication of waveguides through inscription of femtosecond laser light in glass or polymers is a relatively recent development grown with the subject of micromachining for realizing micro- and nano-photonic devices [42–44]. A schematic of the laser-writing or what is sometimes referred to as waveguide formation through laser inscription is shown in Figure 5.12. Due to extremely high intensity and short duration of such pulses, which are usually very tightly focused, sub-micron-scale structural material changes are triggered through multiphoton absorption [42]. As a result, the refractive index in the focused region gets changed, which could be as large as ~10^{-2} [45]. In transparent insulating materials like fused silica, multiphoton absorption is the key absorption mechanism for which the band gap is large. Laser has also been used to write waveguides by applying ideas similar to those used in forming fiber Bragg gratings in which photosensitivity of germania-doped silica is exploited by using UV-emitting lasers (e.g., 157 nm F_2-laser) [46,47]. Photosensitivity implies permanent change in the refractive index of a material due to exposure to radiation at certain UV wavelengths. In fact, formation of fiber Bragg grating is based on exploitation of this very phenomenon. For optical waveguides, at first a planar waveguide is formed by the CVD process in which a GeO_2-doped core layer is sandwiched between two SiO_2 cladding layers. The waveguide is written by focusing and scanning with a CW UV laser beam to a spot size of about 1 µm. For larger refractive index contrast between the core and the cladding, often the waveguide sample is soaked in H_2 or D_2 gas before exposing to the writing laser beam. Exposure to H_2 or D_2 enhances the photosensitivity of the GeO_2-doped silica core.

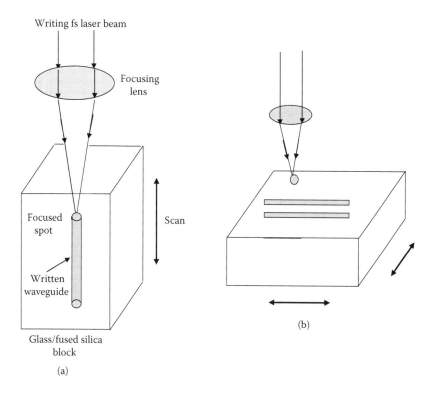

FIGURE 5.12   Waveguide writing through the laser inscription technique in (a) longitudinal and (b) transversal geometries. ((a) After [43] K. Miura, J. Qiu, H. Inouye, and T. Mitsuyu, *App. Phys. Lett.* Vol. 71 (1997), 3329; (b) After [44] C. Florea and K.A. Winick, *J. Lightwave Tech.* Vol. 21 (2003), 246; ©2003 IEEE.)

5.12   CONCLUSION

In this chapter we have attempted a condensed description of the material and design issues connected with fabrication of optical waveguides. After a brief introduction, a tutorial summary of the principles behind wave guidance in an optical waveguide, which is useful from a design point of view, is described. The subsequent sections contain discussions on material issues and fabrication of optical waveguides by various techniques.

ACKNOWLEDGMENT

I would first like to thank Wakaki for inviting me to write this chapter, whom I also thank for his infinite patience and repeated reminders, without which I would not have been able to complete the chapter. Written intermittently over several periods due to my excessive administrative responsibilities at my institute, I am glad that I could finally complete it when I was on leave of absence as a foreign guest scientist of CNRS at Laboratoire de Physique de la Matière Condensée (LPMC), Universite de Nice Sophia Antipolis, CNRS UMR 6622, 06108 Nice Cedex 2, France, during summer 2010. I would like to thank my hosts Gerard Monnom, former director of this laboratory, and current Director Fabrice Mortessagne for inviting and providing me an excellent academic environment and motivation. I would also like to thank Bernard Dussardier, colleague and current group leader of the Fiber Optics group, for the stimulating technical discussions that we often have and for taking good care of me. My graduate student Somnath Ghosh is thanked for plotting Figure 5.2 for me on short notice. This work was partially supported by the UK-India Education and Research Initiative (UKIERI) scheme. Finally I would like to thank my wife Subrata and daughter Parama for their help in editing the manuscript and for pushing me hard to complete the same.

REFERENCES

1. B. P. Pal, Optical Transmission in the book S. S. Jha (Ed.), *Perspective in Optoelectronics*, World Scientific, Singapore (1995).
2. B. P. Pal, Optical Fibers for Lightwave Communication: Evolutionary Trends in Fiber Designs. In B. P. Pal (Ed.), *Guided Wave Optical Components and Devices: Basics, Technology and Applications*, Academic Press/Elsevier, Burlington (2006).
3. R. G. Hunsperger, *Integrated Optics*, 6th edition, Springer, Berlin (2009).
4. B. P. Pal, Silica on Silicon Optical Waveguides: Physics, Technology, and Applications. In E. Wolf (Ed.), *Progress in Optics*, Vol. XXXII, Elsevier, Amsterdam (1993).
5. B. P. Pal and R. R. A. Syms, Guest Editors, Silica on Silicon Optical Waveguides, Proceedings of IEEE Part J: *Optoelectronics*, Vol. 143, issue no. 5 (1996).
6. A. Ghatak and K. Thyagarajan, *Introduction to Fiber Optics*, Cambridge University Press, Cambridge (1998).
7. B. P. Pal, Chapter 1, Optical Waveguides. In M. Butusov (Ed.), *Fiber Optics and Instrumentation* (in Russian), Mashinostroenie, Leningrad (1986).
8. Ajoy Ghatak and K. Thyagarajan, Chapter 3, Integrated Optical Waveguides. In B. P. Pal (Ed.), *Fundamentals of Fiber Optics in Telecommunication and Sensor Systems*, John Wiley/New Age, New York/New Delhi (1992); reprint (2008).
9. K. Thyagarajan and A. Ghatak, Chapter 21, Integrated Optical Waveguide Devices. In B. P. Pal (Ed.), *Fundamentals of Fiber Optics in Telecommunication and Sensor Systems*, John Wiley/New Age, New York/New Delhi (1992), reprint (2008).
10. E. A. J. Marcatili, *Bell Syst. Tech. J.* Vol. 48 (1969), 2071.
11. J. E. Goell, *Bell Syst. Tech. J.* Vol. 48 (1969), 2133.
12. G. B. Hocker and W.K. Burns, *Appl. Opt.* Vol. 16 (1977), 113.
13. A. Kumar, K. Thyagarajan, and A. K. Ghatak, *Opt. Letts.* Vol. 8 (1983), 63.
14. K. Okamoto, *Fundamentals of Optical Waveguides*, Elsevier/Academic Press, Amsterdam/Boston (2005).
15. A. Kumar, A. N. Kaul, and A. K. Ghatak, *Opt. Lett.* Vol. 10 (1985), 86.
16. R. K. Varshney and A. Kumar, *Opt. Lett.* Vol. 11 (1986), 45.
17. M. A. Duguay, Y. Kokubun, T. L. Kock, and L. Pfeiffer, *App. Phys. Lett.* Vol. 49 (1986), 13.
18. Y. Kokubun, T. Baba, T. Sasaki, and K. Iga, *Electron. Letts.* Vol. 22 (1986), 892.
19. R. Tewari, H. Singh, and B. P. Pal, *Microw. Opt. Tech. Letts.* Vol. 3 (1990), 305.
20. A. Ghatak, K. Thyagarajan, and M. R. Shenoy, *J. Lightwave Tech.* LT-5 (1987), 660.
21. K. Thyagarajan, S. Diggavi, and A. Ghatak, *Opt. Quant. Electron.* Vol. 19 (1987), 131.
22. T. Baba and Y. Kokubun, *Photon. Techn. Lett.* Vol. 1 (1989), 232.
23. R. W. Boyd, *Nonlinear Optics*, Academic Press, San Diego (2003).

24. E. L. Wotten, K. M. Kissa, A. Yi-Yan, E. J. Murphy, D. A. Lafaw, P. F. Hallemeier, D. Maack, D. V. Attanasio, D. J. Fritz, G. J. Mcbrien, and D. E. Bossi, *IEEE J. Sel. Topics Quant. Electron.* Vol. 6 (2000), 69.
25. R. V. Schmidt and I. P. Kaminow, *App. Phys.* Vol. 25 (1974), 458.
26. G. P. Agrawal, Chapter 4, *Lightwave Technology: Components and Devices*, Wiley, Hoboken, NJ (2004).
27. G. P. Agrawal and N. K. Dutta, *Semiconductor Lasers*, Van Nostrand Reinhold, New York (1993).
28. S. L. Chuang, *Physics of Optoelectronic Devices*, Wiley, New York (1995).
29. E. Kapon (Ed.), Parts I and II, *Semiconductor Lasers*, Academic Press, San Diego (1999).
30. B. L. Booth, *J. Lightwave Technol.* Vol. 7 (1989), 1445.
31. L. Eldada and L. W. Shacklette, *IEEE J. Sel. Topics Quant. Electron.* Vol. 6 (2000), 54.
32. K. D. Singer, T. C. Kowalczyk, H. D. Nguyen, A. J. Beuhler, and D. A. Wargowski, *Proc. SPIE*, Vol. CR68 (1998), 399.
33. C. R. Kane and R.R. Krchnavek, *IEEE Photon. Technol. Lett.* Vol. 7 (1995), 535.
34. G. Fischbeck, R. Moosburger, C. Kostrzewa, A. Achen, and K. Petermann, *Electron. Lett.* Vol. 33 (1997), 518.
35. R. Yoshimura, M. Hikita, S. Tomaru, and S. Imamura, *IEEE J. Lightwave Technol.* Vol. 16 (1998), 1030.
36. M. C. Oh, H. Zhang, C. Zhang, H. Erlig, Y. Chang, B. Tsap, D. Chang, A. Szep, W. H. Steier, H. R. Fetterman, and L. R. Dalton, *IEEE J. Sel. Topics Quant. Electron.* Vol. 7 (2001), 826.
37. J. L. Jackel, C. E. Rice, and J. Veselka, *Appl. Phys. Lett.* Vol. 41 (1982), 607.
38. A. Himemo, K. Kato, and. T. Miya, *IEEE J. Sel. Topics Quant. Electron.* Vol. 4 (1998), 913.
39. R. M. de Ridler, K. Wörhoff, A. Dreissen, P. V. Lambeck, and H. Alberts, *IEEE J. Sel. Topics Quant. Electron.* Vol. 4 (1998), 930.
40. B. Jalali, S. Yegnanarayanan, T. Yon, T. Yoshimoto, I. Rendina, and F. Coppinger, *IEEE J. Sel. Topics Quant. Electron.* Vol. 4 (1998), 938.
41. B. Schüper, J. Schmidtchen, and K. Petermann, *Electron. Lett.* Vol. 25 (1989), 1500.
42. G. N. Smith, K. Kalli, and K. Sugden, Chapter 15, Advances in Femtosecond Micromachining and Inscription of Micro and Nanophotonics Devices. In B. P. Pal (Ed.), *Frontiers in Guided Wave Optics and Optoelectronics*, In-Tech, Vienna (2010).
43. K. Miura, J. Qiu, H. Inouye, and T. Mitsuyu, *App. Phys. Lett.* Vol. 71 (1997), 3329.
44. C. Florea and K.A. Winick, *J. Lightwave Tech.* Vol. 21 (2003), 246.
45. K. Minoshima, A. M. Kowalevicz, I. Hartle, E. P. Ippen, and J. G. Fujimoto, *Opt. Letts.* Vol. 26 (2001), 1516.
46. G. D. Maxwell and B. D. Ainslie, *Electron Lett.* Vol. 31 (1995), 95.
47. K. P. Chen, P. R. Herman, and R. Taylor, *J. Lightwave Tech.* Vol. 21, 140 (2003).

6 Materials for Optical Thin Films

Takehisa Shibuya

CONTENTS

6.1 INTRODUCTION

The functions of optical thin films are based on the optical phenomena. R. Boyle (1627~1691) and R. Hooke (1635~1703) discovered the optical interference phenomenon independently. This phenomenon was carefully studied by S. I. Newton (1643~1727) [1]. He tried to explain the interference fringes well known as Newton ring by using "fit theory" from the point of view of the particle theory of light. However, the validity of the particle theory was not proved at that time. Newton ring was explained well by the idea of the interference of light waves proposed by T. Young (1773~1829). Taking advantage of his explanation, A. J. Fresnel (1788~1827) and D. F. J. Arago (1786~1853) started to construct the optical wave theory, and this theory became the basic theory for the optical thin film. S. D. Poisson (1781~1840) noticed that the optical interference of the film is caused not only by the interference between reflected lights from front and back surfaces of the film, but also by the repetition reflections of the light on both sides of the film. G. B. Airy (1801~1892) derived a generalized formula that took account of the multireflections on both front and back surfaces of the thin film [2]. L. W. Rayleigh (1842~1919) found out that the reflectance of the surface for a tarnished crown glass plate was lower than a fresh glass plate. It was yielded by the difference of the refractive indices between the surface and inner layers of a bulk glass [3]. This suggested the control of the reflectance on a material surface. This is the start of the study for useful optical applications of thin films.

The optical applications of thin films have come into widespread industrial use. In 1912, mirrors were first made by the evaporation of metal materials [4]. From then, thin films of absorbing and highly reflecting materials have been used widely for the optical applications, typically for instance, as astronomical reflectors. Aluminum, rhodium, and sometimes silver are used for the reflection coatings of mirrors. Rhodium shows 80% reflection in the visible region and has large mechanical strength and chemical resistance. Aluminum possesses the still larger reflectance of 90%, but it is not stable chemically. Silver is modified chemically mainly in the presence of sulfur compounds. Silver shows high reflectance of 97% at the wavelength 550 nm, but very low reflectance of 8% at 320 nm. Platinum, rhodium, and chromium are utilized for semitransparent reflectors in addition to their use for mirrors. The reflectance of these materials remains constant throughout the whole visible region even at various film thicknesses. They may be utilized for the attenuation of the visible light without change in spectral compositions (i.e., for the formation of neutral density filters).

Optical thin films using nonabsorbing materials have been widely applied by utilizing the interference phenomena. Antireflection coatings can be produced on glass surfaces by combining films with suitable thicknesses and refractive indices. The transmittance of the optical system with many interfaces between a glass and air can be controlled by these films.

It is possible to produce various types of interference filters in a similar way. These include transmission-type filters that transmit only some region of the spectrum—narrow-band filters, broad-band filters, and so forth [5–8]. Several examples of transmission and reflection-type filters are illustrated in Figures 6.1 to 6.3. Conventional materials and the film thickness of quarter wavelength were used for the high-pass filters and the narrow band pass filters.

Broadband filters are used for the selection of the colors, which are called dichromatic mirrors or dichromatic filters. These mirrors reflect some components of the spectrum. Reflectance spectra of two such filters are shown in Figure 6.4. One is reflecting practically all blue color and another all red color. By using these mirrors, white light may be decomposed into three colors. This decomposition of color is used in the formation of colored images in films or liquid crystal displays.

The group of materials utilized in thin film optics is formed by weakly absorbing substances. The transmittance in these materials usually increases with the wavelength λ. These materials exhibit a colored hue in transmission, mostly yellow or brown. They are used for the reduction of the light intensity. Most of these are inorganic materials (e.g., SiO, MgF_2, Na_3Al_6, etc.) and have no selective absorption. On the other hand, most organic materials have selective absorption and are used as color filters. Weakly absorbing materials usually have a high refractive index and a high reflectance. The reflection loss effect induced by the property may be overcome

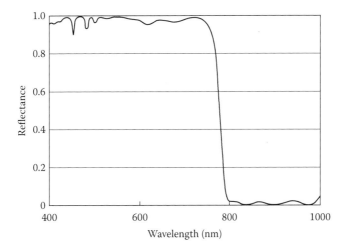

FIGURE 6.1 Spectral characteristics of high-pass filter in visible region composed of 26 layers of SiO_2 and TiO_2.

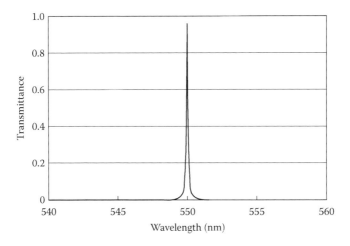

FIGURE 6.2 Spectral characteristics of the band pass filter in visible region composed of 27 layers of $(HL)^6$ H 4L H $(LH)^6$. H and L layers materials are TiO_2 and MgF_2, respectively.

by combining the materials with a layer of nonabsorbing substance having a suitably low index.

A combination of a metal and a nonabsorbing film is also utilized in the protective coatings, for example, of an aluminum mirror coated by a SiO layer. Such a coating introduces, however, a certain degree of spectral selectivity into the reflectance spectra which may be ascribed to interference phenomena. Combinations of nonabsorbing materials with nano-metals are used for optical filters (e.g., a SiO_2 layer dispersed with nano-particles of gold, a ZrO_2 layer with nano-particles of silver, etc.). By combining metals with nonabsorbing films,

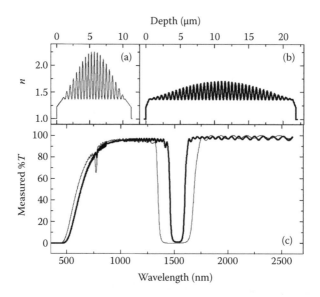

FIGURE 6.3 Comparison of rugate filters made with different index contrasts: (a) nominal profile of a 20-period sample with 1.36 to 2.25 contrast, (b) 40-period sample with 1.36 to 1.7 contrast, (c) transmission spectra of the high-contrast sample (thinner curve) and the low-contrast sample (thicker curve). (From E. Lorenzo, C. J. Oton, N. E. Capuj, M. Ghulinyan, D. N. Urrios, Z. Gaburro, and L. Pavesi: Porous silicon-based rugate filters, *Appl. Optics* 44 (2005) 5415–5421. With permission.)

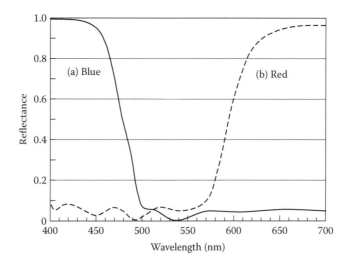

FIGURE 6.4 Reflectance spectra of color filter for: (a) blue; 0.5H (LH)^6 L 0.5H, 16 layers and (b) red; 0.5M (HL)^5 H 0.5L, 13 layers. L, H, and M layer materials are SIO_2, TIO_2, and CeF_3, respectively.

ultra-narrow-band filters can be produced based on the Fabry-Perot interferometer principle, with the difference that a dielectric film is inserted between the semi-transparent mirrors instead of an air gap. Films with the surface on which multi-needles are constructed are expected for anti-reflection filters using some kind of graded index effect.

Evolution of the optical design is also realized for the new film having an arbitrary index range from low index to high index based on the conventional film design.

6.2 BRIEF THEORY OF OPTICAL THIN FILMS [9]

Reflectance from a smooth surface of a material is understood to be caused by the difference of refractive indices between a surrounding medium and the material. Reflectance is described by Fresnel coefficients derived from the refractive index of the material. Reflectance of the material is different in the cases with or without a surface layer on the material. Reflectance also changes by the roughness of the surface. Moreover, the refractive indices of the films are not the same as those of the bulk materials. These values change from starting materials depending on the production processes of thin films. Clear distinction exists between absorbing and nonabsorbing films, and between homogeneous and inhomogeneous films due to optical properties of the materials and the structures of the films.

The refractive index of a homogenous film is the same throughout the film. On the other hand, the optical constants of an inhomogeneous film vary within the film. Some of the incident light is reflected, absorbed, and diffused, and finally the remaining part transmits through single-layer films or multilayer films. The reflectance (R), transmittance (T), and absorptance (A) are related as

$$T + R + A = 1 \qquad (6.1)$$

It is generally approved to set $A = 0$, because the absorption of the optical films is usually very small. Equation (6.1) becomes more simply

$$T + R = 1 \qquad (6.2)$$

There are different methods to evaluate a reflected light from a material composed with films. One example is the solution to the conventional boundary value problem which is the analogy to the electrical theory of transmission lines, and the Fresnel coefficients are used at each interface between optical media. The method utilizing Fresnel coefficients is the most clear physically, and it requires the least demands on mathematical capability.

6.2.1 Fresnel Coefficients for the Reflection and the Transmission

The amplitude, phase, and polarization of the reflected wave in nonabsorbing media can be derived using the optical constants and the incidence angle. The electric vector E_i for the incident light is decomposed into the perpendicular component E_{is} and the parallel component E_{ip} to the plane of incidence as shown in Figure 6.5.

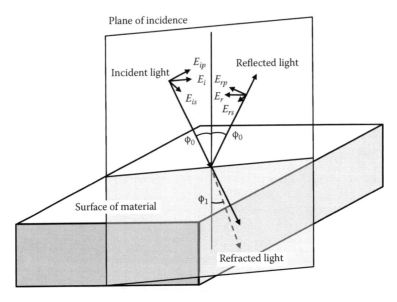

FIGURE 6.5 Decomposition of the incident light E_i into E_{is} and E_{ip} components and of reflected light E_r into E_{rs} and E_{rp} components.

$$E_i = E_{is} + E_{ip} \tag{6.3}$$

The intensity of the incident light is calculated as

$$I_i = E_i^2 = \left| E_{is} + E_{ip} \right|^2 = E_{is}^2 + E_{ip}^2 \tag{6.4}$$

The electric vectors for the reflected light are also defined as E_{rs} and E_{rp} for the perpendicular and parallel components, respectively. The amplitude and phase of the reflected light are derived by Fresnel formulas. The assumption that the amplitudes of E_{is} and E_{ip} are positive at the boundary leads to the Fresnel formulas being written as

$$r_s = \frac{E_{rs}}{E_{is}} = \frac{n_0 \cos\varphi_0 - n_1 \cos\varphi_1}{n_0 \cos\varphi_0 + n_1 \cos\varphi_1} = -\frac{\sin(\varphi_0 - \varphi_1)}{\sin(\varphi_0 + \varphi_1)} \tag{6.5}$$

$$r_p = \frac{E_{rp}}{E_{ip}} = \frac{n_0 \cos\varphi_1 - n_1 \cos\varphi_0}{n_0 \cos\varphi_1 + n_1 \cos\varphi_0} = \frac{(\cos\varphi_0)/n_0 - (\cos\varphi_1)/n_1}{(\cos\varphi_0)/n_0 + (\cos\varphi_1)/n_1} = -\frac{\tan(\varphi_0 - \varphi_1)}{\tan(\varphi_0 + \varphi_1)} \tag{6.6}$$

where n_0 and n_1 are refractive indices of the ambient material and the film, respectively. The sign of these equations relates to the direction of each vector. Generalized refractive indices $(n_0 \cos\varphi_0,\ n_1 \cos\varphi_1)$ and $(\cos\varphi_0/n_0,\ \cos\varphi_1/n_1)$ are replaced by $(N_0,\ N_1)$ for the perpendicular and parallel components, respectively. Taking in the generalized refractive indices, Equations (6.5) and (6.6) are combined to give Equation (6.7).

$$r \cdot \exp(i\delta) = \frac{N_0 - N_1}{N_0 + N_1} \cdot \left(\because \delta = \delta_p - \delta_s \right)$$ (6.7)

The amplitude r_p vanishes in the oblique angle of incidence where $N_0 - N_1$ is equal to zero in Equation (6.7) (i.e., $n_0 \cos\varphi_0$ is equal to $n_1 \cos\varphi_1$. This angle is called the Brewster's angle, and this famous phenomenon is called Brewster's law. It may be noted that the angle between reflected and transmitted lights is perpendicular. When $\varphi_0 + \varphi_1 = \pi/2$, a denominator of Equation (6.7) also diverges to infinity; therefore, r_p becomes zero. Brewster's angle φ_B, or the polarization angle, is calculated from Equation (6.8), and the calculated results are shown in Figure 6.6.

$$\varphi_B = \tan^{-1}\left(\frac{N_1}{N_0}\right)$$ (6.8)

The refracted light beam propagates at an angle φ_1 in the boundary for the light incident from an optically denser medium ($N_0 > N_1$) at φ_0. When φ_0 is increased beyond a certain value, the refracted light emerges parallel to the interface between the media of N_0 and N_1, and φ_1 becomes 90° and $\sin\varphi_1 = 1$. In this case, the total reflection is observed for all incidence angles larger than the critical angle. This angle of the total reflection is given as follows:

$$\varphi_C = \sin^{-1}\frac{N_1}{N_0}$$ (6.9)

The phase of the reflected wave at the total reflection also can be calculated by using the Fresnel formulas. When φ_0 is greater than φ_C, the value of $\sin\varphi_1$ exceeds unity,

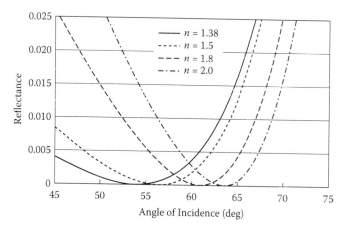

FIGURE 6.6 Angular dependences of reflectance of parallel component for oblique incidence on the medium with different refractive indices.

and φ_1 becomes imaginary. In this case, the equation is transformed slightly to the following equation:

$$\cos\varphi_1 = \sqrt{1-\sin^2\varphi_1} = i\sqrt{\sin^2\varphi_1 - 1} = i\sqrt{\left(\frac{n_0}{n_1}\sin\varphi_0\right)^2 - 1} \quad (\text{law of refraction})$$

The perpendicular component of Fresnel coefficients is given as

$$r_s\exp(i\delta_s) = \frac{N_0 - N_1}{N_0 + N_1} = \frac{n_0\cos\varphi_0 - n_1\cos\varphi_1}{n_0\cos\varphi_0 + n_1\cos\varphi_1}$$

$$= \frac{n_0\cos\varphi_0 - in_1\sqrt{\left(\frac{n_0}{n_1}\sin\varphi_0\right)^2 - 1}}{n_0\cos\varphi_0 + in_1\sqrt{\left(\frac{n_0}{n_1}\sin\varphi_0\right)^2 - 1}} \qquad (6.10)$$

$$= \frac{n_0\cos\varphi_0 - i\sqrt{n_0^2\sin^2\varphi_0 - n_1^2}}{n_0\cos\varphi_0 + i\sqrt{n_0^2\sin^2\varphi_0 - n_1^2}}$$

The numerator and the denominator of Equation (6.10) are the complexes conjugate to each other. The next relation is derived by replacing the denominator of Equation (6.10) to the complex Z:

$$r_s\exp(i\delta_s) = \frac{Z^*}{Z} = (Z^*)Z^{-1}$$

In this case, the total energy of the incident light is reflected at the boundary region. The phase changes continuously at the total reflection. In Equation (6.10), $r_s = (Z^*)Z^{-1} = \exp(i2\alpha) = \exp(i\delta_s)$, because $-2\alpha = \delta_s$:

$$\tan\frac{\delta_s}{2} = \tan\alpha = \frac{\text{Im}(Z)}{\text{Re}(Z)} = \frac{\sqrt{n_0^2\sin^2\varphi_1 - n_1^2}}{n_0\cos\varphi_1} = \frac{\sqrt{\sin^2\varphi_1 - (n_1/n_0)^2}}{\cos\varphi_1} \qquad (6.11)$$

where δ_s is the phase change for the perpendicular component. The phase change for the parallel component is given by a similar calculation:

$$\tan\frac{\delta_p}{2} = \frac{n_0\sqrt{\left(\frac{n_0}{n_1}\sin\varphi_0\right)^2 - 1}}{n_1\cos\varphi_1} = \frac{\sqrt{\sin^2\varphi_1 - (n_1/n_0)^2}}{(n_1/n_0)^2\cos\varphi_1} \qquad (6.12)$$

Thus, the phase difference $\delta = \delta_p - \delta_s$ between the parallel and perpendicular components is derived as follows:

$$\tan\frac{\delta}{2} = \tan\left(\frac{\delta_p - \delta_s}{2}\right) = \frac{\tan\dfrac{\delta_p}{2} - \tan\dfrac{\delta_s}{2}}{1 + \tan\dfrac{\delta_p}{2}\tan\dfrac{\delta_s}{2}} = \frac{\cos\varphi_1\sqrt{\sin^2\varphi_1 - (n_1/n_0)^2}}{\sin^2\varphi_1} \qquad (6.13)$$

These formulas give the amplitude and phase of the wave reflected on the boundary at various angles of incidence if the refractive index of nonabsorbing media is known. The results of such calculations for glasses are shown in Figures 6.7 through 6.10.

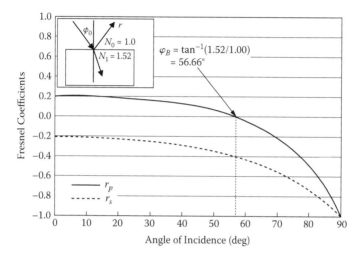

FIGURE 6.7 Angular dependence of Fresnel coefficients of the reflection at the boundary for the light incident from the optically thin medium to dense medium.

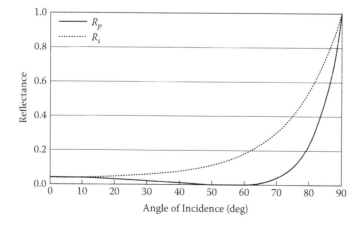

FIGURE 6.8 Angular dependences of reflectance for the light incident from the optically thin medium to dense medium.

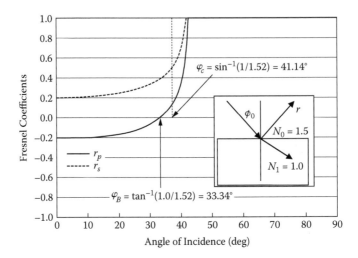

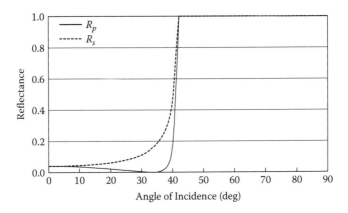

FIGURE 6.9 Angular dependences of Fresnel coefficients of the reflection at the boundary for the light incident from the optically dense medium to thin medium.

FIGURE 6.10 Angular dependence of reflectance for the light incident from the optically dense medium to optically thin medium.

6.2.2   Reflection of Light from a Single Interface

The reflection is calculated using the Fresnel formulas at the boundary region between nonabsorption and an absorbing medium. When the medium is opaque, the refractive index is replaced by the complex refractive index, so that $\widetilde{N}_1 = n_1 - ik_1$. The amplitude and phase of the reflected wave are calculated from Equation (6.7) by replacing N_1 to $\widetilde{N}_1$:

$$r \cdot \exp(i\delta) = \frac{N_0 - N_1}{N_0 + N_1} = \frac{N_0 - \widetilde{N}_1}{N_0 + \widetilde{N}_1} = \frac{N_0 - (n_1 - ik_1)}{N_0 + (n_1 - ik_1)} \qquad (6.14)$$

The reflectance is given by multiplication of the complex conjugate.

$$R = |r \cdot \exp(i\delta)|^2 = \frac{(N_0 - n_1)^2 + k_1^2}{(N_0 + n_1)^2 + k_1^2} \tag{6.15}$$

The moduli of the amplitude of the reflected wave are given by the square root of R, and the phase angle is calculated as

$$\delta = \tan^{-1}\frac{\sin\delta}{\cos\delta} = \tan^{-1}\frac{\mathrm{Im}\{\exp(-i\delta)\}}{\mathrm{Re}\{\exp(-i\delta)\}} = \tan^{-1}\frac{2N_0 k_1}{N_0^2 - n_1^2 - k_1^2} \tag{6.16}$$

The amplitude and phase of opaque metal mirrors are calculated from Equations (6.15) and (6.16) for normal incidence. The reflectance and phase calculated for the absorbing materials are shown in Table 6.1. The reflectance characteristics of a metal surface are almost the same to a nonabsorbing medium with an extremely high refractive index. For instance, the reflectance obtained with aluminum corresponds to that given by a nonabsorbing material with refractive index $n_1 = 40$. For a front surface mirror of aluminum film in the air, typical values are $N_0 = 1$, $n_1 = 0.7$, and $k_1 = 5$, which leads to $R = 90\%$. As a result, the front surface mirrors suggest higher reflectance than back surface mirrors.

An arrangement of three optical media with refractive indices of N_0, N_1, and N_2 have two interfaces where the incident light is partially reflected and transmitted as shown in Figure 6.11. The separation between these interfaces is the order of magnitude of a wavelength of the incident light. It is assumed that a film is plane parallel, homogeneous, and nonabsorbing. Also, the neighboring medium outside of the thin film is assumed to be transparent and extended infinitely. The problem is completely specified by giving the generalized refractive indices N_0, N_1, N_2; the film thickness d_1; the angle of incidence φ_0; and the angle of refraction φ_1. A plane wave is supposedly incident on the surface of medium with N_1. The amplitude reflected at the first boundary between N_0 and N_1 is $r_{01} = (N_0 - N_1)/(N_0 + N_1)$. The wave transmitted into the film has the amplitude of $\sqrt{1 - r_{01}^2}$. At the second boundary between N_1 and N_2, the ratio

---

TABLE 6.1
Reflectance of the Absorbing Materials in the Air and on the Glass

Absorbing Material	Medium	Refractive Index of Metal		Reflectance	tan(δ)	Phase	
	n_0	n_1	k_1			δ	δ'
Al	1.0	0.7	5.0	0.8996	−0.4083	−22.2	157.8
Ag	1.0	0.05	2.87	0.9786	−0.7929	−38.4	141.6
Cr	1.0	2.38	2.97	0.5298	−0.4405	−23.8	156.2
Assumed material	1.0	40	0.0	0.9048	0.0000	0.0	180.0
Al/BK7	1.52	0.7	5.0	0.8578	−0.6557	−33.3	146.7

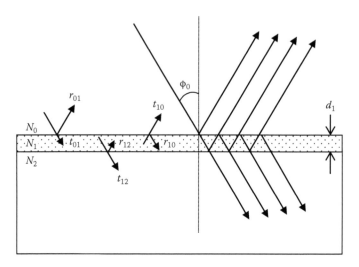

FIGURE 6.11   Fresnel coefficients for the single layer on the glass.

of the reflected light intensity to the incident intensity is $r_{12} = (N_1 - N_2)/(N_1 + N_2)$. A wave with the amplitude of $r_{12}\sqrt{1 - r_{01}^2}$ is reflected back to the first boundary. The wave is reflected at the second boundary with the amplitude $-r_{01}r_{12}\sqrt{1 - r_{01}^2}\exp(i\delta_1)$, and the transmitted light is reduced by the factor of $\sqrt{1 - r_{01}^2}$. The amplitude of the wave becomes $r_{12}(1 - r_{01}^2)$ after the thin film. Compared with the first reflection, this emerging wave has propagated through the film twice. This path difference of $2N_1d_1$ gives the difference of a phase δ_1 relative to the first reflection r_{01}.

$$\delta_1 = \frac{2\pi}{\lambda}2N_1d_1$$

$$r_{01} = \frac{r_{12}\left(1 - r_{01}^2\right)\cdot\exp\left(-i\delta_1\right)}{1 + r_{01}r_{12}\cdot\exp\left(-i\delta_1\right)} \tag{6.17}$$

Reflectance coefficient of the film over a glass is given as follows after calculating r_{01} and writing the sum with a common denominator:

$$r\cdot\exp\left(i\delta_1\right) = r_{01} + \frac{r_{12}\left(1 - r_1^2\right)\cdot\exp\left(-i\delta_1\right)}{1 + r_{01}r_{12}\cdot\exp\left(-i\delta_1\right)} = \frac{r_{01} + r_{12}\exp\left(-i\delta_1\right)}{1 + r_{01}r_{12}\cdot\exp\left(-i\delta_1\right)}$$
$$= \frac{r_{01} + r_{12}\cdot\exp\left(-i\delta_1\right)}{1 + r_{01}r_{12}\cdot\exp\left(-i\delta_1\right)} \tag{6.18}$$

Equation (6.18) is multiplied by the complex conjugate to obtain the measurable quantity R.

$$\left| r_{01} + r_{12} \cdot \exp(i\delta_1) \right|^2 = r_{01}^2 + r_{12}^2 + 2r_{01}r_{12}\cos\delta_1 \tag{6.19a}$$

$$\left| 1 + r_{01}r_{12} \cdot \exp(i\delta_1) \right|^2 = 1 + r_{01}^2 r_{12}^2 + 2r_{01}r_{12}\cos\delta_1 \tag{6.19b}$$

$$R = \frac{r_{01}^2 + r_{12}^2 + 2r_{01}r_{12}\cos\delta_1}{1 + r_{01}^2 r_{12}^2 + 2r_{01}r_{12}\cos\delta_1} \tag{6.20}$$

In Equation (6.18), the numerator is zero or the denominator goes to infinity to derive the expression that $r \cdot \exp(i\delta)$ goes to zero. The latter case is impossible because r_1 and r_2 take values less than or equal to one. As a result, the expression becomes zero in the case $r_{01} + r_{12} \cdot \exp(-i\delta_1) = 0$. This complex equation gives the following two cases: (1) $r_1 + r_2 \cdot \cos\delta_1 = 0$ and (2) $r_2 \cdot \sin\delta_1 = 0$.

In Case (1), $r_1 - r_2 = 0$ for $\delta_1 = \pm\pi, \pm3\pi, \pm5\pi, \ldots$, or the following relation will be required:

$$\frac{N_0 - N_1}{N_0 + N_1} = \frac{N_1 - N_2}{N_1 + N_2}$$

In Case (2), $\sin\delta_1$ is equal to zero for $\delta_1 = 0, \pm\pi, \pm2\pi, \pm3\pi, \cdots$ and the following important conditions are obtained immediately:

$$N_1 = \sqrt{N_0 \times N_2} \tag{6.21}$$

For $\delta_1 = \pi, 3\pi, 5\pi$,

$$N_1 d_1 = \frac{1}{4}\lambda, \frac{3}{4}\lambda, \frac{5}{4}\lambda, \cdots \tag{6.22}$$

The reflectance of a coated surface on a glass becomes zero if the refractive index N_1 of the film on the glass is equal, in the geometric meaning, to the refractive indices of the neighboring media at the target wavelength. Both Equations (6.21) and (6.22) give the amplitude and the phase conditions for the anti-reflection, respectively. For $\pm2\pi, \pm4\pi$, the value is also vanished. A condition that corresponds to $N_1 d_1$ is equal to $\lambda/2, \lambda, 3\lambda/2, \cdots$.

In Case (1), these values give $r_{01} + r_{12} = 0$—that is,

$$\frac{N_0 - N_1}{N_0 + N_1} = -\frac{N_1 - N_2}{N_1 + N_2} \tag{6.23}$$

After a calculation, $N_0 = N_2$ is obtained about the refractive index of the arbitrary film. The refractive index difference between two media becomes zero at the optical thickness of the film $\lambda/2, \lambda, 3\lambda/2, \cdots$, and no reflection occurs at these wavelengths.

In this condition, it is interesting to use Equation (6.18) to calculate the reflection for $\delta_1 = 2\pi$, 4π, $\cdots$, which corresponds to $N_1 d_1 = \lambda/2$, $3\lambda/2$, $\cdots$, assuming arbitrary refractive indices N_0 and N_2. This gives

$$r = \frac{r_1 + r_2}{1 + r_1 r_2} = -\left\{\left(\frac{N_0 - N_1}{N_0 + N_1} + \frac{N_1 - N_2}{N_1 + N_2}\right) \Big/ \left(1 + \frac{N_0 - N_1}{N_0 + N_1}\frac{N_1 - N_2}{N_1 + N_2}\right)\right\} \qquad (6.24)$$

and next relation is obtained:

$$r = \frac{N_0 - N_2}{N_0 + N_2} \qquad (6.25)$$

This formula gives the same amplitude and phase as those without the film. In particular for $N_0 = N_2$, $r = 0$ is obtained as indicated earlier.

No change occurs for the reflectance of a reflecting surface in nonabsorbing films with the optical thicknesses $\lambda/2$, λ, $3\lambda/2$, $\cdots$, for the light at these wavelengths. Now by combining Equations (6.21), (6,22), and (6,25), the following conclusions are deduced. By depositing a nonabsorbing film on a glass substrate with a refractive index chosen to satisfy the amplitude condition, Equation (6.21), zero reflectance is obtained with films of optical thickness $n_1 d_1 = \lambda/4$, $3\lambda/4$, $\cdots$, at those wavelengths, while the reflectance is the same as that obtained with an untreated glass surface for the optical thickness $\lambda/2$, λ, $3\lambda/2$, $\cdots$. The films with the conditions between these extremes in particular cases are treated in most films. If the refractive index of the film does not satisfy the amplitude condition $\delta_1 = 2\pi$, 4π, $\cdots$ or $N_1 d_1 = \lambda/2$, λ, $3\lambda/2$, $\cdots$, the following relation is satisfied in general:

$$R = (r)^2 = \left(\frac{N_0 - N_2}{N_0 + N_2}\right)^2 \qquad (6.26)$$

For any other values of δ_1 or λ, the reflectance is calculated from Equation (6.15) or Equation (6.16). The results of these calculations for $N_0 = 1$ and $N_2 = 1.52$ are shown in Figure 6.12 as a function of δ_1 or $N_1 d_1$, for several refractive indices. These figures offer sufficient information for a nonabsorbing, homogeneous single layer. A film with a low refractive index material between two media with higher refractive indices produces enhanced reflectance. Generally speaking, the films with high refractive index give a high reflectance, and those with low refractive index give anti-reflection coatings. The values of amplitude for r_{01} and r_{12} become unity with silver or gold mirrors in the far infrared (IR). For example, $r_{12} = 1 \cdot \exp(i\delta)$ in Equation (6.18) gives

$$r \cdot \exp(i\delta) = \frac{r_{01} + \exp\{-i(\delta_1 - \varepsilon)\}}{1 + r_{01}\exp\{-i(\delta_1 - \varepsilon)\}} \qquad (6.27)$$

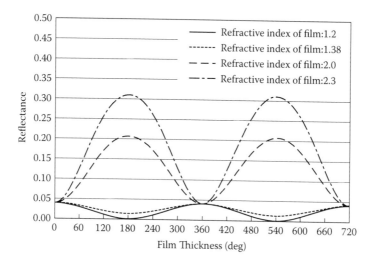

FIGURE 6.12 Reflectance R for homogenous, nonabsorbing single layer on the glass as a function of the phase (film thickness). The refractive index of the glass is 1.5, and the indices of films are changed from 1.2 to 2.3.

where ε is the phase angle corresponding to the value at normal incidence on a gold film. And the reflectance R is written as

$$R = \left| r \cdot \exp(i\delta) \right|^2 = \frac{r_{01} + \exp\{-i(\delta_1 - \varepsilon)\}}{1 + r_{01}\exp\{-i(\delta_1 - \varepsilon)\}} \times \frac{r_{01} + \exp i(\delta_1 - \varepsilon)}{1 + r_{01}\exp i(\delta_1 - \varepsilon)}$$

$$= \frac{r_{01}^2 + \exp\{-i(\delta_1 - \varepsilon)\} \cdot \exp i(\delta_1 - \varepsilon) + r_{01}\exp i(\delta_1 - \varepsilon) + r_{01}\exp\{-i(\delta_1 - \varepsilon)\}}{1 + r_{01}\exp\{-i(\delta_1 - \varepsilon)\} \cdot r_{01}\exp i(\delta_1 - \varepsilon) + r_{01}\exp i(\delta_1 - \varepsilon) + r_{01}\exp\{-i(\delta_1 - \varepsilon)\}} \quad (6.28)$$

$$= 1$$

This relationship is independent of the film thickness or wavelength. If one of the two reflection amplitudes is the unity, there is no interference. No interference is observed with this arrangement, and no wavelength measurement is possible in this way. This is a well-known result to workers in the far IR

6.2.3 REFLECTION OF LIGHT FROM DOUBLE LAYERS OR MULTILAYERS [6–9]

The reflection formula for double layers is derived to add the reflected waves at each interface in amplitude and phase. Fresnel coefficients of multilayers on the glass are defined in Figure 6.13. By using Equation (6.28) for layers of the refractive index of N_2 and N_3, the reflectance r' with phase δ' is written as

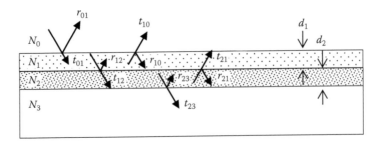

FIGURE 6.13   Fresnel coefficients for multilayers on the glass.

$$r' \cdot \exp(i\delta') = \frac{r_{12} + r_{23}\exp(-i\delta_2)}{1 + r_{12}r_{23}\exp(-i\delta_2)} \qquad (6.29)$$

where $r_{12} = (N_1 - N_2)/(N_1 + N_2)$, $r_{23} = (N_2 - N_3)/(N_2 + N_3)$, and $\delta_2 = \dfrac{4\pi}{\lambda}N_2 d_2$. δ' is the phase change for the single layer at the boundary N_1 and N_2.

For double layers, r and δ are expressed as

$$r \cdot \exp(i\delta) = \frac{r_{01} + r' \cdot \exp(i\delta') \cdot \exp(-i\delta_1)}{1 + r_{01}r' \cdot \exp(i\delta') \cdot \exp(-i\delta_1)} \qquad (6.30a)$$

$$= \frac{r_{01} + \dfrac{r_{12} + r_{23}\exp(-i\delta_2)}{1 + r_{12}r_{23}\exp(-i\delta_2)} \cdot \exp(-i\delta_1)}{1 + r_{01}\dfrac{r_{12} + r_{23}\exp(-i\delta_2)}{1 + r_{12}r_{23}\exp(-i\delta_2)} \cdot \exp(-i\delta_1)}$$

$$r \cdot \exp(i\delta) = \frac{r_{01} + r_{12} \cdot \exp(-i\delta_1) + r_{23}\exp\{-i(\delta_1 + \delta_2)\} + r_{01}r_{12}r_{23}\exp(-i\delta_2)}{1 + r_{01}r_{12} \cdot \exp(-i\delta_1) + r_{01}r_{23}\exp\{-i(\delta_1 + \delta_2)\} + r_{12}r_{23}\exp(-i\delta_2)} \qquad (6.30b)$$

In Equation (6.30b), the first two terms in both numerator and denominator characterize the effect of the single layer. The third term corresponds to the reflection at the $N_0 - N_1$ boundary with the phase of $\delta_1 + \delta_2$ given by the optical path difference through the two films. The fourth term also reflects the single layer with the phase δ_2.

6.3   DESIGN OF THE OPTICAL FILTER [10–13]

The fundamental concept to design the optical thin films was established in the mid of 20th. This theory requires knowledge of the refractive index of the substrate and an optical thickness of the thin films (i.e., the refractive index and a physical thickness of the film). Optical properties for any complex structure with multilayers are obtained by calculating with this method if the structures of the films including the refractive indices of the base material and films are given. The theory of equivalent-layer and the optimization of the film design are explained. Many materials with different values of the refractive index are required for the optical design. However, the index of a thin film is slightly different depending on the deposition processes or conditions, and a desirable magnitude of refractive index cannot be obtained easily.

6.3.1 FUNDAMENTAL METHOD

A conventional optical filter consists of a stack of homogeneous layers of thickness d_j, and refractive indices n_j [14]. The matrix for a single layer is given by using the 2×2 matrix applicable to a homogeneous layer approximation for each layer, such as Macleod [9]. The jth layer is represented by

$$M_j = \begin{pmatrix} \cos g_j & iN_j^{-1}\sin g_j \\ iN_j \sin g_j & \cos g_j \end{pmatrix} = \begin{pmatrix} m_{11} & im_{12} \\ im_{21} & m_{22} \end{pmatrix} \tag{6.31}$$

where $g_j = (n_j d_j \cos\varphi_j)2\pi\lambda^{-1}$, and φ_j is the angle of propagation in the film. $N_j = n_j \cos\varphi_j$ (for s-polarized) and $N_j = n_j \cos^{-1}\varphi_j$ (p-polarized), as the same treatment in Equation (6.31). The matrix for multilayers is described as the products of the matrix for each single layer.

$$M_m = \prod_{j=1}^{k} M_j \tag{6.32}$$

where k is the number of layers. Fresnel coefficients of reflectance and transmittance for multilayers are derived using the elements of the matrix and the indices of an incident medium and a substrate.

$$r = \frac{(m_{11} + im_{12}N_s)N_0 - (im_{21} + m_{22}N_s)}{(m_{11} + im_{12}N_s)N_0 + (im_{21} + m_{22}N_s)} \tag{6.33a}$$

$$t = \frac{2N_0}{(m_{11} + im_{12}N_s)N_0 + (im_{21} + m_{22}N_s)} \tag{6.33b}$$

Reflectance and transmittance for multilayers are also defined as

$$R = rr^* = |r|^2 \tag{6.34a}$$

$$T = \frac{ReN_s}{ReN_0}tt^* = tt^* \frac{N_s \cos\varphi_s}{N_0 \cos\varphi_0} \tag{6.34b}$$

where

$$N_0 = \begin{cases} n_0 \cos\varphi_0 \, (for \ s-polarization) \\ n_0/\cos\varphi_0 \, (for \ p-polarization) \end{cases}, \quad N_s = \begin{cases} n_s \cos\varphi_s \, (for \ s-polarization) \\ n_0/\cos\varphi_0 \, (for \ p-polarization) \end{cases}$$

6.3.2 Optimization of Multilayer Filters [15–18]

Optimization of an optical filter is not easy, or impossible. It is difficult to design an optical filter manually taking consideration of all the specifications for desired applications. The optical filter designers must arrange for targets describing the specifications. The targets can be calculated from the parameters of the filter. It is possible to optimize an optical filter using the information of the $\partial R/\partial t$ computed by the refining program [18]. The $\partial R/\partial t$ indicates how much the transmission or the reflection at various λ depends on the thickness of each layer.

6.3.3 Needle Method [19–21]

The needle method was first described by A. V. Tikhonravov in 1982. However, it began to be widely used only in the middle of the 1990s [19,20]. It consists of adding thin films at the optimal position in the optical filter and adjusting their thickness by refinement [21]. Needles are added until a satisfactory solution is found or the additions of more needles do not improve the filter. The merit function (MF) is defined as

$$MF = \left\{ \frac{1}{N} \sum_{j=1}^{N} \left(\frac{f(\lambda_j) - \hat{f}_j}{\Delta f_j} \right)^2 \right\}^{1/2} \tag{6.35}$$

where f is the actual spectral characteristics (transmittance or reflectance), λ_j is a wavelength point from a given wavelength grid with the total number of N points, $\hat{f}_j$ is the target value, and Δf_j is the specified tolerance at these points. The optimal position to add needles is determined by calculating the derivative of the MF with regard to the thickness of an infinitesimally thin film as the function of the position where it is added. It is favorable to add a needle when the derivative is negative. Usually, one needle at a time is added at the position where the derivative of the MF is the most negative.

6.3.4 Step Method [14]

Materials with a continuous range of the refractive index are available in this method. P. G. Verly proposed to determine both the optimal position and the refractive index of the added film [21]. The step method consists of adding infinitesimal steps in the index profile and refining the refractive index and the thickness of the separated films. Similar to the needle method, the optimal position to add steps is determined by calculating the derivative of the *MF* with regard to the addition of a step as a function of the position where the step is added.

$$MF = \chi^2 = \sum_{i=1}^{m} \left(\frac{B_i - \overline{B}_i}{\Delta B_i} \right)^2 \qquad (6.36)$$

where B_i and $\overline{B}_i$ are properties of interest and the target value for that property. ΔB_i is also the tolerance on that property, and *m* is the number of the target. Refinement of optical filters has a limitation. When a local minimum is found, the filter cannot be further improved. If the optimized filter does not reflect the specifications, refinement must be reinitialized with a better starting design. The step method is a new synthesis method devised specifically for filters made of materials offering a continuous range of refractive indices. As for the needle method, it is based on addition of new parameters to the optimization by the addition of more layers. Instead of adding infinitesimally thin layers, existing layers are separated in many layers by adding infinitesimally small steps in the index profile.

6.3.5 FOURIER TRANSFORM METHOD [22–28]

For the design of the gradient index filter, the Fourier transform method was published by E. Delano in 1967 [22], and J. A. Dobrowolski and D. Lowe also reported later [23,24]. It relates the desired reflection or transmission spectra with the index profile of the filter through the use of amplitude and phase functions, Q and ψ, respectively, called Q function. The index depth profile $n(x)$ is obtained by using the inverse Fourier transform, where *x* is the double-centered optical thickness, $k = 2\pi/\lambda$ is the wave number, and λ is the wavelength.

$$\int_{-\infty}^{\infty} \frac{dn(x)}{dx} \frac{1}{2n} \exp(ikx)dx = Q(k)\exp\{i\varphi(k)\} = f(x) \qquad (6.37)$$

$Q(k)$ is defined as

$$\left(\frac{R}{T} \right)^{1/2} \quad \text{or} \quad Q(k) = \left\{ \frac{1}{2}\left[\frac{1}{T(k)} - T(k) \right] \right\}^{1/2} \qquad (6.38)$$

The rugate filter was reported by P. Baumeister in 1986 [29]. Later, B. G. Bovard [30] and C. S. Bartholomew et al. [31] introduced a Fourier transform–based design technique [32–37]. It can be understood that the Fourier transform of a Q function defined for a single wavelength gives a sinusoidal index profile. A multiband rugate filter can easily be designed by multiplying multiple sinusoidal profiles:

$$\ln \frac{n(x)}{n_m} = \prod_i \overline{Q}_i \sin\left(2\pi x / \lambda_i + \varphi_i\right) w(x) \tag{6.39}$$

$$\overline{Q}_i = \frac{1}{2} \ln \frac{n_m + \Delta n_i / 2}{n_m - \Delta n_i / 2} \tag{6.40}$$

The Kaiser window is defined by

$$w(x) = \frac{I_0\left(\beta\sqrt{1 - 4x^2}\right)}{I_0(\beta)} \Pi(x) \tag{6.41}$$

where λ_i is the wavelength of the band, φ_i is the phase shift that allows one to center the index profile, I_0 is the modified Bessel function of the first kind and zero order, $\Pi(x)$ is a rectangular function, and β is a parameter that controls the amount of apodization.

6.4 MATERIALS FOR OPTICAL THIN FILMS

The optical thin film with an arbitrary refractive index is useful for the design of an optical filter. It is possible to produce such a film by mixing material A and material B which have different refractive indices. Refractive indices of these films are estimated using effective medium theories [38–40].

6.4.1 EFFECTIVE MEDIUM THEORIES

The Maxwell–Garnet theory was developed in 1904 to explain the colors in metallic films [41]. The Bruggeman theory is also used for the film made by the mixed materials [42]. The schematic concepts of this theory are illustrated in Figure 6.14 using effective refractive index n_m. In the Maxwell-Garnet theory, the packing density q of the film material, and dielectric constants ε_m and ε_a of the film and the host (substrate) materials are defined. ε_{eff} is the dielectric constant of the film for the effective medium. The relationship between the effective layer and the islands of film material is defined as

$$\frac{\varepsilon_{eff} - \varepsilon_a}{\varepsilon_{eff} + 2\varepsilon_a} = q \frac{\varepsilon_m - \varepsilon_a}{\varepsilon_m + 2\varepsilon_a} \tag{6.42}$$

Equation (6.42) is described as follows by replacing the dielectric constants by the refractive indices:

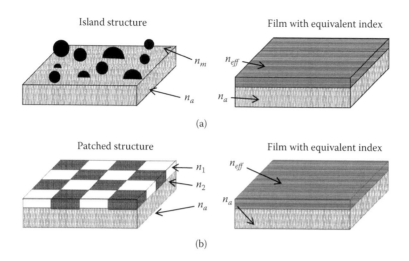

FIGURE 6.14 Effective medium theories: (a) Maxwell-Garnet theory, and (b) Bruggeman theory.

$$\frac{n_{eff}^2 - n_a^2}{n_{eff}^2 + 2n_a^2} = q\,\frac{n_m^2 - n_a^2}{n_m^2 + 2n_a^2} \tag{6.43}$$

The Bruggeman theory for the mixed layer on the substrate is defined as

$$\frac{\varepsilon_{eff} - \varepsilon_a}{\varepsilon_{eff} + 2\varepsilon_a} = (1-q)\frac{\varepsilon_1 - \varepsilon_a}{\varepsilon_1 + 2\varepsilon_a} + q\,\frac{\varepsilon_2 - \varepsilon_a}{\varepsilon_2 + 2\varepsilon_a} \tag{6.44}$$

where ε_1, ε_2, and ε_a are the dielectric constants for materials A, B, and host (substrate), respectively. In this case, ε_a is equal to ε_{eff}, which is the dielectric constant of the mixed layer, and Equation (6.44) is simplified as

$$(1-q)\frac{\varepsilon_1 - \varepsilon_{eff}}{\varepsilon_1 + 2\varepsilon_{eff}} = q\,\frac{\varepsilon_{eff} - \varepsilon_2}{2\varepsilon_{eff} + \varepsilon_2} \tag{6.45}$$

6.4.2 Optical Filters with an Inhomogeneous Structure [43–47]

The problem of designing optical filters using inhomogeneous thin film has been approached by many researchers. The reflectance of an inhomogeneous layer is determined from the boundary conditions of electromagnetic vectors at the two surfaces of the film. An inhomogeneous film with suitable index profile has the intensity reflectance with a low rippled curve at the side robe. The filter with gradient refractive index shown in Figure 6.15 works as a useful antireflection film of a single layer [46,47].

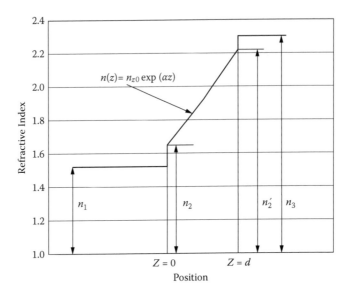

FIGURE 6.15 Dielectric thin film with inhomogeneous structure. The index in the inhomogeneous medium changes exponentially. (From S. F. Monaco: Reflectance of an inhomogeneous thin film, *J. Opt. Soc. Am.* 51 (1961) 280. With permission.)

6.4.3 OPTICAL FILTER USING INTERNAL OR EXTERNAL STRUCTURE OF THE FILM

An antireflective (AR) film on BK7 glass with a refractive index of 1.516 must have the index of ~1.231 from Equation (6.21). This requirement for low index makes it practically impossible to design a single-layer AR film for a glass. The low index materials of MgF_2, NaF, and AlF_3 have values of 1.38, 1.34, and 1.23, respectively, which would not achieve the zero reflection. The refractive index of the material is related to its density. If the density can be lowered by introducing some kinds of porosity, it becomes possible to decrease the refractive index. The effective refractive indices are shown in Figure 6.16 as a function of the packing density for MgF_2, SiO_2, and Al_2O_3.

A packing density of 0.6 is needed to make a low index thin film over the dense material of a glass with the refractive index of 1.52.

Reduction in the reflectivity of freshly polished glass surfaces exposed to sulfuric or nitric acid was first reported by Fraunhofer in 1887. This can be done by leaching a silica-based glass surface to the optical quarter-wave depth [48]. This etched porous silica layer works as an AR coating with its lower refractive index [49].

The reflectance of the surface layer for the high-energy laser was effectively reduced from 8% to 0.5% in the range of 0.35 to 2.5 μm [50].

A gradient index layer for AR coating was also produced by a sol-gel process [51]. An antireflective coating using a porous structured film for the glass plate was carried out by using the sol-gel method [52–55].

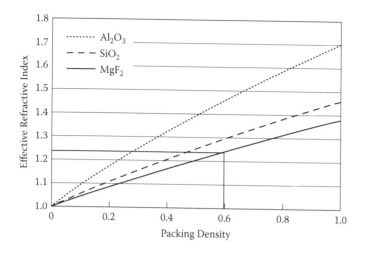

FIGURE 6.16 Effective refractive indices of porous layers for several materials as a function of packing density. The antireflection condition for the glass substrate ($n = 1.516$) is given by the packing density of 0.6 in MgF_2.

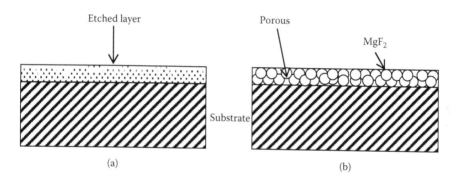

FIGURE 6.17 Low refractive index film produced by etching having a porous structure. (a) Surface-etched film, (b) porous structure.

MgF_2 coatings with an ultra-low refractive index of 1.18 showed high transmittance in the UV region, and the high uniformity of film thickness was achieved over the wide coating area of 300 mm in diameter [56–61].

6.4.4 Characteristics of the Optical Film Materials

Optical characteristics (e.g., the refractive index, reflectance, transmittance, and other properties for conventional optical materials) have been published in other books [62]. Optical properties of the optical materials surveyed in the published scientific journals are summarized in Table 6.2. The materials are classified in alphabetical order [63–87].

TABLE 6.2
Optical Properties of Materials

Materials	Methods	ET: Evaporation Temperature (K) ST: Substrate Temperature (K)	Transmittance Region (μm)	Refractive Index and Extinction Coefficient (μm)	Hardness	Remark
AgS		ST~573K	0.2	2.84, $k = 1.25(0.4)$ 3.04, $k = 0.8(0.52)$ 2.64, $k = 0.27(0.8)$		Ref. 83
AlF$_3$	RE(Mo,W) P-DC-MS R-MS	(F-silica)		$n_1 = 1.22$, $n_2 = 1.38(0.55)$ 1.384(0.193), $k = 0.001(0.193)$, Ref. 67 1.415~1.34, $k = 0.001{\sim}0(0.2{\sim}0.7)$, Ref. 68 1.425~1.375, $k = 0.0015{-}0.0002(0.195{-}0.34)$	S	Ref. 81 Ref. 67, amorphous Ref. 20, 68 Ref. 69
Al$_2$O$_3$	EB, B(W, Ta), ALD	ET2370~3070K ET2323K ST573K(MgF$_2$) ST573K(Si, Soda lime G)	0.2~7	1.54(0.55)ST313, Ref. 81 1.63(0.55)ST573, Ref. 81 $k = 8 * 10^{-6}(1.06)$, $2.3 * 10^{-5}(0.515)$, Ref. 81 1.85(0.12~0.23), $k = 10^{-1}(0.12{-}0.175)$, Ref. 64 1.665~1.63(0.4~1.8), Ref. 70	H	Ref. 64 Ref. 70
BaF$_2$	EB RS	ST473K(MgF$_2$)		1.87~2.0(0.125~0.135), $k = 10^{-2}(0.14{-}0.23)$ 1.657(0.157), Ref. 72		Ref. 64 Ref. 72
BiF$_3$	RE, SPT	STET1173K		1.74(1), 1.65(11)		Ref. 84
BN(Boron Nitride)	EB IAD	423K<ST<523K	0.495~0.69 (less than 2%)	2.8(0.62), 2.6~2.8(0.2~1.7)		Ref. 63 MgF$_2$/BN, AR for semiconductor
DLC(diamond-like carbon)	IP, MS RF, IBP, others			1.8~1.85(1~2), Ref. 88 1.7~2.4(0.633), Ref. 89		AR for Ge Refs. 88, 89
CaF$_2$	RE(W, Ta, Mo)	ET1550K ST523K(MgF$_2$)		$1.3(\lambda > 0.18)$, $k = 10^{-2}(\lambda > 0.18)$: 1.559(0.157), Ref. 72	M	Ref. 64 Ref. 72 Ref. 81, hygroscopic

Material	Method	Temperature	λ range (μm)	Optical properties		Ref./Notes
CdS	B, RE(O~)	ET1073K	0.55~7	2.57~2.27(0.6~7), 2.5(0.6), Ref. 81	S	Ref. 81
CdSe	RE(Ta, Mo)	ET970~1070K		2.45(2)	M	Ref. 81
CdTe	RE(Mo)			2.6(2)	H	Ref. 81
CeF$_3$	EB, RE(W, Mo, Ta)	ET1633K	0.3~5	1.69~1.60(0.25~2), Ref. 71; 1.65~1.58(0.4~2); 1.63(0.55), Ref. 81	H	Ref. 71; Ref. 81
CeO$_2$	EB, RE(W), LA	ET1873K	0.4~12	$k = 1.4 * 10^{-5}(0.633)$, Ref. 81'; 2.2(0.55), Ref. 81; $k = 2.3 * 10^{-5}(0.633)$, Ref. 81'; 2.56~1.91(0.25~2), Ref. 71; 2.52~2.2(0.5~2)	M	Ref. 81; Ref. 71
Er$_2$O$_3$	MO-CVD, EB	ST600(G, Si(100), Al$_2$O$_3$(0001)); ST573K(BK7, Ta)		3.3~2.3, $k = 0.05$~0(0.19~1.65)		Ref. 74; Ref. 91
GdH$_3$	RF		0.32~15,	2.1~1.85, $k = 0.16$~0(0.41~0.83)		Ref. 75
Gd$_2$O$_3$	EB, RE, R-EB	ET~2473K; ST523K, 623K (Lanthanide sesquioxide)	0.4~20	1.8(0.55); 1.97~1.82(0.2~2), $k = 0.13$~0.001(0.2~0.5), Ref. 66; Ref. 66	M	Ref. 81; Ref. 66; Ref. 73
Ge	EB(graphite), RE	ET1873K	2~23	4.4~4.0(2~4)	M	Ref. 81
GeAsTeSe	EB			3.1(10)		Poisonous
Ge$_{30}$As$_{17}$Te$_{30}$Se$_{23}$	EB			3.1(10)		Toxic
	B			3.1(10.6), Ref. 81		
HfO$_2$	EB, RE, LA	ST573K(MgF$_2$)		2.15(0.23); 2.3~1.97(0.25~2), $k = 0.016(0.25)$, Ref. 71	M	Ref. 71

(Continued)

TABLE 6.2 (Continued)
Optical Properties of Materials

Materials	Methods	ET: Evaporation Temperature (K) ST: Substrate Temperature (K)	Transmittance Region (μm)	Refractive Index and Extinction Coefficient (μm)	Hardness	Remark
InAs	DB		3.8~7	4.5(1), Ref. 81		Ref. 81 Toxic Poisonous
InSb	DB		7–16	4.3(1), Ref. 81		Ref. 81 Toxic Poisonous
In$_2$O$_3$	E, R(W)			2.0(0.5)	H	
LaF$_3$	EB, B (Ta, W, Mo)	ET1763K ST523K(MgF$_2$) ST723K(z-quartz)	.25~2	1.63(0.3), 1.57(1) $1.85(0.135-0.23)$, $k = 2.2 * 10^{-2}(\lambda > 0.145)$ 1.65(0.248) 1.65–1.59(0.25~2), $k = 0.001(0.25)$, Ref. 71 1.55(.55), Ref. 81	M	ST < 169 stress cracks; ST > 300 improves durability Ref. 65 Ref. 71 Ref. 71
La$_2$O$_3$	EB RB(W) RE LA	ET1773K	0.3~	1.98(0.33) 1.9(0.55)ST573 $2.1-1.82(0.25-2)$, $k = 0.001(0.25)$, Ref. 71 1.63(0.3), 1.57(1)	H	Ref. 71
LiF	B(Ta)	ET1140K		1.36–1.37(0.55) 1.485(0.157), Ref. 72	S	Highly hygroscopic Ref. 72

Material	Method	Temperature	Range (μm)	Refractive index	Durability	Remarks
MgF$_2$	E, B(Mo, Ta, W)	ET1540~1643K, ST523~575K	0.11~4	1.38(0.55), 1.35(2); 1.7(0.1216), $k = 0.01$(0.1216); 1.46~1.37(0.25~2), Ref. 71; 1.32~1.39(0.55), $k = 6 * 10^{-6}$(1.06), $k = 9 * 10^{-5}$(0.515) Ref. 81	H	thk < 4 um; QWO&T are very durable; low STs are not durable; Ref. 71; Ref. 81
MgO	E	ET2370~3073K	0.2~8	1.7(0.55)ST323, Ref. 81; 1.74(0.55)ST573, Ref. 81; $k = 2.8 * 10^{-5}$(0.633), Ref. 81; 1.75~1.72(0.5~1)	H	Ref. 81; Resistant, Hazy, abs. moisture; Ref. 75
Na$_3$AlF$_2$(cryolite)	E, B(Mo)	ET1070~1470K	0.2~14	1.32~1.35(0.55); 1.35~1.39(0.58)	S	Hygroscopic; Ref. 81
NaF	B(Ta)	ET1260K		1.34(0.58)	S	Hygroscopic
Nb$_2$O$_3$	EB, RB	ET2173K	0.4~	1.79(0.55); 2.15(0.55),ST533	H	Ref. 81
NdF$_3$	E, B(Mo, Ta)	ET1470~1870K, ET1683K, ST573K	0.25~	1.68~1.60(0.25~2),Ref. 71; 1.62~1.58(0.5~1); 1.61(0.55), ST573, $k = 1.2 * 10^{-5}$(0.633), Ref. 81	M	Ref. 71; Ref. 81
Nd$_2$O$_3$ PbCl$_2$	E, B(W, Mo), R B(Pt, Mo) Low evaporation temperature	ET1870~2270K		2.1~2.0(0.5~0.8); 2.3~2.0(0.58~10)	H	
PbF$_2$	B(W)	ET1123K, ST433K	0.25~17	1.98~1.68(0.3~0.8); 1.98(0.3), 1, 75(0.55)	S	Not durable; Ref. 81
PbO	B	ET1172K, ST570K	0.53~	2.6(0.55); 2.6(0.58)	S	

(Continued)

TABLE 6.2 (Continued)
Optical Properties of Materials

Materials	Methods	ET: Evaporation Temperature (K) ST: Substrate Temperature (K)	Transmittance Region (μm)	Refractive Index and Extinction Coefficient (μm)	Hardness	Remark
PbTe	B(Ta)	ET1123K	3.5~20	5.6(5)	S	Ref. 81 Toxic Poisonous
Pr_6O_{11}	RB(W)		0.4	1.92~2.05(0.55) 1.91~1.83(0.4~2)	FH	
Sb_2O_3	B(Mo)	370~670K		2.3~2.0(0.4~0.7)	S	
Sb_2S_3	B	ET643K	0.5~10	3.0(0.58) 3.0(0.55), Ref. 81	S	Ref. 81
Sc_2O_3	E, R	ET2673K	0.35~13	1.9(0.3), Ref. 81 1.89(0.55), Ref. 81 (1.86(0.9)ST523~573, Ref. 81 1.90~1.86(0.4~0.8)	M	Ref. 81
Si	EB B	ET1773K	1~9	3.4(3), Ref. 81 3.4(2)	H	Ref. 81
SiO	RB(Mo, Ta, W)	ET~1570K	0.7~9	2.0~1.85(0.8~2) 2.0(0.7), Ref. 81	H	Howitzer Ref. 81
SiO_2	E, R of SiO, IP	ET~1873K ST573K(MgF_2)	0.2~9	>1.8(0.125~0.15), $k = 10^{-1}$(0.12~0.135), Ref. 64 1.6(0.2) 1.46~1.44(0.58~2) 1.45~1.46(.55), Ref. 81	H	Ref. 64 Ref. 81
Si_2O_3	RB(Ta)	ET~1570K	0.4~9	1.55(0.55)ST303K, Ref. 81 1.55~1.52(0.55)	H	Ref. 81

Material	Method	Range	Index / Properties	Hardness	Reference / Notes
SnO_2	LI		2.0, $k = 0.005(0.4)$; 1.0, $k = 0.0875(1.6)$		Ref. 83
SrF_2	RS, ST433K		1.576(0.157), Ref. 72	S	Ref. 72
Ta_2O_5	EB, ET2373K; RE; RB(W); SPT; IP	0.35~10	2.25(0.4), Ref. 81; 2.1(0.55)ST523, Ref. 81; 2.15~2.05(0.4~2), $k = 0.008(0.3)$, Ref. 71; 2.6~2.1(0.4~0.6)	M–H	Ref. 71; Ref. 81
Te	B(Ta)		$n_1 = 4.93(4)$~4.79(14), $n_2 = 6.37(4)$~6.23(14)		Ref. 90
ThF_4	EB, B ET1373K; ST573K	0.2~15	$n_1 = 1.50$, $n_2 = 1.52$; $k = 2 * 10^{-6}(1.06)$, $5 * 10^{-6}(.515)$, Ref. 81		Ref. 81 durable, Radioactive
ThO_2	E, ET3323K	0.3~	1.95(0.3), Ref. 81; 1.86(0.55)TS523, Ref. 81; 1.95~1.75(0.4~2)	H	Ref.81 Radioactive
TiO_2	EB, B(W), R, IP, anodized Ti, LAiD ET1870–2270K	0.4~3	1.9(0.55)ST303, Ref. 81; 2.3(0.55)ST473, Ref. 81; 2.55(0.55)ST533, Ref. 81; $k = 7.7 * 10^{-5}(1.06)$, $5.5*10^{-4}(0.515)$, Ref. 81; 2.35~2.2(0.5~2)	H	Ref. 81
TiO	B, R ET2020K		1.9(0.58)	H	
TlCl	B(Ta)		2.6(10)		Soft, water-soluble
Y_2O_3	EB, RE, LA ET2673K	0.3~12	1.89(0.33); 1.87(0.55)TS523, Ref. 81; 1.83(0.9), Ref. 81; 1.97~1.75(0.25~2), $k = 0.004(0.25)$, Ref. 71; 1.89~1.87(0.4~0.8)	H	Ref. 71; Ref. 81
ZnO	B, ET1373K	0.4~	2.1(0.55)ST303, Ref. 81; 2.1(0.58)	S	Ref. 81

(Continued)

TABLE 6.2 (Continued)
Optical Properties of Materials

Materials	Methods	ET: Evaporation Temperature (K) ST: Substrate Temperature (K)	Transmittance Region (µm)	Refractive Index and Extinction Coefficient (µm)	Hardness	Remark
ZnS	EB, B(Mo, Ta), SPT	ET1373K	0.4~14	2.38~2.2(0.5~2) 2.3(0.55)ST308, Ref. 81 $k = 4 * 10^{-6}(1.06)$, $2.7 * 10^{-4}(.515)$, Ref. 81 >3.5 * $10^{-5}(.633)$, Ref. 81'	S	Medium compressive stress Ref. 81
ZnSe (Irtran 4)	EB, B(Mo, Ta), SPT	ET1223K	0.55~15	2.5~2.57(0.58) 2.57(0.6)	S	Ref. 81
ZnTe	B	ET1273K		2.79(1.5) 2.8(0.55)	S	Ref. 81
ZrO$_2$	EB, RB(W, Mo), LA, IP	ET2973K	0.34~12	2.05~1.92(0.55) 1.97(0.55)ST303, Ref. 81 2.05(0.55)ST473, Ref. 81 $k = 2 * 10^{-5}(1.06)$, $1.6 * 10^{-4}(0.515)$, Ref. 81 2.3~1.96(0.25~2), $k = 0.01(0.25)$, Ref. 71 2.05~2.2(0.58~2)	H	Ref. 81 Ref. 71 EB-inhomogeneous

6.5   CONCLUSIONS

In this chapter, materials and designs for the optical filters are mainly reviewed. Many new theoretical and experimental approaches may be expected. Symmetric and asymmetric multilayer systems have been designed using personal computers. Nanomaterials and nanostructures are also used for optical and other thin films. These are expected to be applied to various optics that require high angular antireflection performance from the UV to IR region.

REFERENCES

1. I. Newton: *Optiks* 4th Ed. London, (1730) 192.
2. G. B. Airy: On the phenomena of Newton's rings when formed between two transparent substances of different refractive powers, *Trans. Cambridge Phil. Soc.* 4 (1832) 409–424: *The London and Edinburgh Philosophical Magazine and Journal of Science*, Vol. II, No. 4, (1833), 20–29.
3. L. Rayleigh: On the intensity of light reflected from certain surfaces at nearly perpendicular incidence, *Proc. Roy. Soc.* A41 (1886) 275.
4. P. Pringsheim: *Verh. Dtsch. Phys. Ges.*, 14 (1912) 546.
5. L. Young: Synthesis of multiple antireflection films over a prescribed frequency band, *J. Opt. Soc. Am.* 51 (1961) 967–974.
6. L. Young: Multilayer interference filters with narrow step bands, *Appl. Optics* 6 (1967) 297–315.
7. A. Thelen: Equivalent layers in multilayer filters, *J. Opt. Soc. Am.* 56 (1966) 1533–1538.
8. E. Lorenzo, C. J. Oton, N. E. Capuj, M. Ghulinyan, D. N. Urrios, Z. Gaburro, and L. Pavesi: Porous silicon-based rugate filters, *Appl. Optics* 44 (2005) 5415–5421.
9. H. A. Macload: *Thin-Film Optical Filters*, Hilger (1986).
10. J. A. Dobrowolski: Subtractive method of optical thin-film interference filter design, *Appl. Optics* 12 (1973) 1885.
11. G. L. Scheidegger: Circle diagrams applied to the design of thin film optical tunnel filters, *Appl. Optics* 28 (1989) 2061.
12. A. Thelen: Design of optical minus filter, *J. Opt. Soc. Am.* 61 (1971) 365–369.
13. R. R. Willey: Using fence post design to speed the atomic layer deposition of optical thin films, *Appl. Optics*, 47 (2008) C9.
14. S. Larouche and L. Martinu: Step method: A new synthesis method for the design of optical filters with intermediate refractive indices, *Appl. Optics* 47 (2008) 4321.
15. L. Ivan Epstein: The design of optical filters, *J. Opt. Soc. Am.* 42 (1952) 806–810.
16. J. A. Dobrowolski and H. C. Piotrowski: Refractive index as a variable in the numerical design of optical thin film systems, *Appl. Optics* 21 (1982) 1502.
17. J. A. Dobrowolski and R. A. Kemp: Refinement of optical multilayer systems with different optimization procedures, *Appl. Optics* 29 (1990) 2876.
18. P. Baumeister: Design of multilayer filters by successive approximations, *J. Opt. Soc. Am.* 48 (1958) 955–958.
19. A. V. Tikhonravov, M. K. Trubetskov, and G. W. DeBell: Application of the needle optimization technique to the design of optical coatings, *Appl. Optics* 35 (1996) 5493–5508.
20. B. T. Sullivan and J. A. Dobrowolski: Implementation of a numerical needle method for thin-film design, *Appl. Optics* 35 (1996) 5484–5492.
21. P. G. Verly: Modified needle method with simultaneous thickness and refractive-index refinement for the synthesis of inhomogeneous and multilayer optical thin film, *Appl. Optics* 40 (2001) 5718.
22. E. Delano: Fourier synthesis of multilayer filters, *J. Opt. Soc. Am.* 57 (1967) 1529–1533.

23. J. A. Dobrowolski and D. Lowe: Optical thin film synthesis program based on the use of Fourier transforms, *Appl. Optics* 17 (1978) 3039–3050.
24. P. G. Verly, J. A. Dobrowolski, W. J. Wild, and R. L. Butron: Synthesis of high rejection filters with the Fourier transform method, *Appl. Optics* 28 (1989) 2864–2875.
25. P. G. Verly: Fourier transform technique with frequency filtering for optical thin-film design, *Appl. Optics* 34 (1995) 688.
26. P. G. Verly and J. A. Dobrowolski: Iterative correction process for optical thin film synthesis with the Fourier transform method, *Appl. Optics* 29 (1990) 3672–3684.
27. P. G. Verly, J. A. Dobrowolski, and R. R. Willey: Fourier-transform method for the design of wideband antireflection coatings, *Appl. Optics*, 31 (1992) 3836–3846.
28. S. Larouche and L. Martinu: Dispersion implementation in optical filter design by the Fourier transform method using correction factors, *Appl. Optics* 46 (2007) 7436–7441.
29. P. Baumeister: Simulation of rugate filter via a stepped-index dielectric multilayer, *Appl. Optics* 25 (1986) 2644–2645.
30. B. G. Bovard: Derivation of a matrix describing a rugate dielectric thin film, *Appl. Optics* 27 (1988) 1998–2005.
31. C. S. Bartholomew, M. D. Morrow, H. T. Betz, J. L. Grieser, R. A. Spence, and N. P. Murarka: Rugate filters by laser flash evaporation of SiO_xN_y on room-temperature polycarbonate, *J. Vac. Sci. Technol.* A6 (1988) 1703–1707.
32. W. H. Southwell: Spectral response calculations of rugate filters using coupled-wave theory, *J. Opt. Soc. Am.* A5 (1988) 1558–1564.
33. W. H. Southwell and R. L. Hall: Rugate filter sidelobe suppression using quantic and rugated quantic matching layers, *Appl. Optics* 28 (1989) 2949–2951.
34. W. H. Southwell: Using apodization functions to reduce sidelobes in rugate filters, *Appl. Optics* 28 (1989) 5091–5094.
35. B. G. Bovard: Rugate filter theory: An overview, *Appl. Optics* 32 (1993) 5427–5442.
36. P. L. Swart, B. M. Lacquet, A. A. Chtcherbakov, and P. V. Bulkin: Automated electron cyclotron resonance plasma enhanced chemical vapor deposition system for the growth of rugate filters, *J. Vac. Sci. Technol.* A8 (2000) 74–78.
37. A. V. Tikhonravov, M. K. Trubetskov, T. V. Amotchkina, M. A. Kokarev, N. Kaiser, O. Stenzel, S. Wibrandt, and D. Gabler: New optimization algorithm for the synthesis of rugate optical coatings, *Appl. Optics* 45 (2006) 1515–1524.
38. R. Landauer: The electrical resistance of binary metallic mixtures, *J. Appl. Phys.* 23 (1952) 779–784.
39. B. Abeles and J. I. Gittleman: Composite material films: Optical properties and applications, *Appl. Optics* 15 (1976) 2328–2332.
40. T. Shibuya, S. Kawabata, H. Yoshizawa, S. Suzuki, N. Amano, and H. Yokota: Observation of the initial stage of ion assisted deposition films using a rotating-analyzer ellipsometer, 35 (1996) 4556–4560.
41. J. C. Maxwell-Garnett: Colours in metal glasses and in metallic films, *Phil. Trans. R. Soc. London Ser.* A203 (1904) 385–420; A205 (1906) 237–288.
42. D. A. G. Bruggeman: *Ann. Phys. (Leipzig)* 416 (1935) 636–664, 665–679; 417 (1936) 645–672; 421 (1937) 160–178.
43. R. Jacobsson and J. O. Martensson: Evaporated inhomogeneous thin films, *Appl. Optics* 5 (1966) 29–34.
44. P. G. Verly, A. V. Tikhonravov, and M. K. Trubetskov: Efficient refinement algorithm for the synthesis of inhomogeneous optical coatings, *Appl. Optics* 36 (1997) 1487–1495.
45. P. G. Verly: Optical coating synthesis by simultaneous refractive-index and thickness refinement of inhomogeneous films, *Appl. Optics* 37 (1998) 7327–7333.
46. S. F. Monaco: Reflectance of an inhomogeneous thin film, *J. Opt. Soc. Am.* 51 (1961) 280.
47. W. H. Lowdermilk and D. Milam: Graded-index antireflection surface for high power laser applications, *Appl. Phys. Lett.* 36 (1980) 891.

48. F. H. Nicoll and F. E. Williams: Properties of low reflection films produced by the action of hydrofluoric acid vapor, *J. Opt. Soc. Am.* 33 (1943) 434–435.

49. L. M. Cook, W. H. Lowdermilk, D. Milam, and J. E. Swain: Antireflective surface for high energy laser optics formed by neutral-solution process, *Appl. Optics* 21 (1982) 1482.

50. M. J. Minot: Single-layer, gradient refractive index antireflection films effective from 0.35 to 2.5 μ, *J. Opt. Soc. Am.* 66 (1976) 515.

51. S. P. Mukherjee and W. H. Lowdermilk: Gradient-index AR film deposited by the sol-gel process, *Appl. Optics* 21 (1982) 293.

52. B. E. Yoldas: Investigations of porous oxides as an antireflective coating for glass surfaces, *Appl. Optics* 19 (1980) 1425–1429.

53. I. M. Thomas: High laser damage threshold porous silica antireflective coating, *Appl. Optics*, 25 (1986) 1481–1483.

54. I. M. Thomas: Porous fluoride antireflective coatings, *Appl. Optics*, 27 (1988) 3356–3358.

55. L. F. Huang, M. Saito, M. Miyagi, and K. Wada: Graded index profile of anodic alumina films that is induced by conical poles, *Appl. Optics* 32 (1993) 2039–2044.

56. S. Fujihara, H. Naito, M. Tada, and T. Kimura: Sol-gel preparation and optical properties of MgF_2 thin films containing metal and semiconductor nanoparticles, *Scripta Mater.* 44 (2001) 2013–2014.

57. T. Murata, H. Ishizawa, I. Motoyama, and A. Tanaka: Investigation of MgF_2 optical thin films prepared from autoclaved sol, *J. Sol-Gel Sci. Tech.* 32 (2004) 161–165.

58. T. Murata, H. Ishizawa, I. Motoyama, and A. Tanaka: Preparation of high-performance optical coatings with fluoride nanoparticle films made from autoclaved sols, *Appl. Optics* 45 (2006) 1465–1468.

59. H. Ishizawa, S. Niisaka, T. Murata, and A. Tanaka: Preparation of MgF_2-SiO_2 thin films with a low refractive index by a solgel process, *Appl. Optics* 47 (2008) C200–C205.

60. T. Murata, H. Ishizaka, and A. Tanaka: Investigation of MgF_2 optical thin films with ultralow refractive indices prepared from autoclaved sols, *Appl. Optics* 47 (2008) C246–C250.

61. H. Krüger, E. Kemnitz, A. Hertwig, and U. Beck: Transparent MgF_2-films by sol-gel coating: Synthesis and optical properties, *Thin Solid Films* 516 (2008) 4175–4177.

62. M. Wakaki, K. Kudo, and T. Shibuya: *Physical Properties and Data of Optical Materials*, CRC Press, Boca Raton, FL (2007).

63. A. Alemu, A. Freundich, N. Badi, C. Boney, and A. Bensaoula: MgF_2/BN Double layer antireflection coating for photovoltaic application, *Proc. Photovoltaic Specialist Conf.* 978-1-4244-1641-7/08, (2008) IEEE.

64. M. Zukic, D. G. Torr, J. F. Spann, and M. R. Torr: Vacuum ultraviolet thin films, 1: Optical constants of BaF_2, CaF_2, LaF_3, MgF_2, Al_2O_3, HfO_2, and SiO_2 thin films, *Appl. Optics* 29 (1990) 4284–4292.

65. Zs Czigany, M. Adamik, and N. Kaiser: 248nm laser interaction studies on LaF_3/MgF_2 optical coatings by cross-sectional transmission electron microscopy, *Thin Solid Films* 312 (1998) 176–181.

66. N. K. Sahoo, S. Thakur, M. Senthilkumar, D. Bhattacharyya, and N. C. Das: Reactive electron beam evaporation of gadolinium oxide optical thin films for ultraviolet and deep ultraviolet laser wavelength, *Thin Solid Films* 440 (2003) 155–168.

67. C. -C. Lee, M. -C. Liu, M. Kaneko, K. Nakahira, and Y. Takano: Characterization of AlF_3 thin films at 193nm by thermal evaporation, *Appl. Optics* 44 (2005) 7333–7338.

68. C. -C. Lee, B. -H. Liao, and M. -C. Liu: Developing new manufacturing methods for the improvement of AlF_3 thin films, *Optics Express* 16 (2008) 6904–6909.

69. C. -C. Lee, B. -H. Liao, and M. -C. Liu: AlF_3 thin films deposited by reactive magnetron sputtering with Al target, *Optics Express* 15 (2007) 9152–9156.

70. P. Kumar, M. K. Wiedmann, C. H. Winter, and I. Avrutsky: Optical properties of Al_2O_3 thin films grown by atomic layer deposition, *Appl. Optics* 48 (2009) 5407–5412.

71. D. Smith and P. Baumeister: Refractive index of some oxide and fluoride coating materials, *Appl. Optics* 18 (1979), 111–115.

72. J. H. Burnett, R. Gupta, and U. Griesmann: Absolute refractive indices and thermal coefficients of CaF$_2$, SrF$_2$, BaF$_2$, and LiF near 157nm, *Appl. Optics* 41 (2002) 2508–2513.

73. K. Truszkowska and C. Wesolowska: Optical properties of evaporated Gadolinium oxide films in the region 0.2–5 μm, *Thin Solid Films* 34 (1976) 391–394.

74. M. M. Giangregorio, M. Losurdo, A. Sacchetti, P. Capezzuto, and G. Bruno: Metalorganic chemical vapor deposition of Er$_2$O$_3$ thin films: Correlation between growth process and film properties, *Thin Solid Films* 517 (2009) 2606–2610.

75. E. Shalaan, H. Schmitt, and K. -H. Ehses: On the optical properties of gadolinium hydride system, *Thin Solid Films* 489 (2005) 330–335.

76. X. W. Zhang, Y. J. Zou, B. Wang, X. M. Song, H. Yan, G. H. Chen, and S. P. Wong: Optical band gap and refractive index of c-BN thin films synthesized by radio frequency bias sputtering, *J. Mat. Sci.* 36 (2001) 1957–1961.

77. D. W. Shortt, M. L. Jones, A. L. Schawlow, R. M. Macfarlane, and R. F. Farrow: Detection of sharp absorption lines in thin NdF$_3$ films, *J. Opt. Soc. Am.* 8 (1991) 923–929.

78. M. Kennedy, D. Ristau, and H. S. Niederwald: Ion beam-assisted deposition of MgF$_2$ and YbF$_3$ films, *Thin Solid Films* 333 (1998) 191–195.

79. R. Das, K. Adhikary, and S. Ray: Comparison of electrical, optical, and structural properties of RF-sputtered ZnO thin films deposited under different gas ambients, *Jpn. J. Appl. Phys.* 47 (2008) 1501–1506.

80. A. Lehmuskero, M. Kuittinen, and P. Vahimaa: Refractive index and extinction coefficient dependence of thin Al and Ir films on deposition technique and thickness, *Optics Express* 15 (2007) 10744–10752.

81. H. K. Pulker: Characterization of optical thin film, *Appl. Optics* 18 (1979) 1969–1977.

82. D. E. McCarthy: The reflection and transmission of infrared materials: 1, Spectra from 2 50 microns, *Appl. Optics* 2 (1963) 591–595.

83. W. J. Anderson: Optical characterization of thin films, *J. Opt. Soc. Am.* 67 (1977) 1051–1058.

84. T. J. Moravec, R. A. Skogman, and E. Bernal G.: Optical properties of bismuth trifluoride thin films, *Appl. Optics* 18 (1979) 105–110.

85. T. J. Moravec and E. Bernal G.: Automation of a laser absorption calorimeter, *Appl. Optics* 17 (1978) 1938–1943.

86. G. Hass, J. B. Ranseley, and R. Thum: Optical properties of various evaporated rare earth oxides and fluorides, *J. Opt. Soc. Am.* 49 (1959) 116–120.

87. D. E. McCarthy: The reflection and transmission of infrared materials: IV, Bibliography, *Appl. Optics* 4 (1965) 507–511.

88. A. Grill: Electrical and optical properties of diamond-like carbon, *Thin Solid Films* 355–356 (1999) 189–193.

89. S. S. Tinchev, Y. Dyulgerska, P. Nikolova, D. Grambole, U. Kreissig, TZ. Babeva: Optical properties of PECVD deposited DLC films prepared with air addition, *J. Optelectronics and Adv. Mat.* 8 (2006) 308–311.

90. R. S. Caldwell and H. Y. Fan: Optical Properties of Tellurium and Selenium, *Phys. Rev.* 114 (1959) 664–675.

91. M. F. Al-Kuhaili and S.M.A. Durani: Optical properties of erbium oxide thin films deposited by electron beam evaporation, *Thin Solid Films* 515 (2007) 2885–2890.

7 Materials for Nanophotonics

Kiyoshi Asakawa

CONTENTS

7.1   INTRODUCTION

Optical and photonic materials and structures have long been significantly devoted to optical signal transmission, storage, and processing in the optoelectronic community. Since photonic integration technologies were developed in the 1970s, advancement has been made in opto-electronic integrated circuits (OEICs) in the mid 1980s, planar light wave circuits (PLCs) and quantum dot (QD)–based active micro-devices [1] in the 1990s, and photonic crystal (PC) [2] waveguides in the 2000s. Recently, photonic integration technologies have been more advanced since the appearance of nanophotonic materials such as surface plasmon (SP) [3] and negative refractive index material (NIM) [4]. In accordance with these new materials, intensive attention has been paid to application of these nanophotonics that bear ultra-fast and low-energy optical information technologies. In addition, they are expected to break down the limit of conventional electronics or overturn conventional optical principles.

Table 7.1 shows summarized schematic structures (upper level) and unique features (lower level) of these PC, QD, SP, and NIM. Here, the PC column shows a high refractive-index contrast periodic structure with a two-dimensional (2D) air-hole array (upper) and a bend waveguide with a single missing line-defect (lower); the QD column shows a three-dimensional (3D) periodic electronic potential structure (upper) and a density-of-state (DOS) spectrum with a step function (lower); the SP column shows a propagation mode at the metal/dielectric interface (upper) and charge and field localization there (lower); the NIM column shows a fishnet structure originated from a combination of arrayed nano-rods for negative permeability and nano-bar for negative permittivity (upper) and fabricated fishnet structure (lower).

TABLE 7.1
Summarized Schematic Structures (Upper Level) and Unique Features (Lower Level) of Photonic Crystal (PC), Quantum Dot (QD), Surface Plasmon (SP), and Negative Refractive Index Material (NIM)

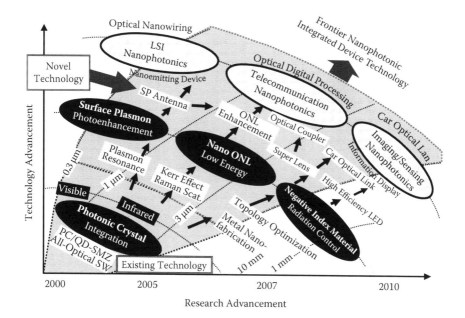

FIGURE 7.1 Nanophotonic technology road map.

Figure 7.1 shows a schematic nanophotonic technology road map that shows key roles of PC, QD, SP, and NIM technologies toward novel nanophotonics targets in LSI and image-sensing fields as well as the telecommunication field. As described in Section 7.3, nanooptical nonlinearity (ONL) represented by QD ensemble is a key element of the PC-based all-optical switch, while SP and NIM structures have potential to enhance the ONL of the QD via optical nano-dipole antenna and strong optical localization in nanostructures. In this chapter, state-of-the-art technologies and applications of these nanophotonics have been described in detail.

7.2 PHOTONIC CRYSTAL MATERIALS

7.2.1 INTRODUCTION

Since a concept of the PC was proposed in 1987 [2], a variety of computer simulations, fabrications, and characterizations on the PC have been performed for aiming at an advanced PIC. Figure 7.2 shows schematic 1D to 3D periodic structures of the PC with alternate high and low refractive index areas. Usage of air for the low index region in an ordinary case is the feature of the PC, which provides a unique wide band gap. For this reason, a semiconductor-based two-dimensional photonic-crystal (2D-PC) slab with a vertical air-hole array is widely used because conventional semiconductor planar epitaxial growth and micro-fabrication technologies such as electron beam lithography and dry etching are available. So far, 2D-PC structures have been most widely studied for applications to ultra-small light sources, optical switches, and waveguide components [5–11].

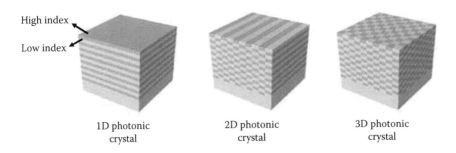

High index

Low index

1D photonic
crystal

2D photonic
crystal

3D photonic
crystal

FIGURE 7.2 One-dimensional to three-dimensional periodic structures of the photonic crystal.

7.2.2 FUNDAMENTAL PROPERTIES OF PHOTONIC CRYSTALS

7.2.2.1 Band Diagram

Behavior of the light wave in the PC material is dominated by the band diagram indicating the relationship between energy and wave number. It is strongly dependent on the crystal lattice and dimension as well as the propagation direction and polarization of the light wave. A typical example is shown in Figure 7.3 [12]. Figures 7.3a and 7.3b show a 2D square lattice configuration and typical propagation directions (Γ-M and Γ-X) in an irreducible Brillouin zone, respectively, while Figure 7.3c shows corresponding band diagrams dependent on the polarization, TE and TM. It is found that a full photonic band gap exists only for TM mode in the normalized frequency range from 0.28 to 0.42 in this case. The existence of the band gap has attracted much attention in the PC materials because of its unique physical interest. The straightforward interpretation is prohibition of photons having the corresponding energy range. This leads to a significantly important property that, if a lattice defect (as shown, for example, by the PC column in Table 7.1) is introduced in the PC, the band gap prohibits the photon from propagating in any direction out of the defect. As a result, the photon energy is localized only in the defect. The result also leads to introduction of the arbitrary configuration of the waveguide by designing the arbitrary lattice defect. In this way, tremendous amounts of theoretical calculation on the band diagram have been performed so far, searching for PC materials/lattice structures exhibiting a band gap as wide as possible and the propagation behavior as close to the design as possible.

7.2.2.2 Classification of One-Dimensional to Three-Dimensional Structures

In general, the band gap width is determined dominantly by a refractive index (n_{ref}) contrast in the PC. A variety of 1D ~ 3D PC structures reported so far is arranged in order for the magnitude of the n_{ref} contrast in Figure 7.4. It is easily imagined that a 3D woodpile structure, as shown in the top-right position, exhibits the strongest band gap effect, while a 2D air-hole array structure, as shown in the middle-right, has a large potential for practical application because a wide range of semiconductor planar processing technologies is available. Among them, the 3D auto-cloned structure at the intermediate n_{ref} contrast, available with a unique fabrication method, attracts much attention for wide practical applicability. Details are discussed in the next section.

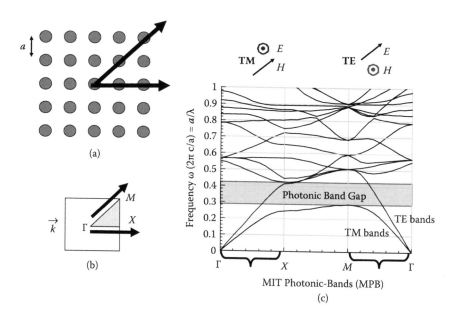

FIGURE 7.3 An example of band diagram in the two-dimensional photonic-crystal indicating the relationship between energy and wave number. (a) Two-dimensional square lattice configuration. (b) Typical propagation directions (Γ-M and Γ-X) in an irreducible Brillouin-zone. (c) Polarization-dependent band diagrams.

7.2.2.3 Applications

The band diagram of the PC has three dominant features: the photonic band gap (PBG), dispersion in the vicinity of the PBG, and strong polarization dependence. Figure 7.5 shows a variety of proposed PC devices classified into these three categories [13]. An optical resonator and micro-laser are typically categorized in the group of "optical confinement," while a dispersion compensator and polarization discriminator are in the groups of "dispersion" and "polarization," respectively. Several waveguide and optical switch devices are extended over three categories. Some of them are shown in detail in the application section.

7.2.3 FABRICATION OF PHOTONIC CRYSTAL (PC) WAVEGUIDES

In this section, the auto-cloned 3D-PC and air-bridge 2D-PC structures are described in detail due to having a large potential ability for practical applications.

7.2.3.1 Auto-Cloned Photonic Crystal (PC) Structure

Figures 7.6a through 7.6c show a fabrication method, schematic picture of the 3D structure and SEM (scanning electron microscopy) image of the cross section in the auto-cloned PC structure, respectively [14]. In Figure 7.6a, high- and low-n_{ref} thin films are alternately sputter deposited on the patterned substrate by rotating the substrate so

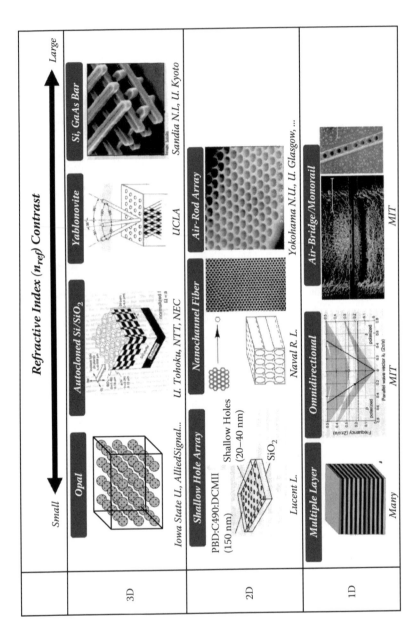

FIGURE 7.4 A variety of one-dimensional to three-dimensional photonic crystal structures reported and arranged in order for the magnitude of the n_{ref} contrast. (See for example, page 144 in "Phototonic Crystals–Application, *Technology and Physics*, (in Japanese)," edited by Shojiro Kawakami, CMC Publishing, Tokyo, March 2002.)

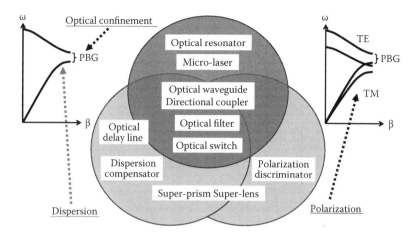

FIGURE 7.5 A variety of proposed photonic crystal devices classified into the three categories: optical confinement, dispersion, and polarization.

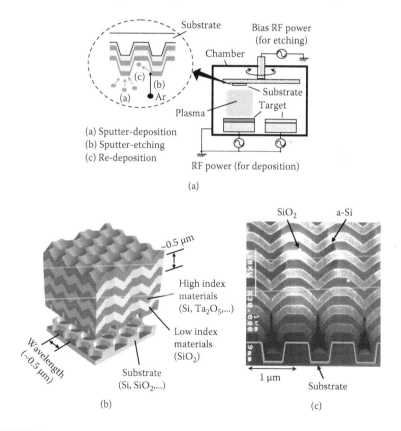

FIGURE 7.6 Auto-cloned three-dimensional photonic crystal structure. (a) Fabrication method. (b) Three-dimensional structure. (c) Scanning electron microscopy (SEM) image of the cross section.

as to face the two different targets alternately in the radio frequency (RF) Ar-plasma machine. A unique feature of this method is the ability to keep the identical 3D structure along the multi-layer deposition up to several tens of layers—that is, auto-cloning of the identical 3D structure. The key to this technique lies in the three elementary plasma processes: sputter deposition on the patterned substrate, sputter etching to form the given tilted angle on the side wall, and re-deposition on the tilted side wall, as shown in the figure. This technique enables fabrication of the periodicity in the ~0.5 μm range, as shown in Figure 7.6b. In Figure 7.6c, it is found that alternate SiO_2 and a-Si layers are auto-cloned up to ~10 layers. The auto-cloned PC has already been applied to several kinds of optical components such as polarization discriminator.

7.2.3.2 Air-Bridge PC Waveguide

A 2D-PC air-bridge-type slab waveguide has a strong optical confinement effect in the vertical direction, so that a large number of excellent transmittance characteristics have been reported so far by the author's group [15]. Figure 7.7 shows a series of fabrication processes of the GaAs-based 2D-PC air-bridge slab waveguide. In the figure, (a) and (b) show a plan-view of the 2D-PC tri-angle lattice pattern and SEM image on the cross section of the MBE-grown GaAs/AlGaAs/GaAs substrate sample, respectively, while (c) to (e) show sequential electron-beam (EB) lithography for patterning the 2D-PC, reactive ion beam etching (RIBE) for forming the air-hole array, and selective wet etching for eliminating the sacrificial AlGaAs layer. The thickness of the GaAs core and AlGaAs sacrificial layers are 250 nm and 2 μm, respectively.

A magnified SEM image on the cleaved edge of the waveguide sample is shown in Figure 7.8a [16]. The waveguide consists of a missing row of air-holes surrounded by 10 rows of air-holes on both sides. A solid-line circle shows a cross section of about 0.7-μm-wide and 0.25-μm-thick single-line defect 2D-PC waveguide. A lattice constant and air-hole diameter are 345 nm and 210 nm, respectively. A single-line defect is formed by leaving a row of perforated air-holes in the Γ–K direction. As shown schematically in Figure 7.8b, the top and bottom surfaces of a line-defect waveguide were observed by SEM. The resultant SEM images of cleaved edges viewed from

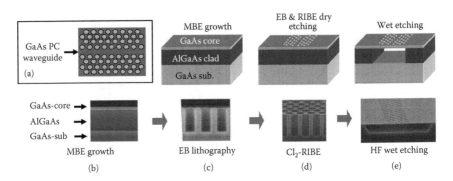

FIGURE 7.7 Series of fabrication processes of GaAs-based two-dimensional photonic crystal (2D-PC) air-bridge slab waveguide. (a) Molecular beam epitaxy–grown epitaxial layers. (b) Resist pattern after electron-beam lithography. (c) Ar-hole array after reactive ion beam etching (RIBE). (d) Cleaved edge of the 2D-PC air-bridge.

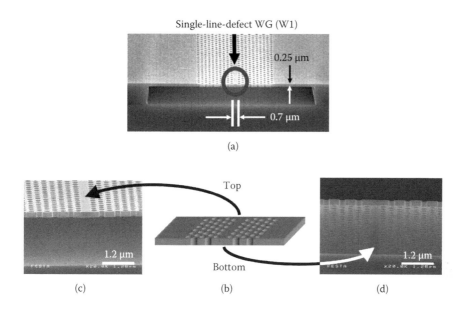

FIGURE 7.8 Magnified scanning electron microscopy (SEM) image of the cleaved edge of the two-dimensional photonic crystal (2D-PC) air-bridge waveguide. (a) Cleaved edge of the 2D-PC air-bridge. (b) 2D-PC slab waveguide. (c,d) Top and bottom surfaces of a line-defect waveguide.

above and below are shown in Figure 7.8c and 7.8d, respectively. The photographs show extremely smooth surfaces. It is also clear that there is no residual material on the rear surface of the slab. From these SEM images, roughness of the top and bottom waveguide surfaces was estimated as less than 10 nm.

7.2.3.3 Design and Characterization of Photonic Crystal Waveguides

7.2.3.3.1 Propagation Loss

In order to derive the propagation loss of these waveguides by a cut-back method, three kinds of straight-line-defect PC waveguides with lengths of 1, 4, and 10 mm were prepared by cleaving both ends of the sample. Then, transmittance values in the measured transmission spectra at the center wavelengths were plotted as a function of the waveguide length, as shown by the empty circles in Figure 7.9a. From a slope of the fitting, as shown by the line in the figure, the propagation loss of (0.7685 ± 0.5) dB/mm was derived. This value is recorded low as long as the GaAs 2D-PC slab waveguide is concerned [16].

The result suggests that the air-bridge structure with the 2D air-hole array has been fabricated with high precision even for the waveguide length up to 10 mm. In addition, surface roughness was measured in more detail by an atomic force microscope (AFM). Figure 7.9b shows an AFM image (top) and a surface-roughness profile (bottom) obtained with the AFM probe line-scanned along the straight line indicated in the AFM image. The result shows that the roughness of the top surface of the 2D-PC slab waveguide is less than 1 nm.

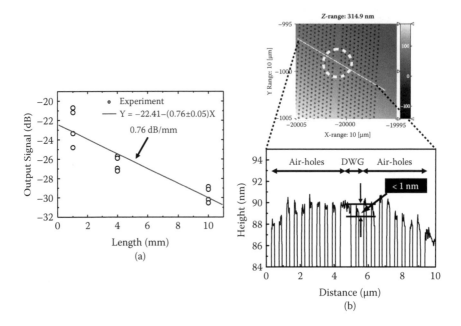

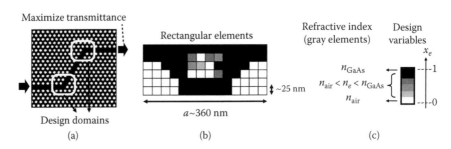

FIGURE 7.9 (a) Transmittance values as a function of the two-dimensional photonic crystal (2D-PC) waveguide length. (b) Atomic force microscopy (AFM) image (top) and a surface-roughness profile (bottom) obtained with the AFM probe line-scanned along the straight line indicated in the AFM image.

FIGURE 7.10 Full computational model of a topology optimization (TO) design method. (a) A computational region. (b) Rectangular elements. (c) Mathematical programming tool.

7.2.3.3.2 Topology Optimization Design

For generic applications, design of the 2D-PC waveguide with a wide bandwidth and transmission spectrum close to the design is important. However, a conventional computer-aided design tool such as FDTD (finite-different time-domain) method has been limited in ability. On the other hand, a topology optimization (TO) design method has been developed for drastic improvement of a bandwidth and transmittance by the group of Denmark [17,18]. A full computational model is shown in Figure 7.10. As shown in designing the bend in Figure 7.10a, the TO procedure is performed to maximize transmittance of the output port by modifying the refractive index distribution in the design domain indicated in the figure. Each unit cell is discretized using 14 × 12 four-noded quadrilateral elements, as shown in Figure 7.10b. An optimization algorithm is based on a 2D finite-element frequency-domain solver. The solver is used repeatedly in an iterative scheme, in which the material distribution is updated every iteration based on an analytical sensitivity analysis and use of a mathematical programming tool, as shown in Figure 7.10c [19].

TO design of the waveguide intersection is shown in Figure 7.11, where the TO procedure is executed to maximize the transmittance in a straightforward line (a through d) by modifying the refractive index distribution in the indicated area, as shown in Figure 7.11a [20]. Continuous design variables x_e ($0 \leq x_e \leq 1$) are used to allow for utilizing a gradient-based optimization strategy. Existence of elements in the final design with values between zero (n_{air}) and one ($n_{dielctric}$), gray elements, is undesirable for fabrication. To remedy this problem, the penalty σ_{pen} and the penalty function $p = 4\,\alpha\,x_e\,(1-x_e)$ are used, where α is a real and positive constant. As shown in Figure 7.11b, the value of p is large in the middle range. Therefore the gray elements will be uneconomical and forced toward either zero or one by applying the parameter α. Calculated characteristics in A to G in Figure 7.11c show snapshots (top) of design during the optimization process and transmittance spectra (bottom) for each design. Many gray elements, appearing in the design shown in C, gradually disappear with increasing α. The final optimized design shown in G is feasible for fabrication and exhibits high transmittance (upper curve) with low crosstalk (lower curve).

Figure 7.12 shows designed intersection patterns and their experimental results as compared between (a) the standard-designed and (b) TO-designed intersections [20]. In the measured transmission spectra (bottom), "ad" means transmittance between the a through d ports, while "ac" means undesirable crosstalk between the a through c ports. It is worth noticing that the crosstalk "ac," which is comparable to "ad" for the standard design, is suppressed by −15 dB for the TO design, and furthermore, the transmission spectra "ad" is identical to that for the straight waveguide in the whole wavelength range. The result clearly shows that the TO method is a desirable tool for designing non-straight 2D-PC waveguides.

Experiments on a TO-designed bend waveguide results in Figure 7.13, where the TO designed snapshot and measured transmittance spectra are shown in Figures 7.13a and 7.13b, respectively [19]. For comparison, the transmittance spectrum of the straight WG is shown by a thin dotted line. Here, the high-transmittance range is realized because the propagation mode is a single-guided mode below the light-line. In case of the standard bend (indicated by STD), a transmission bandwidth is narrowed by 15 nm as compared with the straight waveguide in the longer wavelength range. In contrast, no bandwidth narrowing is observed for the optimized bend (indicated by TO). This is the case for improvement in transmission performance of the TO bend in the vicinity of the band edge.

7.2.3.3.3 Low Group Velocity Design

As mentioned in Figure 7.5, the band diagram in the PC inherently exhibits more dispersive behavior as it approaches the band edge, or in other words, the group

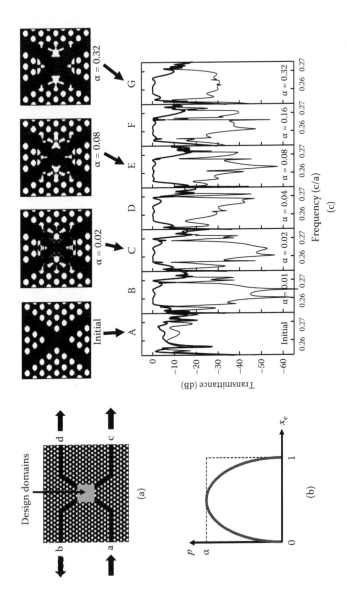

FIGURE 7.11 Topology optimization (TO) design of the waveguide intersection. (a) TO procedure executed to maximize the transmittance in a straightforward line. (b) Modified refractive index distribution in the indicated area. (c) Calculated characteristics that show snapshots (top) of design during the optimization process and transmittance spectra (bottom) for each design.

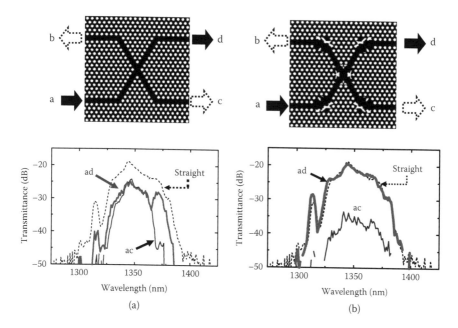

FIGURE 7.12 Designed intersection patterns (upper) and their experimental results (lower) as compared between (a) the standard-designed and (b) topology optimization (TO)–designed intersections.

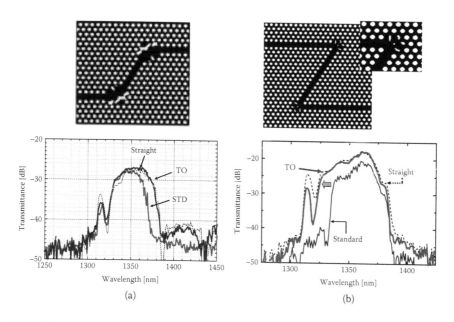

FIGURE 7.13 Topology optimization (TO)–designed bend waveguide. (a) TO-designed snapshot. (b) Measured transmittance spectra.

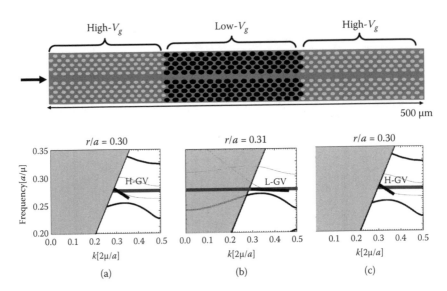

FIGURE 7.14 Design of low V_g waveguide (b) sandwiched by two high V_g waveguides (a) and (c). Upper: Two-dimensional photonic crystal straight waveguides. Lower: Band diagrams.

velocity (V_g) defined by the slope of the band curve decreases and becomes zero at the band edge. Application of the low V_g provides several important optical components such as ultra-small optical delay line and unique slow-wave device. However, design of such a low V_g waveguide needs particular attention in regard to the transmittance. The upper part of Figure 7.14 shows a low-V_g waveguide sandwiched by identical high-V_g waveguides. Usage of the low-V_g waveguide alone exhibits a drastic decrease in transmittance in the vicinity of the band edge if the light wave impinges directly into the edge of the low-V_g waveguide. This is due to a significant reflection of the incident light at the waveguide edge. This problem can be solved by connecting high-V_g waveguides at both ends of the low-V_g waveguide, as shown in Figure 7.14 [21]. In this configuration, the light wave can be transmitted smoothly through the edge of the high-V_g waveguide due to its single-mode design, while the light can also propagate without any reflection at both ends of the low-V_g waveguide because lights in both low-V_g and high-V_g regions exhibit Bloch functions. The author's group has confirmed this principle by theoretical calculation based on the band diagrams and choice of key parameters regarding r/a, as shown in Figures 7.14a to 7.14c, where r and a mean air-hole radius and lattice constant, respectively. The result is shown by the transmission spectra for both low-V_g waveguide alone (Homo-V_g WG) and heterostructure of low-V_g and high-V_g waveguides (Ht-V_g WG) in Figure 7.15, where (a) and (b) show experimental and calculated results, respectively. As shown in Figure 7.15a, the hetero-structure has proved to improve the transmittance by 16 dB over the homo-structure in the low-V_g range (shown here by the high group-index n_g of 17, where n_g is defined by a reciprocal of V_g). The experimental result has also been verified by the calculated result, as shown in Figure 7.15b. The results are useful for designing the PC waveguides including the low-V_g waveguide.

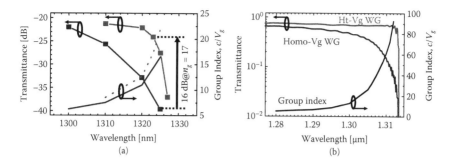

FIGURE 7.15 Transmission spectra for a low-V_g waveguide (WG) alone (Homo-V_g WG) and hetero-structure of low-V_g and high-V_g waveguides (Ht-V_g WG). (a) Experimental results. (b) Calculated results.

7.2.3.4   Applications

2D-PC waveguides have been applied to a variety of applications. Typical examples are shown here. Figure 7.16 shows an ultra-small semiconductor laser using a 2D-PC micro-cavity [22]. An InP-based 2D-PC slab waveguide with four layers of quantum well is formed to be an air-bridge structure to confine the light in the vertical direction, as shown in Figure 7.16a. In the lateral direction, a tri-angle lattice air-hole array with single point defect at the center of the array is perforated to form the micro-cavity, as shown by the upper SEM photograph in Figure 7.16b. A lattice constant a and air-hole radius r are 515 nm and 180 nm, respectively. Radii of two inner air-holes are modified to be 240 nm. When the cavity is optically pumped, it exhibits light emission from the point defect region, as predicted by the calculated electromagnetic wave distributions in the lower part of Figure 7.16b. Optically pumped laser emission has been verified, as shown in Figure 7.16c. This is the first report on the PC-based micro-cavity laser application.

On the other hand, Noda's group has developed two kinds of excellent passive 2D-PC waveguides elements. One is a multiple-wavelength add/drop filter using coupling of a straight waveguide and neighboring output ports composed of three missing point defects with wavelength-dependent sizes, as shown in Figure 7.17 [23]. An upper picture shows a schematic plan-view configuration of the 2D-PC array, single line-defect straight waveguide, and point-defect output ports. Multiple wavelengths in the input light are separated at wavelength-dependent output ports. Experimentally demonstrated results are shown in the lower parts of Figure 7.17. The left side of the photograph in the figure shows near field patterns of the output beams with different wavelengths, while the right-side curves show output-beam intensity peaks with different wavelengths observed from different output ports. It is shown that six to seven wavelengths are clearly separated by this device.

Another example is an ultra-high-Q photonic double-hetero-structure nano-cavity, as shown in Figure 7.18 [24]. In Figure 7.18a, the upper part shows a comparison of schematic 2D-PC waveguides between homo-structure (left) and

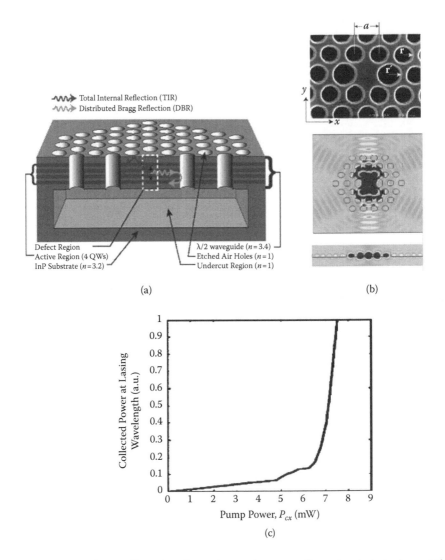

(a)

(b)

(c)

FIGURE 7.16 Ultra-small semiconductor laser using a two-dimensional photonic crystal (2D-PC) micro-cavity. (a) InP-based 2D-PC slab waveguide with four layers of quantum well. (b) Scanning electron microscopy (SEM) pattern of the micro-cavity and calculated electromagnetic wave distributions. (c) Optically pumped laser emission.

hetero-structure (right) in regard to the lattice constant, while the lower part shows a band diagram (left) and its schematic frequency ranges corresponding to transmission and mode gap regions (right). A detailed 2D-PC structure of the proposed double-hetero-structure nano-cavity and its electromagnetic field distribution are shown on the left side of Figure 7.18b, where 2D-PC I has a triangular-lattice structure with a lattice constant of a_1, while 2D-PC II has a deformed triangular-lattice structure with a face-centered rectangular lattice of constant

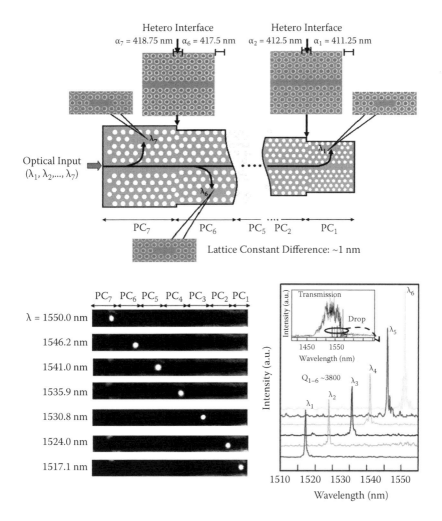

FIGURE 7.17 Multiple-wavelength add/drop filter with a straight waveguide and neighboring point-defect output ports.

a_2 ($>a_1$) in the waveguide direction; it retains the same constant as 2D-PC I in the orthogonal direction in order to satisfy the lattice-matching conditions. In this way, light waves of a specific energy can exist only in the waveguide of 2D-PC II. This results in the desirable calculated Gaussian field distribution, as shown in the left-lower part of Figure 7.18. Right-hand-side pictures in Figure 7.18 show a non-waveguide-type 2D-PC nano-cavity with neighboring three missing point defects reported. This type results in the calculated degraded Gaussian field distribution, as indicated in the figure. As shown in Figure 7.18c, an experimental result exhibits as high a Q-factor as 600,000. This kind of nano-cavity will be applied to a high-sensitivity optical sensor, a part of an optical memory, and so on.

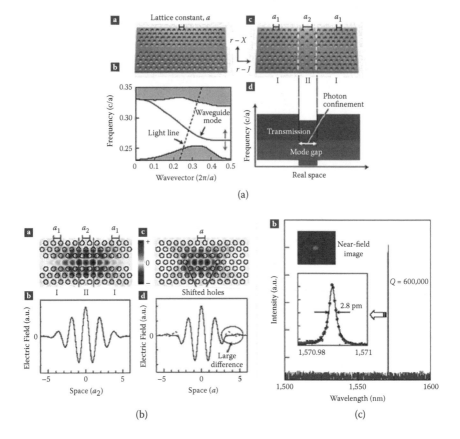

FIGURE 7.18 Ultra-high-Q photonic double-hetero-structure nano-cavity. (a) Upper: Two-dimensional photonic crystal (2D-PC) waveguides between homo-structure (left) and hetero-structure (right). Lower: band diagram (left) and its schematic frequency ranges (right). (b) Left: Detailed 2D-PC structure of the double-hetero-structure nano-cavity and its electromagnetic field distribution. Right: Non-waveguide-type 2D-PC nano-cavity with missing point defects. (c) Experimental transmission spectra with as high a Q-factor as 600,000.

7.3 QUANTUM DOT MATERIALS

7.3.1 INTRODUCTION

Quantum dots (QDs) have been known to have an electronic high density-of-state (DOS) [1]. Figures 7.19a through 7.19c show schematic structures (lower) and corresponding band diagrams (upper) for a quantum well (1D carrier confinement structure), quantum wire (2D one), and quantum dot (3D one), respectively, where the structure shows periodic regions with alternate wider and narrower electronic potential gaps. The high DOS property in the QD is reflected to the approximate step function for the allowed energy state in the band diagram in QD. Large optical nonlinearity (ONL) such as high gain- and absorption-saturation attendant on the QD originates from such high DOS property. In this chapter, QD materials are shown

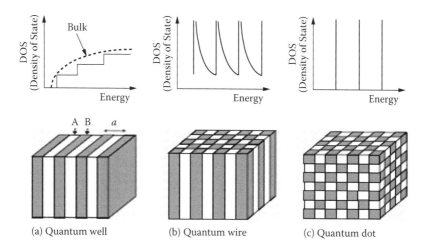

FIGURE 7.19 Structures (lower) of quantum well, quantum wire, and quantum dot, and corresponding band diagrams (upper).

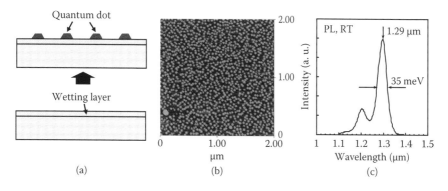

FIGURE 7.20 (a) Stranski–Krastanov (SK) mode. (b) Atomic force microscopy (AFM) image of the quantum dots (QDs). (c) Photoluminescence (PL) spectrum showing a sharp emission peak.

mainly by focusing on their large ONL (absorption saturation) effect and unique application to ultra-small and ultra-fast all-optical switch based on recent work in the author's group.

7.3.2   Fundamental Technologies of Quantum Dot (QD) Growth in the Photonic Integrated Circuit

In this section, fundamental QD growth techniques and properties regarding surface density and uniformity are shown. Figure 7.20a shows schematic pictures of the Stranski–Krastanov (SK) mode widely known as a typical self-assembled QD growth mode. This method is valid only for the strain system with different lattice constants between a substrate and QD material, like a GaAs substrate and InAs QD. In the figure, the QD material is initially two-dimensionally grown on the substrate with the strain accompanied in it. But it instantly changes into three-dimensional (island state) growth mode over the critical thickness. An initial layer (normally within a several mono-layer thickness) under the critical thickness is called a *wetting layer*, while the three-dimensionally grown island is the QD. A typical InAs QD on the GaAs is 30 to 50 nm in diameter and 5 to 10 nm in height. The surface density of the SK-grown QD ranges from 10^9 to 10^{11}/cm^2, depending on the growth temperature, rate, and flux rate. Figure 7.20b shows an AFM image of the QDs with a surface density of 3.5×10^{10}/cm^2, while Figure 7.20c shows a photoluminescence (PL) spectrum that gives a uniformity of 35 meV in terms of emission peak width [25]. These values are available for the device fabrication, as shown later.

Photonic integrated circuits involving optical nonlinear QDs acting as gain and modulation media extend their applicability to devices essential for future telecommunication network systems. An all-optical signal processing device such as an ultrafast optical switch is a suitable target for a QD-containing waveguide device. In such a device, an optical signal pulse is controlled by some sort of ONL phenomena in QD pumped by an optical control pulse without electrical means, thus providing an ultra-high speed that exceeds the limit of the conventional electronic device. In addition, a selective-area-growth (SAG) technique of QDs in the nanophotonic waveguide will accelerate the advancement of the photonic integrated circuit. The author's group has investigated the SAG technologies of the QD and their application based on the scenario of the QD mentioned above. Figures 7.21a and 7.21b show a schematic of an optical flip-flop in a chip and its waveguide configuration showing the principle, respectively [26]. The device consists of two identical SMZ (symmetrical Mach–Zehnder) type all-optical switches based on the QD and 2D-PC waveguide. The SMZ switch is a time-differential phase modulator, while QD-containing PC waveguides form identical ONL (absorption saturation here) arms in the SMZ interferometer. For achievement of such an integrated circuit, several SAG technologies have been developed. They are shown in the next section.

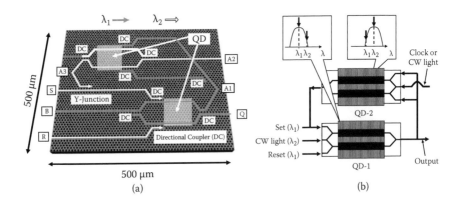

FIGURE 7.21   (a) An optical flip-flop in a chip. (b) Its waveguide configuration showing the principle of the flip-flop function.

7.3.3 SITE-CONTROLLED QUANTUM DOTS

7.3.3.1 Quantum Dots on Patterned Substrate

Site-controlled QDs on a surface with regularly spaced nucleation sites have a possibility for improvement in uniformity of the QD because atoms can evenly migrate to each nucleation site. For this purpose, regular nucleation sites should be formed artificially on a semiconductor surface. In addition, such site-control of QDs can serve as a powerful technique for realizing regular arrays of quantum bits for quantum computers and single-photon emitters for quantum communications, where each QD must be precisely located at its designed position.

Figure 7.22 shows regular arrays of InGaAs QDs on the patterned substrate grown by three steps—that is, a low-temperature deposition of InGaAs, subsequent annealing on nano-holes, and regrowth of InGaAs under the optimized growth condition [27]. The substrate includes nano-hole arrays with periods of 70 nm and 100 nm formed in 50 μm × 50 μm square regions by electron-beam lithography and Cl_2-based reactive ion-beam etching, as shown in Figure 7.22a. Nano-hole diameters are 50 nm and 35 nm for the periods of 100 nm and 70 nm, respectively, and their depths are

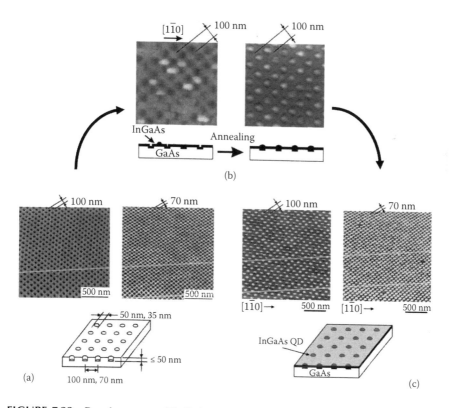

FIGURE 7.22 Regular arrays of InGaAs quantum dots (QDs) on the patterned substrate grown by three steps. (a) Nano-hole arrays formed by electron-beam lithography and Cl_2-based reactive ion-beam etching. (b) Low-temperature deposition of InGaAs, subsequent annealing on nano-holes. (c) Regrowth of InGaAs under the optimized growth condition.

50 nm or less. As shown in Figure 7.22b, an STM (scanning tunneling microscope) image just after the low-temperature deposition shows an irregular surface structure such as empty or half-filled nano-holes and dot-like structures on flat areas. Despite this irregular structure, regular nucleation of QDs in nano-holes was observed after annealing at 450°C. By optimizing the growth condition after the annealing, regular InGaAs QD arrays with both 70 nm and 100 nm periodicity (corresponding to a surface density of $1 \times 10^{10}/cm^2$) have been formed on the nano-hole arrays, as shown in Figure 7.22c. This growth method is a promising selective-area-growth technique of QDs for the advanced photonic integrated circuit.

7.3.3.2 Quantum Dots Grown by Metal Mask Molecular Beam Epitaxy (MBE) Method

Another selective-area-growth method of QDs is a metal-mask (MM) method available for growing QD ensemble selectively in an area of several tens to several hundred μm in size. Figures 7.23a and 7.23b show configurations of the mask in the

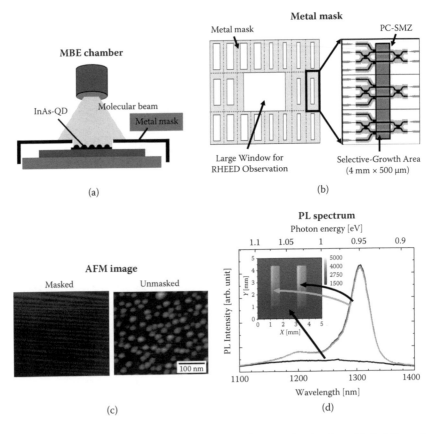

FIGURE 7.23 (a) Configuration of the mask in the growth chamber. (b) Relative relationship between mask open windows and required quantum dot (QD) growth positions on the device. (c) Atomic force microscopy (AFM) image showing selective QD growth. (d) Photoluminescence (PL) spectra for unmasked and masked regions on the GaAs substrate.

growth chamber and relative relationships between an open window in the mask and a required QD-growth position on the device (here, PC-SMZ chip) [28]. The mask has a large open window for real-time observation of reflection-high-energy-electron-diffraction (RHEED) patterns to obtain high-quality InAs QDs as well as an open window array for selective area growth of QDs. Selectivity of the QD growth between the masked and unmasked area has been confirmed by AFM (atomic force microscope) observations, as shown in Figure 7.23c. InAs QDs were characterized with PL spectra. Figure 7.23d shows PL spectra for unmasked and masked regions on the GaAs substrate. A PL-peak width of approximately 38 meV, almost equal to the PL-peak width of the conventionally grown QD, suggests that the QD grown by the new MM method has high optical quality. These results indicate that the MM method is useful for selective area growth of high-quality QDs.

The wavelength of the selective-area-grown QD can be controlled by inserting an InGaAs layer as a stress-reducing layer (SRL) between the InAs QD and GaAs spacer layer. As shown in Figure 7.24a, the strain system between the GaAs spacer and InAs QD induces stress at their interface, while such stress is reduced after insertion of the 3-nm-thick $In_{0.2}Ga_{0.8}As$ SRL and capping of the QD with 3-nm-thick GaAs, as shown in Figure 7.24b. Insertion of the SRL allows red-shift in the PL peak energy. Figure 7.24c shows PL spectra as a function of wavelength, while Figure 7.24d shows PL peak wavelength as a function of SRL thickness [29]. The result indicates that insertion of the SRL is effective for controlling the PL wavelength of QDs without degrading the QD optical quality.

The MM method provides another promising technique useful for selective area growth of QDs. The metal mask mentioned above is designed to be rotatable by 180° on the MBE wafer holder, as shown in Figure 7.25a [30]. After the InAs QD in the first area (QD1) is grown, an $In_{0.2}Ga_{0.8}As$ layer is deposited on the QD and capped with a GaAs layer. Then, the metal mask is rotated by 180° and the second QD (QD2) is grown next to the QD1 position in the same manner except for the thickness of SRL. Figures 7.25b and 7.25c show the schematic picture and observed

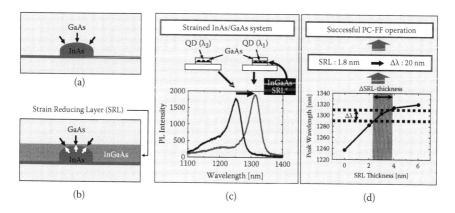

FIGURE 7.24   (a) Strain system between GaAs spacer and InAs quantum dot (QD). (b) Insertion of the $In_{0.2}Ga_{0.8}As$ stress-reducing layer (SRL) for reducing the stress. (c) Photoluminescence (PL) spectra as a function of wavelength. (d) PL peak wavelength as a function of SRL thickness.

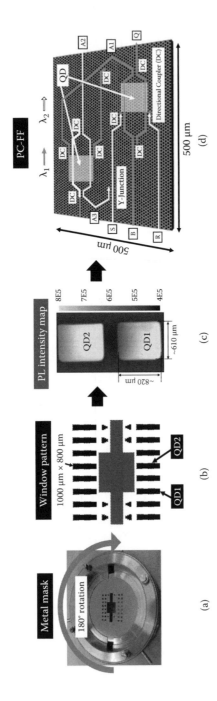

FIGURE 7.25 (a) Metal mask designed as rotatable by 180° on the molecular beam epitaxy (MBE) wafer holder. (b) Window patterns. (c) Observed photoluminescence (PL) intensity mapping. (d) Quantum dots with different wavelengths at the different positions applied to an advanced photonic integrated circuit.

PL intensity mapping, respectively. A SRL thickness is set to 2.7 nm for QD1 and 4.5 nm for QD2. In this experiment, the wavelength difference by 20 nm (1280 nm and 1300 nm) was realized as designed. The results shown here provide a novel selective-area-growth technique of QDs with different wavelengths at the different positions suitable for implementation of the advanced photonic integrated circuit, as shown in Figure 7.25d.

7.3.3.3 Quantum Dots Grown by Nano-Jet-Probe Method

To date, the author's group has developed an STM probe-assisted site-control growth of InAs QDs on the GaAs substrate and demonstrated two-dimensionally arrayed QDs with various as well as constant (50–100 nm) pitches [31]. However, the growth rate of this method was as slow as 0.5 to 1 s/dot and was not high enough for practical application. Alternatively, a nano-jet probe (NJP) method was developed using a specially designed AFM (atomic force microscope) cantilever as a nano-probe capable of available throughput of 1 to 10 ms/dot, as shown in Figure 7.26a, where a series of "in-vacuum" formations of site-controlled InAs QDs is shown [32]. Applying an external voltage to the AFM probe, uniform indium (In) nano-dots are deposited reproducibly at the computer-controlled point. Because the AFM chamber is connected to a molecular beam epitaxy (MBE) chamber via an ultra-high-vacuum (UHV) tunnel, In nano-dots can be directly converted to InAs QDs by subsequent irradiation of arsenic flux in the MBE chamber using a droplet epitaxy technique [33]. Then, a necessary number of QDs in the desired region with high uniformity and high density can be formed by a conventional stacking technique. As shown schematically in Figure 7.26b, a specially designed cantilever has a hollow pyramidal tip with a sub-micron-size aperture on the apex and an In-reservoir tank

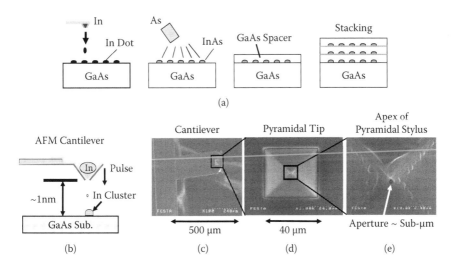

FIGURE 7.26 Nano-jet probe (NJP) method. (a) Sequential process of the NJP method. (b) Specially designed cantilever having a hollow pyramidal tip with a sub-micron-size aperture on the apex and an In-reservoir tank within the stylus. (c–e) Scanning electron microscopy (SEM) images of the cantilever, pyramidal tip, and apex of the pyramidal stylus.

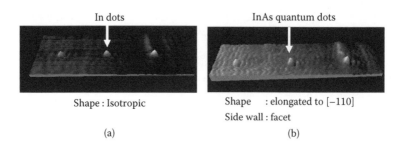

In dots   InAs quantum dots

Shape : Isotropic   Shape   : elongated to [−110]
                                           Side wall : facet

(a)   (b)

FIGURE 7.27   (a,b) Atomic force microscopy (AFM) images of the site-controlled In nano-dots formed at the first step and In As quantum dots grown at the second step in the molecular beam epitaxy growth chamber, respectively.

within the stylus. This cantilever is a piezoelectric type with a hollow pyramidal tip and is used for nano-dot fabrication as well as for sensing the atomic force in an AFM observation mode. By applying a voltage pulse between the pyramidal tip and the sample, In clusters were extracted from the reservoir tank within the stylus through the aperture, resulting in the In nano-dot formation. Figures 7.26c to 7.26e show SEM images of the cantilever, pyramidal tip, and apex of the pyramidal stylus, respectively. Figures 7.27a and 7.27b are AFM images of the site-controlled In nano-dots formed at the first step and InAs QDs grown at the second step in the MBE growth chamber, respectively [34]. Before the conversion process, In nano-dots exhibit a cone shape, while they are changed to anisotropic shape elongated in the [110] direction as an evidence of formation of InAs QDs. This technology has an extremely high selectivity for the QD formation area, because no QD will be formed in the place where the In nano-dots are not deposited. This performance is very important when considering application to the single-photon emitter in the PC high-Q cavity and the advanced photonic integrated circuit, as shown in Figure 7.21.

7.3.3.4 Optical Nonlinearity and Application to All-Optical Switch

7.3.3.4.1 Classification of Optical Nonlinearity

As mentioned in Section 7.1, the QD exhibits a large ONL effect thanks to the high DOS. Table 7.2 shows summarized ONL phenomena and related materials available for optoelectronic applications, categorized by difference of an excitation method. An electro-optic effect is typically known to provide such $LiNO_3$ optical modulator and semiconductor devices, while an absorption saturation pumped by optical electric field is a representative third-order ONL (χ^3) phenomenon available for ultra-fast all-optical devices with a high speed up to a pico-second level. In this section, recent work on the ONL effect (absorption saturation) of the QD which is embedded in the 2D-PC waveguide is shown. Both measured fundamental ONL effects and application to the SMZ-type all-optical switching operation are shown [35].

7.3.3.4.2 Absorption Saturation as Third-Order Optical Nonlinearity

As shown later, an all-optical switch, PC-SMZ is composed of the PC waveguide embedded with the QD, as shown in Figure 7.28a. The QD plays a role of a phase

TABLE 7.2

Summarized Optical Nonlinearity (ONL) Phenomena and Related Materials Available for Optoelectronic Applications, Categorized by Difference of an Excitation Method

Excitation	Phenomena	Materials
• Electrical field	• Electro-optic effect	• $LiNbO_3$, Organics
• Current	• Plasma effect	• InGaAsP, etc.
	• Band filling effect	
• Optical e-field	• Plasma effect	• Semiconductor
	• Band filling effect	• GaAs, InAs, etc.
	• Absorption saturation	• Exciton, quantum dots (QDs), etc.
	• Third ONL polarization effect	• Organics, etc.
	• Thermo-optical effect	• ZnS, ZnSe, etc.
	• Photo-refractive	• $Bi_{12} SiO_{20}$, etc.

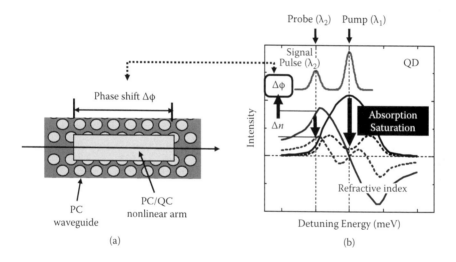

FIGURE 7.28 (a) All-optical switch, photonic crystal symmetrical Mach-Zehnder (PC-SMZ) composed of the PC waveguide embedded with the quantum dot (QD). (b) Refractive index change of the QD, Δn, caused by the absorption saturation pumped by the control pulse.

shifter on account of the ONL-induced refractive index change Δn. Δn is attributed to the absorption saturation pumped by the control pulse, as shown in Figure 7.28b [36]. The phase shift depends on factors such as pumping energy density, control and signal pulse detuning, and inhomogeneous broadening of the QD absorption peak.

Δn is derived from measurement of a transmittance in the QD-containing PC waveguide by changing pump pulse energy in a two-color pump-probe system [37]. As pump pulse energy increases, transmittance increases from an absorptive region to a transparent region due to the absorption saturation of the QD, as shown

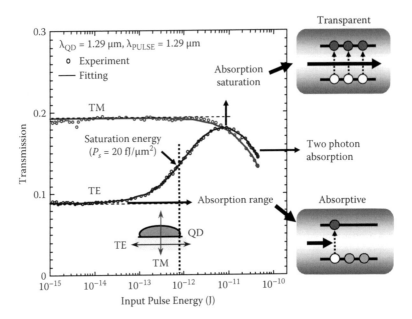

FIGURE 7.29 Transmittance change in the quantum dot (QD)–containing photonic crystal (PC) waveguide as a function of a pump pulse energy in a two-color pump-probe system.

in Figure 7.29. Decrease of the transmittance for further increase of the pump pulse energy is caused by the two photon absorption and has nothing to do with the current absorption saturation. In the figure, empty circles show experimental results, and the transmittance change has been observed only for a TE mode. An absorption saturation power P_s indicated in the figure is a useful criterion for evaluating the pump power needed for causing the absorption saturation. Here, the smaller the absorption saturation is, the lower the switching energy is when applied to an all-optical switch. The experimentally obtained P_s value of 13 fJ/μm² suggests an optical switching energy as low as 100 fJ or less.

For evaluating the quality of the QD as an ONL material, Figure 7.30 shows comparison of a relationship between a saturation intensity (I_s) and energy relaxation time (T_1) among the QD, exciton in the quantum well and bulk quantum well, where $I_s = P_s/T_1$ [38]. Data for excitons in the quantum well of III-V compound semiconductors have been derived from the reported values. In the figure, dotted lines indicate a constant product of I_s and T_1—that is, a constant saturation pulse energy P_s. Regarding the P_s, all the data for the exciton deviate by 1/5 ~ 5 times around the value of 100 fJ/μm². On the other hand, the P_s value of 13.7 fJ/μm² derived for the QD is ~1/7 of the maximum value for the exciton and is comparable to the lowest value. This means that the ONL figure of merit of the QD, defined by $\chi^3/I_s/a_0$ where α_0 denotes a linear absorption coefficient of the QD, is almost an order of magnitude larger than or at least comparable to that of the exciton in the quantum well of III-V compound semiconductors. Taking into account that both the QD and exciton are subject to an atom-like DOS, the result seems qualitatively reasonable.

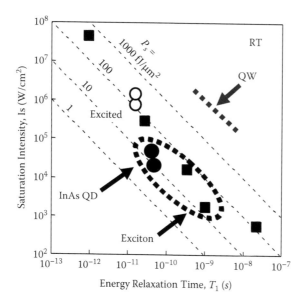

FIGURE 7.30 Comparison of a relationship between a saturation intensity (I_s) and energy relaxation time (T_1) among quantum dots (QDs), exciton in the quantum well, and bulk quantum well.

7.3.3.4.3 Application to All-Optical Switch

Figure 7.31a shows a schematic picture of an ultra-small all-optical switch, PC-SMZ, composed of a 2D-PC waveguide embedded with QDs [39], while Figure 7.31b shows a principle of a time-differential phase modulator as an ultra-fast all-optical switch [40]. A "switch-on" control pulse incident into the upper ONL arm causes Δn, leading to a phase shift $\varphi_1 = 2\pi \, (\Delta n/\lambda) \, l_{ONL}$ for a series of signal pulses, where λ is a wavelength of the signal pulse, and l_{ONL} is an ONL arm length. The phase shift φ_2 is generated similarly in the lower ONL arm by a "switch-off" control pulse. As a result, a phase-shift difference $\Delta\varphi = |\varphi_1 - \varphi_2|$ is generated at the combined Y-junction. Only when $\Delta\varphi = \pi /2$ or π (depending on the junction structure), are the signal pulses switched spatially. Time response of the Δn in the semiconductor is rapid (sub-ps) in rise but slow (sub-ns) in fall depending on a carrier lifetime in the semiconductor. However, because the Δn is excited time-differentially by the "switch-on" and "switch-off" control pulses, the Δn in the tailing slow component is cancelled, as shown in Figure 7.31b.

Prior to switch operation of the PC-SMZ, influence of a low group (V_g) velocity in the 2DPC waveguide on an enhancement effect in the ONL phase shift has been investigated using a 500-μm-long, QD-embedded straight 2DPC waveguide. In the single-mode band diagram for the 2D-PC waveguide sample used in the experiments, a low group velocity range lies in the vicinity of the band edge, as shown in Figure 7.32a. On the other hand, Figure 7.32b shows a phase shift in the ONL arm as a function of a wavelength of the single signal pulse [41]. A desirable drastic increase for the phase shift and group index (reciprocal of the V_g), as shown by a solid line

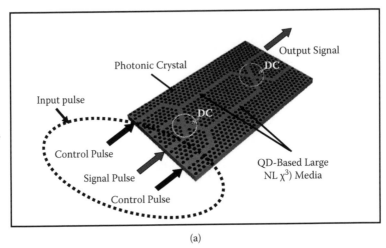

(a)

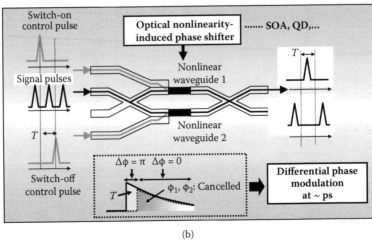

(b)

FIGURE 7.31 (a) The photonic crystal symmetrical Mach–Zehnder (PC-SMZ). (b) Principle of a time-differential phase modulator as an ultra-fast all-optical switch in the PC-SMZ.

curve, is due to a long light/matter interaction on account of the low V_g for the SP. Figure 7.33 shows an ONL-induced phase shift as a function of the control pulse energy when the signal pulse wavelength is set to 1325 nm [41]. Lower circles show a similar dependence for a 1-mm-long non-PC ridge-waveguide with similar QD. Importantly, the control pulse energy necessary for a $\pi/2$ phase shift in the 2D-PC waveguide is as low as 100 fJ and more than three orders of magnitude lower than that in the non-PC ridge waveguide. The result means that the PC/QD waveguide contributes to low-energy all-optical switch.

An optical switching demonstration of the PC-SMZ resulted in rapid rise and fall times of ~2 ps and switching window width of 20 ps. Afterward, high repetition-rate operation of the PC-SMZ was also demonstrated by using the four-pulse train with

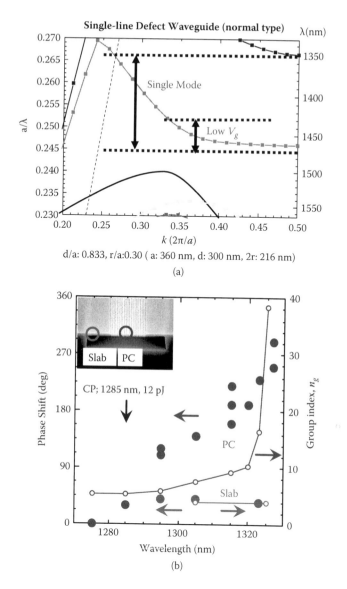

FIGURE 7.32 (a) Single-mode band diagram for the two-dimensional photonic crystal (2D-PC) waveguide and low group velocity range in the vicinity of the band edge. (b) Measured phase shift in the ONL arm as a function of a wavelength of the single signal pulse.

the period of 25 ps. Figures 7.34a and 7.34b show a schematic waveguide configuration and plan-view SEM image of the PC-SMZ sample, respectively. As shown in Figures 7.34c and 7.34d, sequential operations with repetition frequencies of 20 and 40 GHz were demonstrated, respectively [42]. Although a switching ratio in the current experiment is as low as 17%, the result suggests that the PC-SMZ is capable of a switching speed of 20 to 40 GHz. The problem of the switching ratio is now under

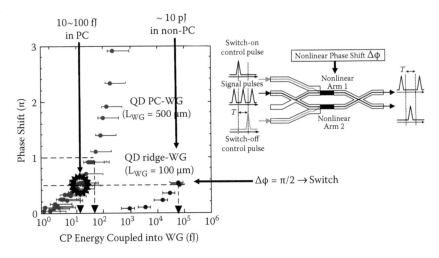

FIGURE 7.33 Optical nonlinearity (ONL)–induced phase shift as a function of the control pulse energy.

improvement. Because technologies of the QD growth and PC waveguide fabrication are advancing rapidly, performance of the integrated nanophotonics represented by the PC-SMZ and its modified devices will also improve rapidly in the near future.

7.4 SURFACE PLASMONIC MATERIALS

7.4.1 BASIC PROPERTIES OF SURFACE PLASMONICS

It is well-known solid-state physics that when it is incident on a metal, an electromagnetic wave is reflected on it below the plasma frequency ω_p. For most of the metals, ω_p lies in the visible light wave range (300 nm ~ 500 nm in wavelength). At the boundary between the metal and air (dielectric materials in general), a wave function of the light penetrates into the metal from the surface with an evanescent wave, and a dielectric constant (or permittivity) of the metal, ε_m, is negative. An electric field of the evanescent wave whose penetration depth is much less than the wavelength, in this case, induces longitudinal oscillation of the collective electron cloud in the metal surface. Surface plasmon (SP) is defined by such a light wave–coupled electron motion that propagates along the surface interface. Figure 7.35 shows behavior of the SP. It propagates along the surface as accompanying the alternate positive- and negative-induced charges, while its energy distribution is localized at the interface, from where it decays exponentially into both metal and dielectric materials. In this sense, the evanescent wave can be regarded as a near field in an optics community. In the microwave frequency range, energy propagation loss due to the electronic collective motion can be ignored, and the metal can be regarded as a perfect conductor. However, the loss in the light wave range cannot be ignored, and the propagation distance of the SP, dependent on the metal, ranges from several μm to several tens of μm on the metals such as aluminum, silver, and gold.

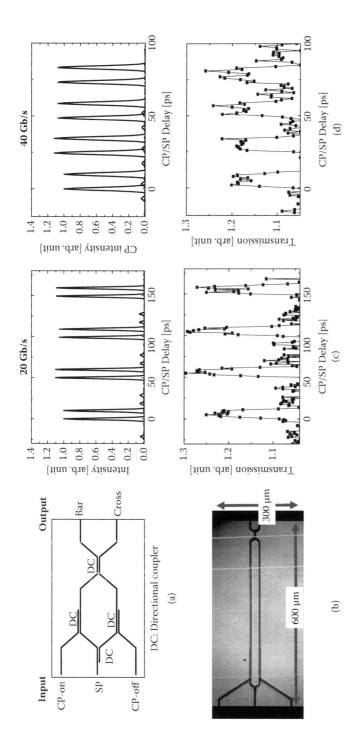

FIGURE 7.34 Optical switching demonstration of the photonic crystal symmetrical Mach-Zehnder (PC-SMZ). (a,b) Waveguide configuration and plan-view scanning electron microscopy (SEM) image of the PC-SMZ, respectively. (c,d) Sequential operations with repetition frequencies of 20 and 40 GHz, respectively.

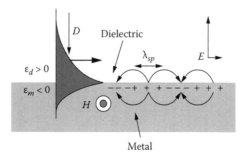

FIGURE 7.35 Behavior of the surface plasmon (SP)—the boundary between metal and dielectric materials.

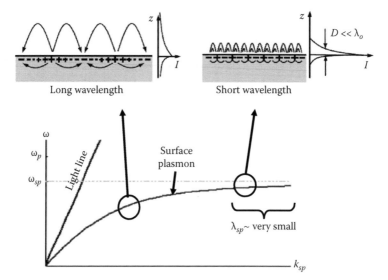

FIGURE 7.36 Band diagram of the surface plasmon (SP) showing large wave-number state with much shorter wavelength and stronger energy localization at the interface small and small wave-number state with longer wavelength and weaker energy localization.

Figure 7.36 shows an energy band diagram of the SP as well as a light line. Because the SP originated from coupling with the bulk plasmon, as mentioned above, its band diagram exhibits split into branches—that is, uniquely dispersive curves around the plasma frequency ω_p and cut-off frequency, ω_{sp}, below the ω_p. The figure shows a lower-energy branch only and a higher-energy branch lies above ω_p. Two unique features of the SP can be found from the band diagram. One is that since the band diagram of the SP does not cross the light line in the entire frequency range, it does not couple with the free-space light wave. This fact needs a particular means for exciting the SP. It will be described later in detail. The other feature is that the large wave-number state in the vicinity of ω_{sp} provides much shorter wavelength and stronger energy localization at the interface than the small wave-number state, as shown by the upper parts of Figure 7.36 [43]. This fact suggests that a size of an SP-based optical device can be compatible with that of an electronic device. Resultantly, it might be said that the SP device has excellent advantages of both ultra-fast operating speed (GHz~THz) comparable to that of the conventional photonics and also ultra-small critical dimension (10 nm ~ 100 nm) comparable to that of conventional electronics, as shown in Figure 7.37 [43]. For this reason, the SP device is a good candidate of silicon photonics in the future LSI. Here, particular attention should be paid to the SP device application, because a large wave-number range on the SP energy band is accompanied by a low group velocity. This means that a propagation speed of the ultra-fast SP short pulse is restricted to some extent. Below a critical operating speed, the SP device will attract much attention for promising application to the silicon photonics field. In the silicon LSI community, such an SP device will be expected in the era more than Moore or beyond complimentary metal-oxide-semiconductor (CMOS).

As mentioned above, the SP does not couple with the free-space light wave. Nevertheless, several means for exciting the SP mode have been reported so far. Figure 7.38 shows a typical structure and principle of coupling from high refractive index materials such as a glass (SiO_2) prism, named the Kretchmann method [44]. A metal film is deposited on the prism surface. The light wave is incident from the lower-left direction and excites the SP wave at the interface between the metal and air via a near field in the metal film (not at the interface between the metal and SiO_2). Several band diagrams in the lower picture show energies versus wave-number components parallel to the interface. Because the refractive index of the glass prism (SiO_2) is larger (~1.5) than that of the air, band curves of the light line in the glass and the SP at the metal/glass interface lie below those of the light line in air and the SP at the metal/air interface. As a result, only the band curves of the light line in the glass and the SP at the metal/air interface alone cross to each other. In this way, the SP at the metal/air interface can be excited by the light in the glass (i.e., the light incident

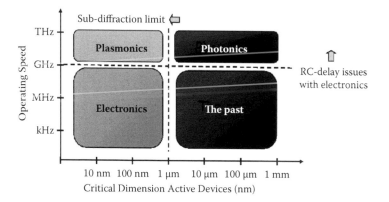

FIGURE 7.37 Excellent advantages of the surface plasmon (SP) device in terms of both ultra-fast operating speed comparable to conventional photonic devices and ultra-small critical dimension comparable to conventional electronic devices.

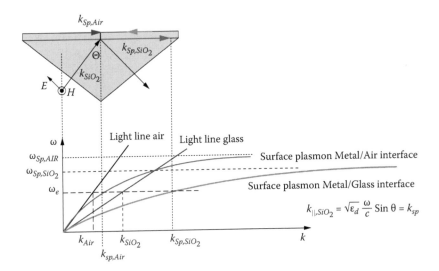

FIGURE 7.38 Typical excitation method of the surface plasmon (SP), Kretchmann method, showing structure and principle of coupling from high refractive index materials such as a glass (SiO$_2$) prism.

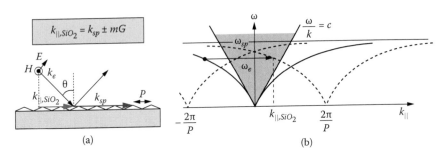

FIGURE 7.39 (a) Surface plasmon (SP) excitation method using a grating surface. (b) Its band diagram showing the SP mode as a Bloch wave.

into the glass prism). This coupling configuration has been widely used for application such as in a gas sensor on the metal, as shown later.

Another coupling method of the SP is known to use a periodic structure like a grating coupler. Figures 7.39a and 7.39b show structure and principle of the grating coupler, respectively [43]. Light incident from the top-left excites the SP on the grating surface, leaving an uncoupled reflected light from the surface. Because the grating has a periodic structure (G as an inverse lattice vector), the SP on the grating surface behaves as a Bloch wave, as shown by the extended band diagrams in Figure 7.39b. With the aid of this phenomenon, a wave-number component parallel to the surface of the light in air, $k_{//,SiO_2}$ can be coupled to that of the SP within the light cone (zone surrounded by $\omega = \pm c \cdot k$). That is why the light in air can excite the SP on the grating surface.

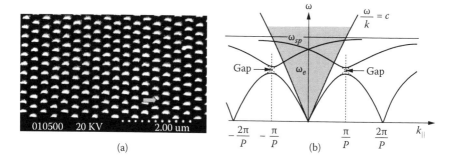

FIGURE 7.40 (a,b) Scanning electron microscopy (SEM) image of the two-dimensional (2D) periodic metal and band diagram of the surface plasmon (SP) in the extended Brillouin zone, respectively.

The SP on the surface with a 2D periodic structure behaves like a light wave in the 2D-PC structure. Figures 7.40a and 7.40b show an SEM image of the 2D periodic metal and band diagram of the SP in the extended Brillouin zone, respectively [43]. It is noted that an SP band gap appears at the zone boundary in the same manner as the PC. This means that an analogy similar to the PC is almost valid for the SP except for necessity of a special consideration on the propagation loss in the metal even in the geometrically perfect periodic structure. One of the analogies from the PC is a means of an SP guiding along a missing line defect, and, actually, several experimental demonstrations have been reported so far. A technique of the SP waveguide is a base of a metal optics proposed afterwards, as shown in Figure 7.46 [45]. Another important property of this structure is the capability to couple with light in the free space. In the band diagram in Figure 7.40b, the light in the free space (within the light cone) can couple with the SP with wave-numbers less than ω/c, where c is a velocity of light in air. In this way, the light is emitted upward in the air or absorbed downward into the metal film.

7.4.2 Photonic Applications of Surface Plasmonics

To date, the SP has already been proposed to a variety of applications. SPs can be categorized into four regions as follows: The first is a highly sensitive gas/molecule sensor using a flat metal surface on the glass [44]. A coupling condition between an incident light and SP on it is sensitively modulated by a small amount of adsorption of the gas/molecule, which induces a shift in the band diagram due to a strong energy localization of the SP on the surface. The second is an application using an extraordinary transmission of light incident on a nano-scale via-hole on a metal/dielectric material [46]. A single nano-hole on the periodically corrugated metal/dielectric material enables conversion of the incident light beam into the regenerated fine beam beyond a diffraction limit via the SP effect, thus applying to a highly sensitive photodetector whose size is compatible in the LSI [47]. An array of nano-scale air-holes on a metal/dielectric material can be applied to an optical filter with high precision. The third is a channel waveguide of the SP on the substrate surface

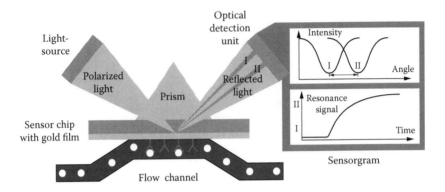

FIGURE 7.41 A gas/molecule sensor chip using the surface plasmon (SP).

with strip metal [48] or grooved metal patterns [49]. The fourth is an optical dipole nano-antenna with a nano-scale feed gap [50]. Light wave incident on the metal dipole at a resonant condition induces an extremely high electric field within the feed gap on the dipole antenna. It will induce strong light/matter interaction (i.e., large optical nonlinearity in the organic or inorganic nano-materials inserted into the feed gap). In this section, a typical example in each application is shown.

Figure 7.41 shows a schematic picture of a gas/molecule sensor chip using the SP [51]. A flow channel of the gas/molecule is attached to the surface of a gold film deposited on the prism. Polarized light is obliquely incident on the gold film from the back of the prism, while intensity of the reflected light is detected as a function of an oblique incident angle. When the incident light is coupled to the SP, an intensity of the reflected beam shows a sharp absorption peak as the incident angle is changed. The sensor chip uses the high sensitivity of the absorption peak shift caused by a small amount of adsorption of the gas/molecule on the gold film. The chip has already been commercialized in the medical and other industrial communities.

Figure 7.42 shows an example of an extraordinary transmission of the light incident on the nano-scale via-hole on the metal/dielectric materials [46,47]. As shown in Figure 7.42a, a light wave incident to a nano-scale via-hole in a dielectric slab decays exponentially at the outside of the slab. However, if the outside surface is covered with a corrugated metal film, or the slab itself is composed of a metal with a corrugated surface, as shown by the SEM photograph in Figure 7.42b, the incident light propagates through the hole with an extraordinary transmission rate [52]. This is explained by reradiation of the light as a result of the SP on the grating (corrugation) excited by the transmitted light through the hole, as shown in Figure 7.42c. The reradiation of the light is based on a principle of coupling between the free space light and the SP on the periodic structure surface, as shown in Figure 7.40b. As shown by transmission spectra in Figure 7.42d, a peak height depends on the shape of the apex in the corrugation. A sharp apex exhibits a strong radiation, while a round apex shows a weak one [53]. A corrugated metal film with concentric circulars is called *bull's eye* and has been applied to a coupler of the incident light, called *plasmon antenna*, into a silicon photodiode in the LSI, as shown in Figure 7.43 [47].

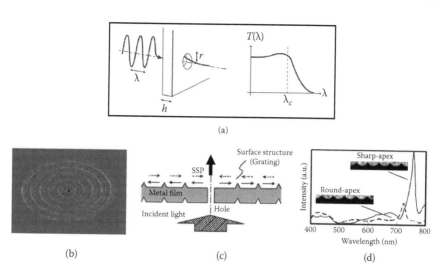

(a)

(b)   (c)   (d)

FIGURE 7.42 Extraordinary transmission of incident light due to surface plasmon (SP) mode excitation on the metal/dielectric materials. (a) Light wave incident to a nano-scale via-hole, decaying exponentially at the outside of the slab. (b) Metal with a corrugated surface for SP excitation. (c) Reradiation of the light as a result of the SP on the grating. (d) Peak height depending on the shape of the apex in the corrugation.

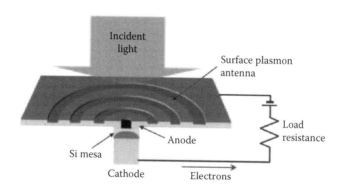

FIGURE 7.43 Corrugated metal film with concentric circulars called *bull's eye*, applicable to a coupler of incident light into a silicon photodiode in the Large Scale Integrated Circuit (LSI).

The photodiode is attached with the Ag/Cr metal film with a SiO_2 spacer. It is noted that an air-hole diameter as small as 300 nm, which is otherwise unable to transmit the light due to the diffraction limit, enables transmission of the light on account of the reradiation by means of the SP effect mentioned above. The radiation mechanism is the same as the case in Figure 7.42c. As a result, a light with a broad beam incident from upward is converted into a narrow beam with extremely high intensity, leading to the high sensitivity and high operation speed photo-detector.

One of the promising applications of the SP is an SP waveguide. Figure 7.44a shows a variety of the wedge-shaped groove waveguide patterns on the metal

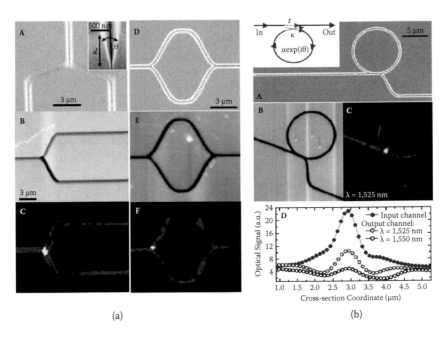

FIGURE 7.44 A variety of the wedge-shaped groove-type surface plasmon (SP) waveguide patterns on the metal film. (a) Y-branch and Mach–Zehnder-type interferometric waveguides. (b) Ring-coupled double-bent SP waveguide.

film [49]. The SP propagates along a bottom edge of the groove, as shown by a cross-sectional view of the wedge in the SEM photograph, due to a high effective refractive index effect. The light beam can be observed to propagate along the Y-branch and Mach-Zehnder-type interferometric waveguides. Figure 7.44b shows a similar observation of light propagating along the ring-coupled double-bent waveguide. The results suggest the possibility of an SP-based ultra-small integrated circuit on the metal film.

As an application of the SP in the fourth category, a concept of an optical dipole nano-antenna has been proposed [50]. Figure 7.45 shows a fundamental structure (left) and several items for possible applications (right). In the left figure, a split metal bar with a narrow feed gap at the center and total length of a half-wavelength plays a role of a dipole antenna, similar to the Yagi antenna. The dipole induces negative and positive charges at each end of the elements, as shown in the figure. When the dipole absorbs the light with a resonant condition, the energy of the incident light is almost entirely localized in the small feed gap so that a field in the gap is extremely enhanced. If several kinds of materials are inserted into the feed gap, strong light/ matter interaction due to the high field produces unique applications. Insertion of DNA of protein or semiconductor produces a highly sensitive biosensor or photodetector, while insertion of nonlinear medium or quantum emitter produces a low-energy optical switch or high-efficiency single-photon source. Their experimental verifications are under way.

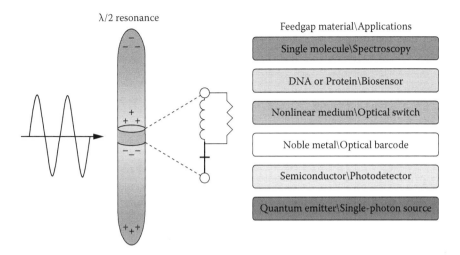

FIGURE 7.45 Fundamental structure (left) of an optical dipole nano-antenna and several items for possible applications (right).

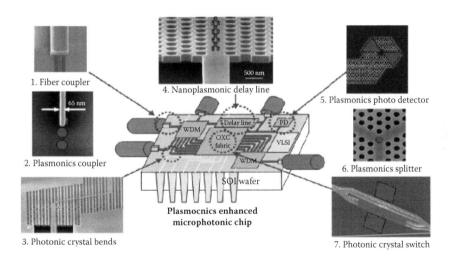

FIGURE 7.46 A surface plasmon–enhanced microphotonic chip.

As shown by the fundamental properties and examples of applications in SP mentioned above, it is predicted that the SP will situate a solid position of a promising photonic integrated device. Figure 7.46 shows a schematic picture of an SP-enhanced microphotonic chip [54], where a conventional photonic crystal-based Y branch, coupler, optical switch, detector, and other components are improved in performance by partial insertion of metal units for the SP components, thus providing a photonic integrated circuit much smaller than ever, comparable with an electronic circuit, as shown in Figure 7.37. There are still potentials for advancement in the photonic integrated technologies.

7.5 NEGATIVE INDEX MATERIALS

7.5.1 BASIC PROPERTIES OF NEGATIVE INDEX MATERIALS

A negative index material (NIM) is one of the three categories specified in the meta-material. Metamaterials are defined by artificial composites tailored for specific electromagnetic properties that are not found in nature and are not observed in constituent materials. Figure 7.47 shows a Veselago classification with four territories involving a metamaterial territory regarding electrical permittivity (ε) and magnetic permeability (μ). The first quadrant involves abundant naturally occurring materials ($\varepsilon > 0$, $\mu > 0$) represented by common transparent dielectrics, where electromagnetic (EM) waves propagate in a sinusoidal function. The second quadrant involves the electrical plasma state ($\varepsilon < 0$, $\mu > 0$) like in a metal at an optical wavelength, where an EM wave behaves in an evanescent function within its penetrate depth. They are also called *electrical metamaterials*. These two materials have no magnetic response—that is, $\mu = 1$ in the optical range. They are dominated electromagnetically by the right-handed system regarding an electric field (E), magnetic field (H), and wave vector (k), so that they are called right-handed materials (RHMs). On the other hand, materials in the third and fourth quadrants are called left-handed materials (LHMs) where they are dominated by the left-handed system in E, H, and k relations. The negative index materials are involved in the third quadrant and exhibit very artificial properties—that is, negative ε and μ. Finally, the fourth quadrant shows magnetic metamaterials where $\varepsilon > 0$, $\mu < 0$. In this section, the negative index materials and their behaviors in the optical range are shown.

One of the physical models for easily understanding the generation of the negative ε or μ is to consider vibration of a spring with a weight subject to a Lorenz-type resonance, as shown in Figure 7.48 [55]. The figure in the center shows spectra of ε or μ as a function of a normalized angular frequency near the resonance frequency of

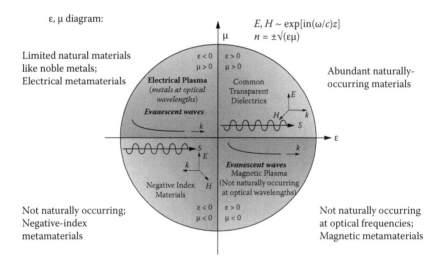

FIGURE 7.47 Veselago classification with four territories regarding electrical permittivity (ε) and magnetic permeability (μ).

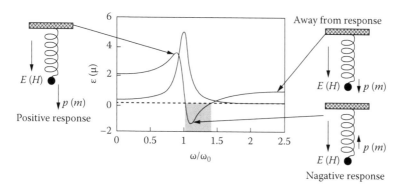

FIGURE 7.48 Spring with a weight subject to a Lorenz-type resonance, showing spectra of ε or μ as a function of a normalized angular frequency near the resonance frequency of the system.

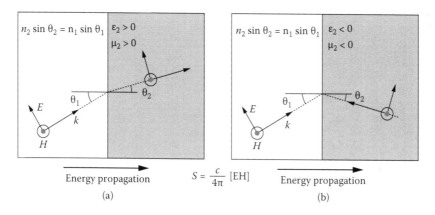

FIGURE 7.49 Strange behaviors between phase and group velocities in the frequency range of the negative ε and μ. (a) Ordinary behavior in the conventional positive index material. (b) Extraordinary behavior in the negative index material.

the system. Typical three-motion points are depicted by the spring vibration models. In a frequency range just below the resonance, a motion (ρ) of the weight oscillates in phase with an external field (E or H) resulting in a positive response (ε or $\mu > 0$), while it oscillates out of phase with the external field (E or H) just above the resonance, resulting in the negative response (ε or $\mu < 0$). Far away from the resonance, the weight moves in phase with E or H but in a small amplitude. In the actual NIM, development of a simultaneous condition for both negative ε and μ in as wide a range as possible is the subject in an optical range.

In the frequency range of the negative ε and μ, strange behaviors occur between phase and group velocities, as shown in Figure 7.49 [55]. In the NIM, a direction of a group velocity determining the energy propagation differs by 180° from that of the phase velocity with an extraordinary phenomenon of a negative refractive angle, as shown in Figure 7.49b. Here, Figure 7.49a shows an ordinary behavior in the conventional positive index material (PIM). With deeper insight into the refraction/reflection surface in the PIM/NIM system, as shown in Figure 7.50, a calculated electromagnetic distribution shows that energy of an incident optical beam is refracted mostly and reflected a little. This is one of the unique features in the NIM. The result leads to the proposal of an application to the cloaking materials, as will be mentioned later.

Then, what kind of structures could be the NIM? Typical structures for the NIM proposed in the range from microwave to optical regions are shown here. Figure 7.51

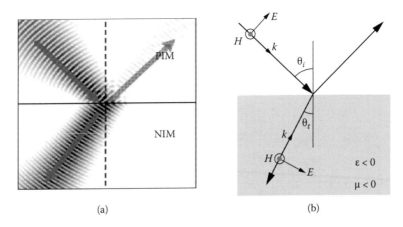

(a)   (b)

FIGURE 7.50 Calculated electromagnetic distribution showing that energy of an incident optical beam is refracted mostly and reflected a little.

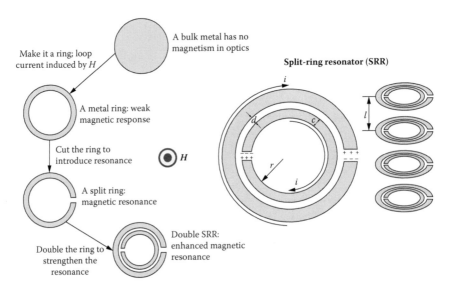

FIGURE 7.51 A split-ring resonator (SRR) as the negative index material (NIM) and a series of metallic structures; bulk metal having no magnetism in optics, metal ring having weak magnetism response, split ring having larger magnetic response, and double split-ring resonator (SRR) exhibiting enhanced magnetic resonance.

shows schematic pictures of a split-ring resonator (SRR) as the NIM proposed theoretically by Pendry [4] and demonstrated experimentally by Smith et al. [56]. A series of metallic structures have changed from a bulk metal having no magnetism in optics, via a metal ring having weak magnetism response, and a split ring having a larger magnetic response, up to a double SRR exhibiting enhanced magnetic resonance. The split ring forms a resonator with a coupling of an inductance along the ring-shaped rod and a capacitance at a gap in the ring. Experimentation of the SRR with ~mm size resulted in the calculated negative μ in microwave [56]. Simultaneous achievement of $\varepsilon < 0$ and $\mu < 0$ requires a combination of metal rods for negative ε and SRR for negative μ, as shown in Figure 7.52. Sizes of both structures are optimized for matching two negative-ε and -μ regions, as shown in the figure.

Interpretation of NIM structures from the microwave to optical region was achieved by introducing what is called a *fishnet* structure, as shown in Figure 7.53 [57]. In the figure, bars ($\varepsilon < 0$) and rods ($\mu < 0$) both composed of metal-oxide-metal sandwich elements are combined in-plane to form a 2D periodic network like a fishnet, as shown in Figures 7.53a through 7.53c. Using this structure, negative index n' of -2 is theoretically calculated at a 1.45-μm-wavelength, as shown in Figure 7.53d. Length and width of bars and rods are on the order of several hundred nm.

Figure 7.54 shows summarized photographs and schematic pictures of some periodic structures for NIMs reported so far. 2D and 1D arrays of SRR in the microwave range are shown in Figures 7.54a and 7.54b, respectively. In Figures 7.54c and 7.54d, two kinds of 2D arrays of NIM, both modified from the fishnet structure, are shown. In Figure 7.54c, a nano-rod element with an aluminum oxide (Al_2O_3)

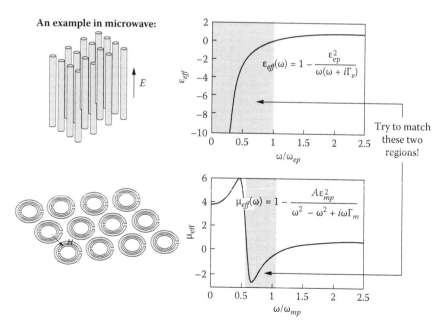

FIGURE 7.52 Simultaneous achievement of e < 0 and m < 0 with a combination of metal rods for negative ε and split-ring resonator (SRR) for negative μ.

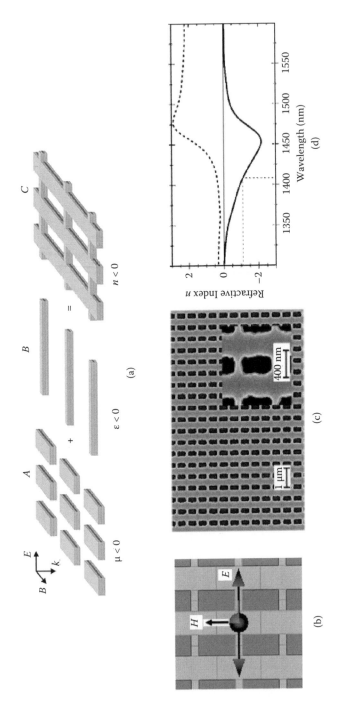

FIGURE 7.53 Fishnet structure with mixture of bars (ε < 0) and rods (μ < 0) both composed of metal-oxide-metal sandwich elements, combined in-plane to form a two-dimensional (2D) periodic network like a fishnet.

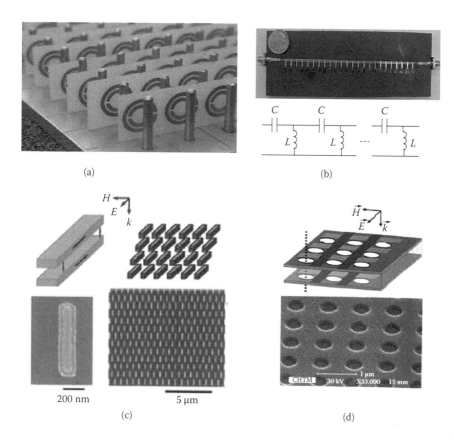

FIGURE 7.54 Summarized photographs and schematic pictures of some periodic structures for negative index materials (NIMs). (a,b) Two-dimensional (2D) and one-dimensional arrays of split-ring resonator (SRR) in the microwave range, respectively. (c,d) Two kinds of 2D arrays of NIM, both modified from the fishnet structure.

thin film sandwiched by a gold (Au) metal film is periodically arrayed to form the modified fishnet [58]. In Figure 7.54d, on the other hand, a similar Au/Al$_2$O$_3$/Au composite thin film is perforated with a periodic round-void array to form another modified fishnet [59]. Representative experimental results using these samples are summarized in Table 7.3 [55]. At a wavelength of 1.8 μm, a negative refractive index n' of −4 was reported by using a nano-fishnet with round voids, while n' of −1 was reported at a wavelength of 1.4 μm by using a nano-fishnet with rectangular voids.

7.5.2 Photonic Applications of Negative Refractive Index Materials

Representative applications of the NIM reported so far are shown in this section. One of the most important applications of the NIM is a perfect lens, as shown in Figure 7.55 [60]. The figure shows a comparison of geometrical ray optics (upper) and simulated optical field distributions (lower) from an object to its image with (a) a conventional convex lens using the PIM and (b) perfect flat lens with the NIM.

TABLE 7.3

Representative Experimental Results Using Negative Index Material (NIM) Samples in the Light Wave Range

Year and Research Group	First Time Posted and Publication	Refractive Index, n'	Wavelength λ	Figure of Merit $F = nV''$	Structure Used
2005					
Purdue	April 13 (2005) arXiv:physics/0504091 Opt. Lett. (2005)	–0.3	1.5 µm	0.1	Paired nano-rods
University of New Mexico and Columbia	April 28 (2005) arXiv:physics/0504208 Phys.Rev. Lett. (2005)	–2	2.0 µm	0.5	Nano-fishnet with round voids
2006					
University of New Mexico and Columbia	J. of OSA B (2006)	–4	1.8 µm	2.0	Nano-fishet with round voids
Karlsruhe and Iowa State University	Opt. Lett. (2006)	–1[a]	1.4 µm	3.0	Nano-fishnet with rectangular voids
Karlsruhe and Iowa State University	In press	–0.6	780 nm	0.5	Nano-fishnet with rectangular voids

[a] Minimum n' is –2 occurring at 1.45 µm with $F = 1.5$.

A classical imaging system shows that special resolution on the image is determined by how many optical Fourier components emitted from the object will be reproduced on the image. In the conventional system in Figure 7.55a, only propagating wave components are focused on the image through the convex lens, and an evanescent wave component is omitted because it is decayed much farther away from the lens, as shown in the figure. As a result, the object cannot be restored on the image completely. The perfect lens using the NIM, on the other hand, enables us to transmit all the optical components attached to the object into the imaging point because the perfect lens is composed of a flat lens. Therefore, it can be closed within the evanescent distance near the object. In this way, the object can be perfectly restored on the image, as shown in Figure 7.55b.

In reality, the decay distance of the evanescent wave is on the order of less than 100 nm, so that the perfect lens is effective in the near-field optics only. Some examples of the perfect lens applications are shown in Figures 7.56a through 7.56c [55]. Figure 7.56a is an application to the remote SERS (surface-enhanced Raman

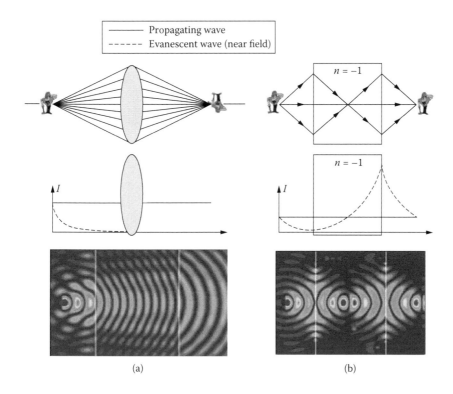

(a)   (b)

FIGURE 7.55 Explanation of perfect lens using geometrical ray optics (upper) and simulated optical field distributions (lower). (a) Conventional convex lens using the positive index material (PIM). (b) Perfect flat lens with the negative index material (NIM).

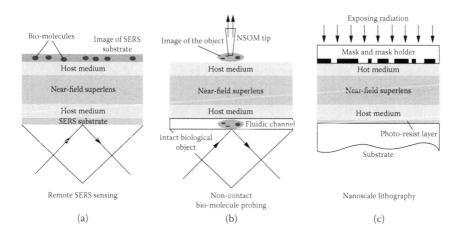

(a)   (b)   (c)

FIGURE 7.56 Application of a perfect lens. (a) Remote SERS (surface-enhanced Raman scattering) sensing. (b) Intact biological object in the fluidic channel between the prism and the flat lens. (c) Nano-scale lithography using the flat lens, exhibiting focused nano-patterns in the photo-resist layer between the flat lens and the substrate.

scattering) sensing. Here, a SERS substrate on the prism surface is focused on top of the flat lens where a biomolecule is involved, while Figure 7.56b shows that the intact biological object in the fluidic channel between the prism and the flat lens is focused on top of the flat lens where the image of the object is probed with a Near-Field Scanning Optical Microscope (NSOM) tip as noncontact biomolecule probing. On the other hand, Figure 7.56c is nano-scale lithography using the flat lens. Optical beam incident on the mask separately focuses nano-patterns in the photo-resist layer between the flat lens and the substrate. Some of them have already been experimentally verified.

Another unique application of the NIM is a concept of cloaking. An object is cloaked when it is wrapped with the NIM because an optical beam incident on the NIM is not reflected but is transmitted completely through the NIM. The result means that the object is not observed by the outer observer. Figures at the right side in Figure 7.57 show calculated optical field distributions in and around the object wrapped with the ring-shaped NIM [61]. The top, middle, and bottom show an ideal case, case for reduced parameters, and experimental data in the microwave range, respectively. For the ideal case, incident beam flux remains almost undisturbed after propagating through the NIM. This means the NIM-wrapped object is not observed, so that it is transparent.

The last example is an application of the NIM to beam steering in a microwave antenna system with a wide steering angle. Figure 7.58a shows a schematic of an antenna configuration with a periodic array electrode with 17 unit cells [62]. As shown in Figure 7.58b, the substrate involves multilayers of bottom glass substrate, middle porous Teflon film impregnated with liquid crystal, and top Teflon film printed with antenna patterns. The top antenna pattern is designed to exhibit PIM or NIM

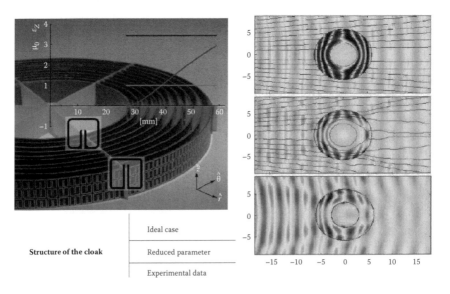

FIGURE 7.57 Left: Structure of the cloak. Right: Calculated optical field distributions in and around the object wrapped with the ring-shaped negative index materials (NIM).

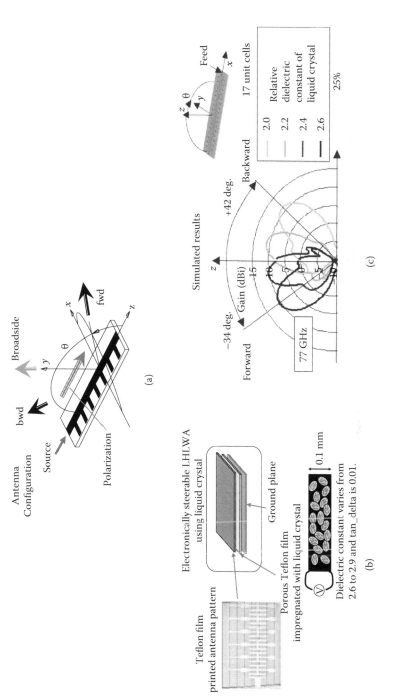

FIGURE 7.58 Application of negative index material (NIM) to beam steering in a microwave antenna system with a wide steering angle. (a) Antenna structure. (b) Cross-sectional structure of antenna. (c) Radiation patterns as a refractive index is changed between 2 and 2.6 in the antenna system.

depending on the refractive index of the liquid crystal changed by an external electric field. As a result, an incident beam is steered forward or backward within a steering range of about ± 40° due to the principle of Cherenkov radiation. Figure 7.58c shows radiation patterns as a refractive index is changed between 2 and 2.6. This technique is practically applied to an antenna system in the car safety system.

REFERENCES

1. Y. Arakawa and H. Sakaki: Multidimensional quantum well laser and temperature dependence of its threshold current, *Appl. Phys. Lett.*, 40, 939 (1982).
2. E. Yablonovitch: Inhibited spontaneous emission in solid-state physics and electronics, *Phys. Rev. Lett.*, 58, 2059–2062 (1987).
3. T. W. Ebbesen, H. J. Lezec, H. F. Ghaemi, T. Thio, and P. A. Wolff: *Nature*, 391, 667–669 (1998).
4. J. B. Pendry, A. J. Holden, W. J. Stewart, and I. Youngs: *Phys. Rev. Lett.*, 76, 4773 (1996). J. B. Pendry, A. J. Holden, D. J. Robbins, and W. J. Stewart: *IEEE Trans. Micr. Theory and Techniques*, 47, 2075 (1999).
5. T. F. Krauss, R. M. De La Rue, and S. Brand: Two-dimensional photonic-bandgap structures operating at near-infrared wavelengths, *Nature*, 383, 699–702 (1996).
6. T. Baba, N. Fukaya, and J. Yonekura: Observation of light propagation in photonic crystal optical waveguides with bends, *Electron. Lett.*, 35, 654–655 (1999).
7. S. Noda, A. Chutinan, and M. Imada: Trapping and emission of photons by a single defect in a photonic bandgap structure, *Nature*, 407, 608–610 (2000).
8. S. G. Johnson, S. Fan, P. R. Villeneuve, and J. D. Joannopoulos: Guided modes in photonic crystal slabs, *Phys. Rev. B*, 60, 5751–5758 (1999).
9. M. Loncar, T. Doll, J. Vuckovic, and A. Scherer: Design and fabrication of silicon photonic crystal optical waveguides, *J. Lightwave Technol.*, 18, 1402–1411 (2000).
10. M. Notomi, A. Shinya, K. Yamada, J. Takahashi, C. Takahashi, and I. Yokohama: Structural tuning of guiding modes of line-defect waveguides of silicon-on-insulator photonic crystal slabs, *IEEE J. Quantum Electron.*, 38, 736–742 (2002).
11. Y. Sugimoto, Y. Tanaka, N. Ikeda, H. Nakamura, K. Kanamoto, S. Ohkouchi, Y. Watanabe, K. Inoue, and K. Asakawa: Fabrication and characterization of photonic crystal based symmetric Mach-Zehnder (PC-SMZ) structures based on GaAs membrane slab waveguides, *IEEE J. Selected Areas in Comm.*, 23, 1308–1314 (2005).
12. For example, *Photonic Crystals—Physics, Fabrication and Applications*, edited by K. Inoue and K. Ohtaka, Springer Series in Optical Sciences Berlin, Heidelberg, Germany (2004).
13. Courtesy of H. Yamada originated from Figure 2 in Chapter 15, *Photonic Crystal Technology—Scenario of Industrialization*, edited by S. Kawakami, CMC Books (2005) (in Japanese).
14. S. Kawakami, T. Kawashima, and T. Sato: Mechanism of shape formation of three-dimensional nanostructures by bias sputtering, *Appl. Phys. Lett.*, 74, 463 (1999).
15. N. Ikeda, Y. Sugimoto, Y. Tanaka, K. Inoue, Y. Watanabe, and K. Asakawa: Studies on key nano-fabrication processes for GaAs-based air-bridge-type two-dimensional photonic-crystal slab waveguides, *Semiconductor Sci. Technol.*, 22, 1–9 (2007).
16. Y. Sugimoto, Y. Tanaka, N. Ikeda, Y. Nakamura, K. Asakawa, and K. Inoue: Low propagation loss of 0.76 dB/mm in GaAs-based single-line-defect two-dimensional photonic crystal slab waveguides up to 1 cm in length, *Opt. Express* 12, 1090–1096 (2004).
17. P. I. Borel, L. H. Frandsen, A. Harpøth, J. B. Leon, H. Liu, M. Kristensen, W. Bogaerts, P. Dumon, R. Baets, V. Wiaux, J. Wouters, and S. Beckx: Bandwidth engineering of photonic crystal waveguide bends, *Electron. Lett.*, 40, 1263–1264 (2004).

18. J. S. Jensen, O. Sigmund, L. H. Frandsen, P. I. Borel, A. Harpøth, and M. Kristensen: Topology design and fabrication of an efficient double 90-degree photonic crystal waveguide bend, *IEEE Photonics Technol. Lett.*, 17, 1202–1204 (2005).
19. Y. Watanabe, N. Ikeda, Y. Sugimoto, Y. Takata, Y. Kitagawa, A. Mizutani, N. Ozaki, and K. Asakawa: Topology optimization of waveguide bends with wide, flat bandwidth in air-bridge-type photonic crystal slabs, *J. Appl. Phys.*, 101, 113108 (2007).
20. Y. Watanabe, Y. Sugimoto, N. Ikeda, N. Ozaki, A. Mizutani, Y. Takata, Y. Kitagawa, and K. Asakawa: Broadband waveguide intersection with low crosstalk in two-dimensional photonic crystal circuits by using topology optimization, *Opt. Exp.*, 14, 9502–9507 (2006).
21. K. Inoue, N. Kawai, Y. Sugimoto, N. Carlsson, N. Ikeda, and K. Asakawa: Observation of small group velocity in two-dimensional AlGaAs photonic crystal slabs, *Phys. Rev. B*, 65, 121308 (2002).
22. O. Painter, R. K. Lee, A. Scherer, A. Yariv, J. D. O'Brien, P. D. Dapkus, and I. Kim: Two-dimensional photonic band-gap defect mode laser, *Science* 284, 1819–1821 (1999).
23. B. -S. Song, S. Noda, and T. Asano: Photonic devices based on in-plane hetero photonic crystals, *Science*, 300, 1537 (2003).
24. B. -S. Song, S. Noda, T. Asano, and Y. Akahane: Ultra-high-Q photonic double heterostructure nano-cavity, *Nature Materials*, 4, March (2005).
25. S. Ohkouchi, Y. Nakamura, H. Nakamura, N. Ikeda, Y. Sugimoto, and K. Asakawa: Selective growth of high quality InAs quantum dots in narrow regions using in situ mask, *J. Crystal Growth*, 293, 57–61 (2006).
26. S. Nakamura, A. Watanabe, X. Wang, N. Ikeda, Y. Sugimoto, N. Ozaki, Y. Watanabe, and K. Asakawa: Optical flip-flop based on coupled ultra-small Mach-Zehnder all-optical switches, *Proc. of International Conference on Optical Fiber Communication 2008*, San Diego, CA, February (2008).
27. Y. Nakamura, N. Ikeda, S. Ohkouchi, Y. Sugimoto, H. Nakamura, and K. Asakawa: Two-dimensional InGaAs quantum-dot arrays with periods of 70–100nm on artificially prepared nanoholes, *Jpn. J. Appl. Phys.*, 43, L 362–L 364 (2004).
28. N. Ozaki, Y. Takata, S. Ohkouchi, Y. Sugimoto, Y. Nakamura, N. Ikeda, and K. Asakawa: Selective area growth of InAs quantum dots with a metal mask towards optical integrated circuit devices, *J. Crystal Growth*, 301–302, 771–775 (2007).
29. S. Ohkouchi, Y. Nakamura, H. Nakamura, and K. Asakawa: Control of InAs quantum dot emission wavelengths in narrow regions by selective formation of GaInAs covered layers grown with in situ mask, *Jpn. J. Appl. Phys.*, 44, 5677–5679 (2005).
30. N. Ozaki, Y. Takata, S. Ohkouchi, Y. Sugimoto, N. Ikeda, and K. Asakawa: Selective-area-growth of InAs-QDs with different absorption wavelengths via developed metal-mask/MBE method for integrated optical devices, *Appl. Surf. Sci.*, 254, 7968–7971 (2008).
31. S. Kohmoto, H. Nakamura, T. Ishikawa, and K. Asakawa: Site-controlled self-organization of individual InAs quantum dots by scanning tunneling probe-assisted nanolithography, *Appl. Phys. Lett.*, 75, 3488–3490 (1999).
32. S. Ohkouchi, Y. Sugimoto, N. Ozaki, H. Ishikawa, and K. Asakawa: Molecular beam epitaxial growth of site-controlled InAs quantum dot arrays using templates fabricated by the nano-jet probe method, *Physica E*, 40 1794–1796 (2008). S. Ohkouchi, Y. Nakamura, H. Nakamura, and K. Asakawa: InAs nano-dot array formation using nano-jet probe for photonics applications, *Jpn. J. of Appl. Phys.*, 44, 5777–5780 (2005).
33. N. Koguchi, K. Ishige, and S. Takahashi: *J. Vac. Sci. & Technol. B*, 11, 787 (1993). T. Mano, S. Tsukamoto, N. Koguchi, H. Fujioka, and M. Oshima: *J. Cryst. Growth*, 227–228, 1069 (2001).
34. S. Ohkouchi, Y. Sugimoto, N. Ozaki, H. Ishikawa, and K. Asakawa: Selective growth of stacked InAs quantum dots by using the templates formed by the nano-jet probe, *Appl. Surf. Sci.*, 254, 7821–7823 (2008).

35. K. Asakawa, Y. Sugimoto, Y. Watanabe, N. Ozaki, A. Mizutani, Y. Takata, Y. Kitagawa, H. Ishikawa, N. Ikeda, K. Awazu, X. Wang, A. Watanabe, S. Nakamura, S. O. K. Inoue, M. Kristensen, O. Sigmund, P. I. Borel, and R. Baets: Photonic crystal and quantum dot technologies for all-optical switch and logic device, *New J. Phys.*, 8, 1–26 (2006).

36. H. Nakamura, S. Nishikawa, S. Kohmoto, K. Kanamoto, and K. Asakawa: Optical non-linear properties of InAs quantum dots by means of transit absorption measurements, *J. Appl. Phys.*, 94, 1184–1189 (2003).

37. H. Nakamura, K. Kanamoto, Y. Nakamura, S. Ohkouchi, H. Ishikawa, and K. Asakawa: Nonlinear optical phase shift in InAs quantum dots measured by a unique two-color pump/probe ellipsometric polarization analysis, *J. Appl. Phys.*, 96, 1425–1434 (2004).

38. H. Nakamura, S. Nishikawa, S. Kohmoto, K. Kanamoto, and K. Asakawa: Optical non-linear properties of InAs quantum dots by means of transit absorption measurements, *J. Appl. Phys.*, 94, 1184–1189 (2003).

39. Y. Sugimoto, N. Ikeda, N. Carlsson, K. Asakawa, N. Kawai, and K. Inoue: Fabrication and characterization of different types of two-dimensional AlGaAs photonic crystal slab, *J. Appl. Phys.*, 91, 922–929 (2002).

40. K. Tajima, All-optical switch with switch-off time unrestricted by carrier lifetime, *Jpn. J. Appl. Phys.*, 32, L1746–L1748 (1993).

41. H. Nakamura, Y. Sugimoto, K. Kanamoto, N. Ikeda, Y. Tanaka, Y. Nakamura, S. Ohkouchi, Y. Watanabe, K. Inoue, H. Ishikawa, and K. Asakawa: Ultra-fast photonic crystal/quantum dot all-optical switch for future photonic network, *Opt. Exp.*, 12, 6606–6614 (2004).

42. Y. Kitagawa, N. Ozaki, Y. Takata, N. Ikeda, Y. Watanabe, A. Mizutani, Y. Sugimoto, and K. Asakawa: Measurements of optical non-linearity induced phase-shifts of signal pulse with repetitive control pulses in photonic crystal/quantum dot waveguide, *Proceedings of LEOS Annual Meeting*, October (2007).

43. Courtesy of V. M. Shalaev at Purdue University from 2006.10.05-ece695s-I09.

44. E. Kretschmann: *Z. Phys.*, 216, 313–324 (1971).

45. Cary Gunn, CMOS photonics for high-speed interconnects, *IEEE MICRO* (2006).

46. T. Ishii, J. Fujikata, and K. Ohashi: Large optical transmission through a single sub-wavelength hole associated with a sharp-apex grating, *Jpn. J. Appl. Phys.*, 44, L 170–L 172 (2005).

47. T. Ishi, J. Fujikata, K. Makita, T. Baba, and K. Ohashi: Si nano-photodiode with a surface plasmon antenna. *Jpn. J. Appl. Phys.*, 44, L364–L366 (2005).

48. E. Verhagen, A. L. Tchebotarev, and A. Polman: Erbium luminescence imaging of infra-red surface plasmon polaritons, *Appl. Phys. Lett.*, 88, 121121 (2006).

49. S. I. Bozhevolnyi1, V. S. Volkov, E. Devaux, J. -Y. Laluet, and T. W. Ebbesen: Channel plasmon sub-wavelength waveguide components including interferometers and ring resonators, *Nature*, 440, 508–511 (2006).

50. M. L. Brongersma, Engineering optical nanoantennas, *Nature Photonics*, 2, 270–272 (2008).

51. Institute of Chemistry, Hebrew University of Jerusalem, http://chem.ch.huji.ac.il/~eugeniik/spr.htm#reviews.

52. C. Genet1 and T. W. Ebbesen: Light in tiny holes, *Nature*, 445, 39–46 (2007).

53. D. E. Grupp, H. J. Lezec, T. W. Ebbesen, K. M. Pellerin, and T. Thioa: Crucial role of metal surface in enhanced transmission through sub-wavelength apertures, *Appl. Phys. Lett.*, 77, 1569–1571 (2000).

54. Courtesy of Y. A. Vlasov at IBM, T. J. Watson Research Center, Yorktown Heights, New York, USA.

55. Courtesy of V. M. Shalaev at Purdue University.

56. D. R. Smith, J. B. Pendry, and M. C. K. Wiltshire: Metamaterials and negative refractive index, *Science*, 305, 788–792 (2004).

57. G. Dolling, C. Enrich, M. Wegener, C. M. Soukoulis, and S. Linden: Simultaneous negative phase and group velocity of light in a metamaterial, *Science,* 312, 892–894 (2006). S. Linden, C. Enrich, G. Dolling, M. W. Klein, J. Zhou, T. Koschny, C. M. Soukoulis, S. Burger, F. Schmidt, and M. Wegener: Photonic metamaterials, magnetism at optical frequencies, *IEEE J. Selected Topics in Quantum Electron.,* 12(6), November/ December (2006).

58. V. M. Shalaev, W. Cai, U. K. Chettiar, H. -K. Yuan, A. K. Sarychev, V. P. Drachev, and A.r V. Kildishev: Negative index of refraction in optical metamaterials, *Opt. Lett.,* 30, 3356–3358 (2005).

59. S. Zhang, W. Fan, N. C. Panoiu, K. J. Malloy, R. M. Osgood, and S. R. J. Brueck: Experimental demonstration of near-infrared negative-index metamaterials, *Phys. Rev. Lett.,* 95, 137404 (2005).

60. J. B. Pendry: Negative refraction makes a perfect lens, *Phys. Rev. Lett.,* 85, 3966–3969 (2000).

61. D. Schurig, J. J. Mock, B. J. Justice, S. A. Cummer, J. B. Pendry, A. F. Starr, and D. R. Smith: Metamaterial electromagnetic cloak at microwave frequencies, *Sci. Expr.,* Page 1, October (2006) /10.1126/science.1133628.

62. A. Sanada, C. Caloz, and T. Itoh: Planar distributed structures with negative refractive index, *IEEE Trans. on Microwave Theory Tech.,* 52, 1252–1263 (2004).

Index

Printed and bound by CPI Group (UK) Ltd, Croydon, CR0 4YY

21/10/2024

01777089-0011